THE TURBAN FOR THE CROWN

THE TURBAN
FOR THE CROWN

The Islamic Revolution in Iran

Said Amir Arjomand

New York   Oxford
OXFORD UNIVERSITY PRESS
1988

OXFORD UNIVERSITY PRESS

Oxford   New York   Toronto
Delhi   Bombay   Calcutta   Madras   Karachi
Petaling Jaya   Singapore   Hong Kong   Tokyo
Nairobi   Dar es Salaam   Cape Town
Melbourne   Auckland

and associated companies in
Berlin   Ibadan

Published by Oxford University Press, Inc.,
200 Madison Avenue, New York, New York 10016

Library of Congress Cataloging-in-Publication Data
Arjomand, Said Amir.
The turban for the crown. (Studies in Middle Eastern history)
Bibliography p.   Includes index.
1. Islam and politics—Iran.   2. Iran—Politics and government—20th century.
3. Iran—Politics and government—1979–   .   I. Title.
II. Series: Studies in Middle Eastern history (New York, N.Y.)
DS316.6.A74   1988   955'.05   87–15231
ISBN 0–19–504257–3

1 3 5 7 9 8 6 4 2

Printed in the United States of America
on acid-free paper

To my mother and the memory of my father

Shah raft, Imam amad
(the Shah has gone, the Imam has come)

Preface

This work was conceived at the onset of the revolutionary upheaval in Iran in 1978 to 1979. In addition to the cited historical sources, documents and publications, which include a number of important recent memoirs, it draws on a number of interviews I have conducted since that time with the key personalities of the old and the new regime. The first of these was with Grand Ayatollah Ruhollah Khomeini in Neauphle-le-Chateau, France, on January 2, 1979. I have subsequently interviewed the former prime minister, Shahpour Bakhtiar, the former president, Abo'l-Hasan Bani-Sadr; two of the surviving highest ranking generals who took over the army after the Shah's departure on January 16, 1979; the former British ambassador to Iran, Sir Anthony Parsons; the former American ambassador to Iran, William Sullivan; the former director of the Iranian National Oil Company, Hasan Nazih who was the first important nationalist figure to break with Khomeini, and many other Iranians most of whom prefer to remain anonymous. I was also present during an exclusive interview with the former Empress, Farah Pahlavi, conducted by my wife, Kathryn Arjomand.

Furthermore, at the time of the outbreak of the revolution I was working on a project on religion and the state in Shi'ism (*The Shadow of God and the Hidden Imam,* published by the University of Chicago Press in 1984). In 1977 and 1978, the research on the project brought me to encounters with "The Signs of God" (literal translation of *ayatollah*), including the late Grand Ayatollahs Kazim Shari'at-madari and 'Abdollah Shirazi and the Grand Ayatollah Shehab al-Din Najafi Mar'ashi, who were then quite accessible and could talk fairly objectively and openly about the issues that were suddenly to become capital with the establishment of Islamic government. I am grateful to all the above for their assistance with this project.

The crucial period for the writing of the manuscript was the year I spent at the Institute for Advanced Study, Princeton (1984–85). I am most grateful to the Institute's faculty of the School of Social Science, to the fellow members with whom I discussed my work, and to the staff, especially Lucille Allsen whose editorial suggestions while typing the manuscript I came to value.

I also wish to express my gratitude to the State University of New York at Stony Brook for providing me with a grant-in-aid to start the project at an early stage, and to Professors Lewis Coser, John Gagnon, Dick Howard, and James Rule for their comments on an earlier draft of this book at a symposium organized by the Department of Sociology.

Professor A. K. S. Lambton read and commented on an earlier draft of the manuscript. I am very grateful to her.

Finally, I wish to thank the editors of *Government and Opposition* and *Middle Eastern Studies* for their kind permission to use materials previously published in their journals. My article, "Iran's Islamic Revolution in Comparative Perspective," *World Politics* 38, no. 3 (April 1986) Copyright © 1986 by Princeton University Press, is also reprinted here in a modified form with permission of Princeton University Press.

Needless to say, none of the persons named above bear any responsibility for the ideas and opinions expressed in this book or for its shortcomings. For these, I alone am responsible.

Stony Brook, Long Island
June 1987   S. A. A.

Contents

Chronology of Significant Events
in Iranian History Since 1500

1501	Foundation of the Safavid empire and establishment of Shi'ism as the state religion of Iran.
1587–1629	Reign of 'Abbas the Great; centralization of Safavid state.
1722	Afghan conquest of Isfahan and the overthrow of the Safavid Dynasty.
1736–47	Reign of Nader Shah Afshar; subversion of Shi'ism.
1747– early 1760s	Anarchy and the dominance of tribal warlords.
mid-1760s–1779	Pacification of the tribes and the reign of Karim Khan Zand.
1779–94	Tribal anarchy and the rise of the Qajars.
1770s–1800	Independent growth of the influence of the Shi'ite hierocracy.
1796	Coronation of Aqa Mohammad Khan Qajar.
1797–1834	Reign of Fath 'Ali Shah Qajar; concord between the state and the Shi'ite hierocracy.
1834–48	Reign of Mohammad Shah Qajar.
1848–96	Reign of Naser al-Din Shah Qajar.
1848–51	Centralizing reforms of Mirza Taqi Khan, Amir Nezam.
1891–92	Nationwide protest against the tobacco concession is led by the Shi'ite religious leaders and results in its repeal.
1896–1907	Reign of Mozaffar al-Din Shah Qajar.
Aug. 5, 1906	Iran is granted a parliament (Majles) in response to popular agitation led by the Shi'ite religious leaders.
1907–09	Reign of Mohammad 'Ali Shah Qajar.
Oct. 25– Dec. 19, 1907	Reforming cabinet of Naser al-Molk, who also serves as finance minister.
Feb. 29, 1908	Sani' al-Dawleh, minister of public works since Oct. 1907, also takes over the ministry of finance.
June 23, 1908	Bombardment of the Majles and restoration of autocracy.
July 1909	Conquest of Tehran by the Constitutionalists and restoration of constitutional government.
1909–25	Reign of Soltan Ahmad Shah Qajar.

Oct. 30, 1910– Mar. 11, 1911	Reforming (second) Cabinet of Mostawfi al-Mamalek; Sani' al-Dawleh serves as finance minister until his assassination on Feb. 6, 1911.
Nov.–Dec. 1911	Occupation of Northern Iran by Russian troops and aborting of the Constitutionalists' reforms.
Oct. 1925	Abolition of the Qajar Dynasty.
Dec. 1925	Reza Khan is declared Shah and monarchy is transferred to the Pahlavi Dynasty.
1925–41	Reign of Reza Shah Pahlavi; formation of a centralized bureaucratic state.
1941–79	Reign of Mohammad Reza Shah Pahlavi.
1944–53	Nationalization of oil, masterminded by Mohammad Mosaddeq, dominates Iranian politics.
1963–79	Mohammad Reza Shah's programs of reform and modernization, officially designated the "White Revolution" and the "Revolution of the Shah and the People."
Feb. 1979	Overthrow of the Pahlavi Dynasty and end of monarchy.
Dec. 1979	Ratification of the Constitution of the Islamic Republic of Iran by national referendum.

THE TURBAN FOR THE CROWN

Introduction

The objective of this book is to explain the Islamic revolution of 1979 in Iran and to assess its significance in world history. It is clear that the Islamic revolution in Iran is a cataclysm as significant and as unprecedented in world history as the French revolution of 1789 and the Russian revolution of 1917. It has indeed a good claim to being considered the surprise of the century, a century not devoid of surprises. Few considered the rise of theocracy in a modernized state a possibility, and even fewer thought it might result from a popular revolution. What added to the wry poignancy of the event was that the Ayatollah succeeded in raising the banner of Islamic theocracy in the land of the Shah, the last King of Kings who, until the day of his departure, commanded a large army equipped with the latest weapons.

The unfolding of the Islamic revolution ran completely counter to the conventional wisdom about revolutions, and more generally, about the philosophy of history. A White House aide who had to face some of its most unpleasant consequences expresses the surprise caused by this revolution very well when he writes: "The notion of a popular revolution leading to the establishment of a theocratic state seemed so unlikely as to be absurd." And further, "what was truly 'unthinkable' . . . was not the Shah's demise but the emergence of a clerical-dominated Islamic republic."[1] The inability to understand the unfolding of the Islamic revolution is by no means confined to American observers. It also characterizes many of the key actors in the revolution who paid dearly for their incomprehension and consequent misreadings of the events and trends. In fact, the widespread inability to comprehend the Islamic revolution persists in the Iranian émigré communities and accounts for a mushrooming of the most fantastic conspiracy theories among them.

The fact that a phenomenon is surprising and defies understanding in terms of stereotypes does not mean that it is incomprehensible. What it does mean is that there is something wrong with the conventional wisdom of the scholars and with the common sense notions of the participants in the revolutionary events. Both of these beliefs are actually quite close as they both stem from the modern myth of revolution. Revolution is one of the most powerful myths of modern politics. During the nineteenth century, revolutions came to be seen as glorious milestones in the liberation of humanity from oppression and its inexorable march toward progress and freedom. Under the impact of twentieth century Marxism, the modern political myth of revolution was fused with the notion of the struggle and inevitable victory of the rising classes. Equipped with such preconceptions, most observers and many participants did not make much headway in comprehending the unfolding revolution in Iran, a revolution that was neither "bourgeois" nor "proletarian" and whose slogans emphasized neither democracy nor progress. Those who were disposed to see the revolution of 1979 in Iran from the perspective of the wars of national liberation and the Third World anti-imperialist struggle could persist longer in their misconceptions, but their dismissal of Islam as a mere camouflage for nationalism or socialism can now clearly be seen as no less erroneous. For the observers and participants who adhered to this last view, Ayatollah Khomeini would variously be a Gandhi to be succeeded by a Nehru or—for those of a more extreme persuasion—"the Chiang Kai-Chek of Iran" but a Chiang (and here we are forced to mix China and Afghanistan in our metaphor) who would pave the way for an Iranian Babrak Karmal. To the chagrin of these observers, the Ayatollah was not to be "Bazarganized"[2] as was the third-worldist president of the Islamic Republic of Iran, Bani-Sadr.

Discard the conventional wisdom and dispassionately discount the modern political myth of revolution, and you will have no difficulty understanding Iran's Islamic revolution. I hope the following pages will demonstrate that the Islamic revolution, though unprecedented in some respects, can indeed be understood in terms of familiar categories and concepts. They should make the Islamic revolution comprehensible in two ways: by putting it in the context of the relationship between Shi'ism and political order in the history of Iran, and by comparing it to other revolutions.

Revolution can be defined as the collapse of the political order—the prevalent system of authority—and its replacement by a new one. Modern revolutions occur in political orders that are dominated by the state. However, in addition to the state, political order comprises other institutions and corporate entities that have some measure of autonomous authority in the religious, judiciary, or economic spheres. The most important of these is usually the religious institution, that is, the church or its equivalent.

As will be seen in Chapter 1, a distinctively Shi'ite dualistic system of authority became prevalent in Iran in the nineteenth century. The modernizers of the twentieth century sought to destroy this dualistic political order and to replace it by one exclusively dominated by a centralized bureaucratic state. They succeeded in erecting such a monolithically secular political order, but only to see it taken over by the surviving forces of theocracy in 1979.

Modern revolutions occur not in stagnant societies, but in those undergoing considerable social change. In Iran, as in the rest of the Middle East and the Third World, the centralized bureaucratic state has been the chief agent of social change in the twentieth century. Social change results in social dislocation and normative disturbance. The dislocated groups and individuals need to be reintegrated into societal community[3] and may also demand inclusion in political society. Modern states try to meet these needs by increasing controlled political participation but often fail. The rival integrative social and political movements that consequently arise to meet these demands have often been a major contributing factor to the occurrence of revolutions. In Iran, the most effective of these movements drew its inspiration from Shiʻite Islam and was led by Shiʻite men of religion.

The collapse of the political order in revolutions is caused by two sets of factors: its internal weaknesses and vulnerabilities, and the concerted action of the social groups and individuals opposing it. Assessing the vulnerability of a political order obviously brings the state to the forefront of discussion. So does understanding the motives of the supporters of revolution, albeit less obviously and less directly. The groups and individuals who support a revolution can have political motives for opposing the regime, notably resistance to loss of power and demands for political participation and citizenship, which are often set in motion by modernization of the state. They can also have moral motives. These often require the preconditions of social dislocation and normative disturbance, which are in turn largely or partly the consequence of the economic and developmental policies of the state. An adequate explanation of the causes and preconditions of the Islamic revolution in Iran must therefore begin with an account of the prior rise of the modern bureaucratic state and the impact of its policies.

It should also be stated at the outset that the theory of revolution implicit in this book distinguishes between the preconditions and causes of revolutions, on the one hand, and their consequences, on the other. It assumes that one set of factors bears on the preconditions and causes of a revolution; a separate set of factors determines its direction and can predict its outcome. According to this theory, the rise of the modern state and its consequences, concentration of power in a centralized bureaucratic state and increasing integration of the various segments of society into a unified nation-state, constitute the most important preconditions of modern revolutions; the factors making for a temporary paralysis of the state compose these revolutions' most important immediate causes. By contrast, the direction and consequences of modern revolutions is to be sought in the international development of the general myth of revolution and the specific ideologies on which it has been grafted, the latest of these being the contemporary Islamic ideology.

A historical perspective is necessary to understand both the causes and the consequences of the revolution in Iran. Regarding its causation, the Iranian revolution of 1979 is distinctly modern in that its occurrence is inconceivable without the prior rise of the centralized state, its transformation of Iranian society, and finally, its paralysis after the onset of crisis in 1978. At the same time, Islamic ideology, which has determined the direction of the Iranian revolution—though novel in that it incorporates the modern myth of revolution, could not have succeeded without the unusual institutional assets of its proponents, the Shiʻite men of religion. The story of the Islamic revolution in Iran must therefore begin with the establishment of Shiʻism as the state religion in 1501, and it should include a brief account of the consolidation of Shiʻite clerical power and the prevalence of a dual system of authority until the onset of centralization and the expansion of the power of the state.

This historical background is indispensable for explaining why, when a modern political revolution occurred in Iran in 1979, it was destined to be an Islamic revolution with the goal of establishing a theocracy.

Part I covers the historical background of the Islamic revolution with special attention to the rise of the modern state and its impact on the hierocracy-state relationship. As the book is addressed not only to the specialist but also to the general reader, those who are not particularly interested in the sociology of state formation or in the finer points of Iranian history may prefer to begin with Parts II and III and then return to Part I for the amplification of the historical context.

Part II deals not only with the immediate causes of the revolution of 1979 but also with its consequences. Chapter 5 traces the emergence of an Islamic revolutionary movement under Khomeini's leadership in its immediate sociopolitical context. The collapse of the Pahlavi political order and the fall of the Shah are explained in Chapter 6. Chapter 7 deals with the revolutionary power struggle after the collapse of monarchy. It underscores the determination and consummate skill with which Khomeini and his militant followers, as the bearers of the Islamic ideology and advocates of theocratic clericalism, triumphed over the other contending groups. Finally, the consolidation of Islamic theocracy and its distinctive institutional structure are treated in Chapter 8. Together, these chapters give an analytical account of the transition from one monistic political order to another, from the temporal absolutism of the Shah to the theocratic absolutism of Khomeini.

The reader will find the measured intrusion of my value judgments in Chapters 6 and 7 at variance with the detached and dispassionate tone of the rest of the book. It would have been misleading for me to present my account under the guise of scholarly objectivity and to hide my deep emotional, if not active, involvement in the events I analyze in these chapters. Nor was it necessary. But a brief comment regarding my judgment on the new middle class, which lost the revolutionary struggle between September 1978 and June 1981, may be in order because the reader may find this judgment too severe. Perhaps it should be considered together with the backhanded compliment I pay the new middle class in the epigram to Part III.

Although the analysis of Part II is self-sufficient in accounting for the course of revolutionary events in causal terms, it is hardly adequate for a full understanding of the Islamic revolution in Iran. More is needed. We need to assess the *significance* of the Islamic revolution, both in the history of Shi'ism and in world history. This is the objective of Part III. Chapter 9 assesses the paradoxical impact of the Islamic revolution on the Shi'ite tradition, which amounts to nothing short of a revolution *in* Shi'ism. Last but by no means least, in Chapter 10 I have sought to bring out the significance of the Islamic revolution in Iran as the last of the "great revolutions" by comparing it to other revolutions. No major event in world history can be understood without, at the same time, throwing new light on earlier events of the same kind. In fact, in making the Islamic revolution intelligible in comparative perspective, quite a few surprises regarding the neglected or misinterpreted aspects of other revolutions are brought to light, thereby enhancing our understanding of revolution in general. As I hope to persuade the reader, the Islamic revolution in Iran forces us to reassess the balance of progressive and reactionary aspects, of traditionalism and modernism, in all revolutions.

A glossary of Persian words is supplied with the appendices to make the reading easier and terms are translated when they first appear in the text. No diacritical marks are used in the text. When dual dates are given, the first figure refers to the common era (A.D.) and the second to the Islamic calendar (A.H.). A chronological table of significant events in the history of Iran since the sixteenth century is also provided at the beginning of the book. I have tried to avoid using technical terms if possible. The one important exception is the use of the term "hierocracy" with reference to the Shi'ite religious institution. Neither "clergy" nor "church" would have conveyed the requisite sense of a system of authority.[4] I have also used the term "teleology" to refer to the direction and intended consequences of revolutions. Whenever possible, the evidence for the trends described and the arguments put forward has been summarized in table form. The tables, all of which contain important information, are referred to in the text to illustrate or substantiate a point but placed in the appendix in order not to impede the flow of argument and ease of reading.

I

RISE OF THE
MODERN STATE AND THE
TRANSFORMATION OF SOCIETY

Seek not faith in covenant from this world, weakly stayed
This ancient crone has a thousand bridegrooms betrayed
Hafiz of Shiraz (d. 1389–90)

1

Religion, Government, and the Social Structure in the Eighteenth and Nineteenth Centuries

Establishment of Religious Uniformity and Consolidation of the Shi'ite Clerical Power in Iran

The various Shi'ite branches of Islam have their nucleus in "the Party" (*shi'a*) of 'Ali, the son-in-law of Mohammad, who became the fourth and last universally recognized "rightly guided" Caliph in 656 and died in 661. Of these, a group organized into a religious sect by the mid-eighth century was to survive various crises of succession and become known as the Imami Shi'a on account of their doctrine of Imamate (divinely inspired leadership). The sect was also referred to as the "Twelvers," because of their belief in a line of twelve Imams as divinely inspired, infallible leaders of a community of believers and teachers in religion. The Twelfth Imam, Mohammad al-Mahdi, is believed to have gone into hiding in the year 874. He is considered to be the Lord of the Age, to reappear at the End of Time.

Twelver Shi'ism bore the permanent imprint of the doctrine of the Imamate formulated under the leadership of the Sixth Imam, Ja'far al-Sadeq (d. 765). To assure the lasting sectarian organization of the Shi'a as a disciplined sect under institutionalized religious authority, he dissociated supreme religious authority from actual political rule and rested it on divine inspiration and on *'ilm* (knowledge). Later generations of Shi'ite scholars, the *'ulama* (the learned), derived their religio-legal authority from this basic premise without any reference to reigning political authority. Throughout the medieval period, the Twelver Shi'a reached an accommodation with non-Shi'ite rulers and were therefore considered the moderate branch of Shi'ism. Extremism (*gholovv*) in other Shi'ite sects was denounced, and its recurrence among the Twelvers considered anathema. In contrast to the mainstream, Sunni Muslims, however, the Twelver Shi'a refrained from granting the ruling powers any religious authority, which was invested exclusively in the *'ulama*. In this way, the *'ulama* were able to persist, supported by the community, and to flourish independent of government and of political vicissitudes.

The establishment of Shi'ism as the state religion of Iran dates from the foundation of the Safavid Empire in 1501. The empire of the Safavids was created by the military force of a millenarian Sufi warrior order. At that time, the population of Iran was overwhelmingly Sunni, but Sufism (Islamic mysticism) was widespread, as was the devotion to the House of the Prophet that included the Imams of the Twelver Shi'a. For most strata of society, religious life was organized by the Sufi orders and controlled by the Sufi shaykhs. The members of the Safavid Sufi order were Turkman tribesmen who had adopted certain "extremist" Shi'ite notions such as the divinity of the first Imam, 'Ali. They considered their shaykh the reincarnation of 'Ali and of the holy Imams, the incarnation of God and the Lord of the Age, the Mahdi. The Shi'ism of these tribesmen was indeed extremist, and their millenarian zeal in the battlefield a most useful means of conquest. Once the conquest of Iran was completed, however, millenarianism lost its political utility to the new dynasty and became more of a liability than an asset. The Safavids invited a number of Arab Shi'ite theologians to their kingdom to spread the orthodox creed of the moderate Twelver Shi'ism among the predominantly Sunni population of Iran. The inflow of Shi'ite theologians and jurists from the Arab lands, notably from the Jabal 'Amil region in present-day southern Lebanon, continued for two centuries, and they soon trained a large number of Iranian *'ulama* with whom they intermarried. Meanwhile, despite their origins in Sufism, or perhaps because of it, the Safavid rulers took strong measures to suppress rival Sufi Orders, which dominated the religious life of the masses in the fifteenth and sixteenth centuries. The Shi'ite *'ulama,* for their part, incorporated many of the features and practices of popular Sufism into the official belief system during the seventeenth century. These developments eliminated the rivalry of the Sufi shaykhs as popular religious leaders and enabled the emergent Shi'ite hierocracy in Iran to control the daily religious life of the masses to an extent unknown in other Islamic lands.

There remained, however, important obstacles to the consolidation of clerical power under the Safavids. These limited the possibilities open to the Shi'ite *'ulama* for exploiting their domination of the masses. First, the Safavid state remained "caesaropapist" to its last day, with the monarch enjoying religious legitimacy as the descendant of the holy Imams and the Lieutenant of the Lord of the Age. The monarch possessed the prerogative of making clerical appointments and controlled the endowments (*awqaf*) that constituted the source of financial support for institutions of religious learning. Second, the Shi'ite jurists had to face the rivalry of an estate of notables who constituted the corps of Safavid clerical and local administrators—the *sayyeds*. This group possessed a clerical or quasi-clerical status and could boast of considerable *religious* charisma of lineage as the descendants of the Prophet and the Imams. The collapse of the Safavid Empire as a result of the Afghan invasion in 1722 put an end both to the religious pretensions of the rulers and the political dominance of the *sayyeds*. Thus obstacles to the consolidation of Shi'ite clerical power in the form of religious or quasi-religious pretension by royalty and from the notables also disappeared in the eighteenth century.

The seventeenth and eighteenth centuries witnessed the rise of a distinct group of Shi'ite religious professionals as the guardians of the Sacred Law. The religious institutions under the control of this Shi'ite hierocracy were still heteronomous—still neither administratively nor financially independent of the state. The ruler appointed the chief religious official, the *shaykh al-Islam,* of the cities. Nevertheless, the Shi'ite hierocracy became increasingly differentiated from other branches of the state, and the last Safavid monarch appointed a special functionary, the *Mollabashi,* as its head. After the collapse of the Safavid empire in 1722, the members of the Shi'ite hierocracy were forced to subsist on their own, totally independent of the state. The state, furthermore, assumed a ruthlessly hostile posture toward Shi'ism and its custodians under Nader Shah (1736–48) who was hoping to build a pan-Islamic empire after his conquests in the east and India. The rigors of forced self-subsistence resulted in an immediate and drastic decline in religious learning; but in the long run, it bore fruit in the form of a religious/intellectual movement known as the *Osuli* movement.

It is interesting to note that the final obstacle to the growth of Shi'ite clerical power in Iran was a doctrinal one. Ever since the disappearance of the Twelfth Imam in 874, the transfer of his authority to the *'ulama* had been problematic. For over a century, the prevailing attitude was staunchly traditionalist: no jurist or rather no human authority could derive any legal norms other than what was explicitly transmitted through the Traditions (*akhbar*) of the Imams. In the eleventh century, Akhbari traditionalism was overcome by the proponents of rational theology and jurisprudence. The science of jurisprudence and the practice of deriving legal norms developed progressively until, around the beginning of the fourteenth century, the principle of *ejtehad* (the competence of the jurists to derive new legal norms from the sources of the Sacred Law) was explicitly adopted. The adoption of the principle of *ejtehad* constituted an intellectual revolution in medieval Shi'ism. It greatly enhanced the juristic authority of the Shi'ite *'ulama,* the most prominent of whom were henceforth given the designation of *mojtahed* (he who practices *ejtehad*). The orthodox Shi'ism that the Safavid rulers brought into Iran and sponsored in order to replace the extremism of the period of conquest was the Shi'ism of the *mojtaheds.* Nevertheless, the traditionalist orientation had not disappeared in the Shi'ite scholarly community and was vigorously revived in seventeenth century Iran. The revival of Akhbari traditionalism during the second half of the seventeenth century was connected with the resistance of the *sayyeds* to the growth of hierocratic power in Iran. The Akhbaris rejected the principle of *ejtehad* and denied the authority of the *'ulama* to engage in jurisprudence and to derive new legal norms. Akhbari traditionalism, which rejected jurisprudence and advocated simply the study and collection of the Traditions of the Imams, became a major force in the second half of the seventeenth century and completely dominant during the decades following the collapse of the Safavid empire.

The *Osuli* movement flourished outside Iranian territory in the holy cities of Ottoman Iraq and consisted of a revival of Twelver Shi'ite jurisprudence that dominated the last decades of the eighteenth and the whole of the nineteenth centuries. It consisted in the vigorous reaffirmation of the principle of *ejtehad* and thereby resulted in very considerable enhancement of the power and the independence of the Shi'ite hierocracy. The revival of religious jurisprudence greatly augmented the prerogatives of the Shi'ite *mojtaheds* (doctors of jurisprudence) as the authoritative interpreters of the Sacred Law. The failure of Nader's policy to promote Sunnism and restrict certain Shi'ite practices proved that Iran's conversion to Shi'ism was irreversible. Henceforth, their unrivaled dominion over the religious life of the masses was not just the result of the absence of rival Sufi shaykhs, but had a firm doctrinal basis, which gained virtually universal acceptance in the nineteenth century. By the end of the century, the Shi'ite hierocracy controlled the religious life of over 90 percent of the population of Iran.

The *Osuli* movement assured the independence of religious authority from the political authority of the ruler and, consequently, the autonomy and autocephaly of the Shi'ite hierocracy. Furthermore, it assured a large measure of financial autonomy for the hierocracy through the authorization of the collection of religious taxes on behalf of the Hidden Imam. As this last point is of some consequence, I should perhaps point out that a novel interpretation of a part of the religious tax of *khoms,* known as "the share of the Imam" (*sahm-e imam*), was introduced by the *Osuli* jurists, and that an instance of cooperation between the hierocracy and the state during the first Perso-Russian War (1811–13) enabled the Shi'ite *mojtaheds* to secure the tacit consent of the state to their right to its collection. From the nineteenth century onward, the Shi'ite hierocracy collected this voluntarily discharged religious tax as the lawful recipients of "the share of the Imam." Finally, the independence of the hierocracy from the Iranian government was enhanced by the fact that the leading *mojtaheds* resided in the holy cities of Iraq, which were under Ottoman rule.

The upper rank of the Shi'ite hierocracy consisted of the *mojtaheds,* who enjoyed tremendous power and prestige during the nineteenth century. By virtue of their judicial and religious authority, they heard the complaints of the people against injustice and tyrannical misrule and, on occasion, took up their cause. However, even though it is useful to conceive of the Shi'ite *'ulama* as a hierocracy, the extent of their organization into a system of authority should not be exaggerated. There was no formal hierarchy of authority among the *mojtaheds,* and the congeries of clerics at the lower ranks could be differentiated only by the patronage of different *mojtaheds* and by a nonformalized scale of prestige and learning. Because of their amorphous internal organization, therefore, the Shi'ite *'ulama* could not act as a unified body except on rare occasions, such as the wars against Russia in 1811 to 1813 and 1826 to 1827. This inability was made painfully clear in the face of the millenarian heresy of the Bab who put forth the "extremist" claim of being the Lord of the Age in the 1840s. The hierocracy was unable to suppress the heretics or prevent the younger *tollab* (seminarians) from joining them. It was the state that stemmed the tide of Babism, suppressing a series of millenarian insurrections and an attempt to assassinate the king (1848–52).

During the long reign of Fath 'Ali Shah (1797–1834), both the Qajar dynasty and the Shi'ite hierocracy consolidated their respective political and religious authority. Safavid caesaropapist monism had given way to a hierocracy-state dualism. A rough division of the political and the religious functions of government was worked out, and the Shi'ite hierocracy assumed independent control of the latter. These comprised the religious, the judicial, and the educational institutions. At the same time a number of tracts on political ethics and political theory written by the Shi'ite jurists legitimized this division. These tracts stressed the interdependence of religion and kingship, and the importance of concord and collaboration between the state and the hierocracy.[1]

Close ties between the hierocracy and urban petty bourgeois strata, that is, merchants and craftsmen, are probably more the rule than the exception in preindustrial societies. As Weber explains, these ties could give rise to enduring alliances against the patrimonial and feudal powers. On the basis of "opposition to political charisma,"

> the elective affinity between bourgeois and religious powers, which is typical of a certain stage in their development, *may grow into a formal alliance* against the feudal powers; this happened rather frequently in the Orient and also in Italy at the time of the struggle over lay investiture [the Investiture Contest of the eleventh century].[2]

The formation of such an alliance was hindered by the heteronomy of the religious institution and its subordination to the state during the Safavid period. Under the Qajars, from the last decades of the eighteenth century onward, the autonomy of the Shi'ite hierocracy made an alliance with civil society—with urban guildsmen, merchants, and shopkeepers—possible and likely. In the last quarter of the nineteenth century, an enduring alliance against the state between mosque and bazaar came into being. Against the background of noticeable growth of the economic power of civil society, this alliance was cemented by the common opposition of the two parties to foreign penetration that resulted from the political privileges and economic concessions granted to imperialist powers by a servile state. It was an alliance in defense of the respective cultural and economic interest of the hierocracy and the bourgeois strata. Popular bourgeois support in this period greatly enhanced the power and prestige of the great *mojtaheds* who began to assume the lofty title of *ayatollah* (sign of God).[3]

A very important factor in determining the capacity of a hierocracy to act against the state or upon society is the character and extent of autonomy of its organization. As we have pointed out, the autonomy of the Shi'ite hierocracy became established in principle in the early years of the nineteenth century, although the appointment of the *imam jom'eh* (prayer leaders of the cathedral mosques) and *shaykh al-Islams* of the major cities remained with the ruler. By contrast, the Shi'ite hierocracy was still very loosely organized in 1900 and lacked the elaborately hierarchical structure characteristic of the Catholic Church. The principle of juristic authority in Shi'ism, that is, *ejtehad,* only made possible the division of the clerical professionals into the *mojtahed* (jurist) and the non-*mojtahed*. The divisions within these two categories were still informal and not legally or doctrinally based. The Ranking among the *mojtahed*s had come to depend on the number of their followers (*moqalled*s); those with the larger followings were referred to singly as the *marja'-e taqlid* (source of imitation) and occupied the positions of leadership in the Shi'ite hierocracy. There was another informal but all-important division within the Shi'ite hierocracy around 1900. The prominent *'ulama* could be divided into two groups: "those who have no relations with the functionaries of the state, do not administer the Sacred Law and avoid luxuries," and "those whose quest for worldly dominion (*riyasat-ma'abi*) outweigh their religious aspect," who intermingled with the functionaries of the state, held religious courts, and usually amassed large fortunes. This latter group issued verdicts in legal disputes and acted as intermediaries between government and people. For this group, judiciary function constituted a source of income whose enlargement resulted in the issuance of a plethora of contradictory verdicts. Ranking well below these categories in prestige and authority were the preachers (*ahl-e menbar; vo"az*) who had close ties with the nobility because of their role in the latter's religious ceremonies (*rawzeh-khani* and *ta'zieh*). To these observations we should add that the *imam jom'ehs* and *shaykh al-Islams* were the appointees of the state, thus belonging to the group of *'ulama* seeking worldly dominion, and that some members of this latter group possessed extensive landholdings and controlled endowed (*vaqf*) properties. When the holy cities of Arab Iraq eclipsed Isfahan and Tehran as centers of religious learning in the second half of the nineteenth century, the dichotomy between the "worldly" *'ulama* and those who piously withdrew from political engagements, found a corresponding geographical dimension.[4]

Collapse of the Safavid State and Tribal Domination of Urban Society in the Eighteenth Century

The eighteenth century is among the most turbulent in Iran's long history. It witnessed the collapse of the centralized Safavid empire in 1722, followed by seven decades of internecine tribal warfare, and finally, the establishment of Qajar monarchy in the 1790s. Tribalism subsided in the nineteenth century, but generally speaking, the central government remained weak and the kingdom minimally integrated. Iran entered the twentieth century with the Qajar governmental system on the verge of complete breakdown and bankruptcy.

Like most dynasties in the history of Iran, the Safavid House established an empire at the beginning of the sixteenth century through a tribal confederation. The power of the Qizilbash tribes, which constituted the military backbone of Safavid state in the sixteenth century, was drastically reduced by 'Abbas the Great (1587–1629), the number of Qizilbash troops being cut from 60,000 to 30,000. Nevertheless, the Qizilbash tribes remained successfully integrated into the centralized Safavid state through the subsequent century. The Qizilbash troops constituted one of the four branches of the Safavid army, the Qurchis, and their leaders held governorships. Indeed, one can argue that in 1722 they constituted the most important branch of the army. They received over one-third of the army budget and their head, His Excellency the Qurchi-bashi, the Pillar of Royalty (*rokn al-saltaneh*), was ranked the most important *amir* (commander) and given precedence over the three *amirs* who headed the other branches of the Safavid army.[5]

The most important component of 'Abbas the Great's centralization policy was the creation of new troops: a corps of musketeers (*tofangchi*) numbering 12,000 and a corps of royal slaves (*qollar* or *gholaman*). The revenue from the royal domains (*khasseh*) was used to pay the new troops. Furthermore, 'Abbas inaugurated the policy of converting the state provinces (*mamalek*) into royal (*khasseh*) provinces, governed by intendants appointed directly by the king, in the latter part of the 1590s. From 1600 onward, Allahverdi Khan, the *qollar-aqasi* (commander of the *qollar*) and the first royal slave to attain equality of status with the Qizilbash commanders, began the reorganization of the army in conjunction with Sir Robert Sherley. The slave corps under his command was increased from 4,000 to 25,000 men.[6] Shah Safi (1629–42) and 'Abbas II (1642–66) continued 'Abbas the Great's policies of centralization. Safi's grand vazir, Saru-Taqi, reformed and rationalized the administration of royal domains, extending it to cover the government of the inner provinces not threatened by war. 'Abbas II abolished more governorships and placed them under the *khasseh* administration.[7] Although the size of the slave corps might have become smaller under 'Abbas II (down to 10,000 according to Chardin), their importance as the tool for centralization of power was undiminished. Both 'Abbas I and 'Abbas II relied on the royal slaves for provincial administration, as the Mamluks had been relied upon by the Muslim rulers in the medieval period. By the time of 'Abbas I's death, eight of the fourteen major provinces were administered by the royal slaves. During 'Abbas II's reign some twenty-three to twenty-five of the thirty-seven governorships were held by royal slaves.[8]

To gain some perspective on the extent and nature of the centralization of the Safavid state, comparisons with Western European states during the age of absolutism are helpful. The first point to note is that the crucial importance of the royal domains as a source of revenue for the central government and for the maintenance of the nontribal regular army makes the Safavid state more similar to the Prussian than to the French variant of European absolutist state. *Khasseh* revenue under 'Abbas II was estimated by Chardin at some 44 percent of the total government revenue, and the amount collected by the *khasseh* department in 1722 is just under one fourth (the real revenue from the royal domains may well have been higher, some of it being reported in other categories). The Ottoman Register of 1727/1140 for Tabriz, conquered after the fall of the Safavids and almost certainly reflecting the *status quo ante,* shows a larger figure for royal domains (*khass-e shahi*) than for all other land assignments (9.5 as against 8.4 million aspers).[9] This is similar to the fiscal pattern in Prussia where, by 1740, the revenue from the royal domains was almost equal to public taxes.[10] Beyond this, however, differences between Iran and Western Europe are much greater than similarities. The first important difference is that the tribal component of the Safavid army, the Qurchis, which remained numerically preponderant, finds no counterpart in the armies of the absolutist states of Europe. The concentration of financial resources and the fiscal power of the Safavid state was also considerably lower than its European counterparts. Given the comparable levels of economic prosperity as attested to by such perceptive travelers as the Protestant French knight, Chardin, and the Ottoman bureaucrat, Katib Chelebi,[11] the weaker fiscal basis of the central state in Iran is demonstrated by Table 1 (see Appendix).

From the figures given in Table 1 we may conclude that both in absolute terms and relative to the size of population, the concentration of fiscal resources under the state is much greater in France and even Spain and England than in eighteenth century Iran at the height of centralization under the Safavids.[12] The comparatively modest size of the central bureaucracy of the Safavid state in Isfahan corresponds to this difference. We know that in France, "when the Court and the Government installed themselves at Versailles, the *bureaux* occupied exclusively the two long wings on either side of the forecourt."[13] The Safavid Shah's bureaucracy was considerably smaller: it consisted of some forty-eight functionaries and well over one hundred scribes (see Appendix, Table 2).

The remuneration of the functionaries of the central government consisted of fixed salaries supplemented by specified levies on land grants and, occasionally, on religious endowments.[14] Multiple office holding was common, and there was a tendency for the commanders and other high functionaries of central government to hold specific governorships, with offices connected in a stereotyped manner.[15]

In sharp contrast to the growth of absolutist states in Europe, the history of Iran in the eighteenth century is marked by the resurgence of the tribalism to which the Safavid state succumbed. According to Lambton, the decades preceding the collapse of the Safavid empire witnessed a tribal reassertion not only among the Afghan tribes, which were eventually to overthrow the empire in 1722, but also among other tribes, the Baluchis, the Kurds, the Lors, and the Bakhtiyaris.[16] The tribal resurgence became overwhelming after the collapse of Safavid central authority. Nader-qoli, who began his career as a general in the service of the Safavid crown prince by raising troops to recapture the capital, expelled the Afghans in 1729 and ruled Iran as Nader Shah after 1736. Though from a tribal background, he was not a tribal chief and could not create a typical tribal confederation. Nevertheless the forces he amassed to reconquer Iran and to conquer India were overwhelmingly tribal. His militarist government collapsed with his assassination in 1747, and Iran became the arena of a devastating struggle among the tribal warlords. Karim Khan Zand was successful in creating a tribal confederation and in unifying and pacifying the country, ruling it from Shiraz by the mid-1760s. Tribal anarchy returned after his death. The devastation of Iranian cities and countryside by the tribal warlords continued until Aqa Mohammad Khan, chief of the Qajar tribe, created a tribal confederacy along the traditional lines and succeeded in establishing the Qajar monarchy in the 1790s. His coronation took place in 1796.

Eighteenth century chronicles have preserved harrowing accounts of the devastation of the major cities, especially Isfahan, Qazvin, Shiraz, and Yazd, as a result of repeated sieges, sacks, and depredations.[17] During the seven decades between Safavid and Qajar rule, the main cities lost at least two-thirds of their inhabitants.[18] The last major catastrophe befell the city of Kerman. It was sacked by Aqa Mohammed Khan Qajar in October 1794. Hearing of the flight of the last Zand ruler who had held the city, "he grew so angry that he ordered the slaughter of the inhabitants of Kerman. About 8,000 women and children were distributed as slaves among the army; all the men were either killed or deprived of their sight."[19]

Nader's military campaigns and the internecine struggle of the tribal warlords resulted in the imposition of a crushing tax burden on the countryside. During his last years, Nader confiscated religious endowments (*awqaf*) throughout the country[20] and raised the taxes fantastically, executing the tax collectors who pointed out that the new figures were unrealizable. We have a graphic eyewitness account of the encounter between Nader Shah and the tax collectors of the province of Fars. Having ordered his designated tax commissioner for Fars to choose the ten least corrupt tax agents, Nader had the remaining seventy-three decapitated together with some other wrongdoers. (The executioners were ordered to build two towers with the severed heads and, as some of the tax collectors had fled overnight, killed other persons "so that the tower of [severed] heads should not be defective".)[21] Nader Shah then sent off the surviving collectors to Fars, giving the tax commissioner the following instruction: "You are permitted to decapitate up to fifty persons from the army if their accounts are irregular. As for beating and [cutting off] nose and ears in cases of irregularity, there is no limit [to your authority]."[22] The crushing tax increases caused a series of rebellions (1744–47). Mutilated and blinded victims of Nader's fiscal severity were still to be seen in 1765 and later.[23] Mirza Mohammad Kalantar also documents the tax exactions of the Bakhtiyari and Zand warlords of the interregnum[24] and points to instances of the abandonment of agrarian land as a result of excessive taxation.[25]

The years of anarchy intervening between the death of Karim Khan and the supremacy of Aqa Mohammad Khan (subsequently Shah) also demonstrated the military weakness of the cities and their inability to defend themselves against tribal forces. In these years sundry armed cohorts in the cities did come into existence and mayors or urban notables did assume control of the cities for brief periods.[26] In February 1785, with most tribal elements already expelled from Isfahan by the Zand ruler who had died near the city, the mayor Baqer Khan summoned four to five thousand musketeers from the surrounding area and some Lezgi troops resident in Isfahan. Styling himself Shah Baqer, he "ascended the throne with royal splendor."[27] His rule lasted four-and-a-half days, when he was attacked and wounded by Zand soldiers and had to flee the city. Reinstated as governor by Aqa Mohammad Khan Qajar, he was put to death four months later by the returning Zand.[28] In one remarkable instance in 1791, the citizens of Shiraz under the mayor Haji Ebrahim ingeniously divided the tribal army loyal to the Zands into small groups after a pay parade, disarmed them, and even took away their clothes. "By this means the whole tribal army, naked and disarmed, was removed from the city. Nobody knew what happened to the others [i.e., those who had not appeared at the parade]. Then their families were expelled from the city."[29] Haji Ebrahim and the citizens of Shiraz successfully defended the city against Lotf 'Ali Khan Zand and opened its gates to the Qajar monarch. Nevertheless, the military weakness of the cities and their penetration by tribal military contingents is evident. Even though he was proved wrong in that particular instance, it was not without reason that the last Zand ruler Lotf 'Ali Khan contemptuously dismissed the rebellious mayor of Shiraz, Haji Ebrahim, with the following remark: "This traitor is only a citizen and his followers are merchants and dealers. They cannot compete with the victorious army, which is trained in handling musket and sword only."[30]

If the eighteenth century began with the resurgence of tribalism, it ended with the reassertion of central monarchical authority. After the unification of Iran, Aqa Mohammad Khan, who claimed Safavid descent through his grandmother, set out to revive the Safavid empire. In January 1794, he conferred the governorship of the provinces of Fars, Kerman, and Yazd upon his chosen successor, Crown Prince Baba Khan, and made Haji Ebrahim, the mayor of Shiraz, his grand vazir. *"After the pattern of the Safavids,* the Shah conferred upon Haji Ebrahim the honorific title *E'temad al-Dawleh"* (Trust of the State).[31] However, if the Safavid central bureaucracy was of modest size by French standards, Aqa Mohammad Khan's was excessively small by Safavid standards. No chart or table is necessary to represent it; it can be depicted more elegantly and just as rigorously in prose. The grand vazir apart, Aqa Mohammad Khan had two financial officers, a *mostawfi* in charge of taxation and a *lashkar-nevis* in charge of army accounts. Both traveled constantly with him. He himself administered justice and supervised accounts. He "had no official private secretary, no minister of justice, no minister of court. He ran a highly personal administration and, finances aside, paper work was almost non-existent"[32]

The Qajar Government and the Social Structure of Iran in the Nineteenth Century

In 1797, the year after his coronation, Aqa Mohammad Khan was assassinated. His nephew Baba Khan succeeded him as Fath 'Ali Shah after eliminating the competing pretenders. The long reigns of Fath 'Ali Shah (1797–1834) and later of Naser al-Din Shah (1848–96) freed Iran from frequent bloodshed over succession to the throne and offered the population of Iran a century of general internal peace and relative economic prosperity. The population of Iran increased from some six million in 1800 to under ten million in 1900. As is indicated in Table 3 (see Appendix), the depopulation of cities was arrested and there was a slight increase in the rate of urbanization. The tribal and nomadic population probably declined, though it still remained substantial.

The assertion of central authority during the first decade of Fath 'Ali Shah's reign was very impressive. The tribes were subdued and some transplanted. Many tribal chiefs were executed, others or their close relatives were forced to reside at the royal court as hostages.[33] Fath 'Ali set up a sumptuous court in the new capital, Tehran. The administrative staff of the Zand dynasty were co-opted eagerly and the bureaucracy was expanded considerably. The two financial officials in charge of taxation and army accounts each became the head of a department and their ranks were upgraded to *mostawfi al-mamalek* (finance minister) and *vazir* of the army respectively. Other ministerial posts were created for a controller-general (*saheb divan*) and the head of the state secretariat (*monshi al-mamalek*). The royal treasury was divided into the public treasury and the privy purse, though in practice the two were not kept strictly separate. Haji Ebrahim of Shiraz was confirmed as prime minister (*sadr-e a'zam*) and presided over the central bureaucracy. The Garden of Lalehzar near the Shah's palace became the residence of prime ministers, though the ministers did not acquire fixed offices and conducted government business from their homes or other places.[34]

In the first decade of the nineteenth century, state revenues reached the Safavid level.[35] It is variously estimated at 2 million *tumans* (£2,000,000 [1807]) for 1807 and 3 million for 1808–1809, as compared to roughly the same revenue for the Ottoman Empire, with a population twice as large as that of Iran, and to a revenue of 7 million *tumans* in the previous decade from the territories of the East India Company, with a population five times as large.[36] However, the Qajar state's newly acquired vigor and fiscal strength were wasted in two major wars against Russia (1811–13 and 1826–27), in which Iran was defeated, suffering significant loss of territory and population. From the second quarter of the nineteenth century onward, the picture is one of irreversible decline in the fiscal, and military, resources of central government.

As is shown in Table 4 (see Appendix), during the nineteenth century, state revenue in cash (about 85 percent of total revenue) and kind (about 15 percent)[37] declined in real terms while the population and economy of Iran grew considerably. The figures in Table 4, though they reveal the general trend, do not adequately reflect the true fiscal weakness of Qajar government; nor do they convey the fiscal chaos and financial insolvency of central government. Even in the best of years, central government probably controlled less than half of this revenue.[38] Taxes were frequently in arrears and were often either never remitted or partly or wholly remitted under the threat or actuality of military expeditions.[39] The finance minister was forced to issue drafts against provincial revenue. These drafts (*barat*s) were rarely honored upon presentation and often realized no more than half of their nominal value. In bad times, they are reported to have been traded for as little as one-tenth of their nominal value.[40]

The fact that state revenue from taxes was low did *not* mean that the burden of taxation on the subjects was light. On the contrary, the inefficiency of the central government in collecting taxes greatly increased the tax burden.[41] Governors were expected to collect above the tax assessment, the difference being officially termed the *tafavot-e 'amal* (difference of the operation), and invariably did so. Sale of office was a marked feature of Qajar government,[42] and governors had to recuperate the considerable sums they had paid for their offices and make a large profit to maintain their clients. With the exception of the most important men of the realm who enjoyed longer tenure, governors or deputy governors were rotated fairly frequently—on average, perhaps once every two or three years.[43] Some of the governors amassed huge fortunes. For instance, the Georgian eunuch Manuchehr Khan Mo'tamed al-Dawleh, who held the governorships of Isfahan, Khuzestan, and Fars, died in 1847, leaving an estate of 3.5 million *tumans* or nearly one-and-a-half times the theoretical revenue of the state.[44] Furthermore, tax farming was the regular practice: collection of taxes in different areas was farmed out to the highest bidders, and tax farms were negotiable and transferrable. Taxes could be arbitrarily increased by tax farmers.[45]

The Iranian economy in the nineteenth century was, needless to say, overwhelmingly agrarian. As in all agrarian societies, land itself—which could be assigned temporarily or permanently—constituted the major fiscal resource of the state other than taxes in cash and kind. In addition to using the state revenue, Fath 'Ali Shah granted *tuyul* (a land assignment or a grant to collect the taxes of a given area, or a specific tax) in lieu of salary, as benefice for service to the king or in exchange for the obligation to provide military contingents. According to Lambton, the dominant type of *tuyul* was a grant of land belonging to the royal domains. Furthermore, there was a strong tendency for land or taxation rights alienated by the state as *tuyul* to become hereditarily transmitted, even alienable by sale, and converted into private property. This tendency, which prevailed unchallenged in the second half of the century, was manifest in the vehement opposition to the forcible assertion of the theoretical right of the state to repossess *tuyul* by Haji Mirza Aqasi—referred to by one contemporary writer as "the destroyer of the notable"—in the late 1830s and 1840s.[46]

Mention has been made in passing of Nader Shah's order for confiscation of all religious endowments. Under the Safavids, these endowments were very extensive, and much of the distribution of benefices to the local notables and Shi'ite clerics, as well as the financing of the institutions of learning, drew on this important resource, which as far as I can tell, is not included in the figures from the *Tazkirat al-Muluk* that I have used for government revenue. The extent of Nader's actual confiscation of the endowments is not clear. However, as it occurred in the last year of his reign, it resulted in much usurpation of religious endowments by their administrators during the ensuing period of anarchy, at least in the Isfahan area.[47] This can only have meant an increased claim by benefices and pensions on state revenue from taxation. As endowed land under state control decreased considerably, the royal domains (now referred to as *khaleseh* instead of *khasseh*) assumed increasing importance among the resources of central government. Thanks to large-scale confiscations by Nader and Aqa Mohammad, Fath 'Ali Shah began his reign with extensive royal domains, comparable as a source of revenue to the *khasseh* under the Safavids.[48] Furthermore, the cornerstone of the centralizing policy of Haji Mirza Aqasi, the prime minister of Mohammad Shah (who reigned 1834–48), consisted in the extension and improved cultivation of the royal domains.[49]

In crucial contrast to the Safavids, the Qajars failed to centralize and regularize the administration of the royal domains. Under Naser al-Din Shah in the second half of the nineteenth century, the royal domains were "for the most part in a state of decay and made little contribution to the revenue."[50] During the last ten years of his reign, Naser al-Din resorted to sale of *khaleseh* land on an appreciable scale, a practice that was transformed by his successor into a means of dispensing favors during the closing years of the nineteenth century.[51] At the beginning of the twentieth century, less than 5 percent of government revenue came from the royal domains;[52] the percentage could not have been much higher in 1876.[53]

In addition to its greater fiscal weakness, the Qajar polity differed from the Safavid in regard to military organization. Throughout the nineteenth century, Qajar government intermittently but constantly pursued two goals: to break the military power of the tribes and to create a standing army; they succeeded in the first but failed in the second. Fath 'Ali Shah's success in subduing the tribes has already been mentioned. Under his successors Mohammad and Naser al-Din, "the power of the tribes was further reduced and the authority of the central government asserted."[54] The tribes remained under the obligation to supply the state, on demand, with irregular cavalry whose number was theoretically about 20,000 in the middle of the nineteenth century.[55] However, we find no systematic integration of tribes into the Qajar political and military organization comparable to the Safavid pattern. On the other hand, several attempts were made to create a standing army. These foundered because of the fiscal weakness of Qajar government. Rawlinson, who was himself instrumental in pursuing both the goals of the Qajar state, commented on the consequences of the success of the first and the reasons for the failure of the second:

> To a nation devoid of organization in every other department of government a regular army was impossible. . . . [In the attempt to create a standing army] the resources of the country were lavished on the army to an extent which grievously impoverished it; above all, the tribes, the chivalry of the Empire, the forces with which [Nader] overran the East, and which, ever yielding but ever present, surrounded, under [Aqa] Mohammad Khan, the Russian armies with a desert, were destroyed.[56]

As was the case with taxation, the century opens with a rosy picture but darkening gloom sets in after Iran's defeat by Russia in 1813. Although the supply of slaves was somewhat replenished by Aqa Mohammad Khan's invasion of Georgia, the Safavid slave corps of *qollar* could not be reconstituted. Furthermore, the use of royal slaves as governors was rare and had practically disappeared in the second half of the nineteenth century. However, in the first decade of the nineteenth century, the slave soldiers, *gholaman,* figure prominently in the Shah's guard, which according to Malcolm, was composed of 3,000 to 4,000 men from "Georgian slaves, and sons of the first nobles of Persia."[57] In mid-century, the royal bodyguard consisted of 2,500 "excellent horsemen" and in 1890 of 2,000 men.[58] In the first decade of the nineteenth century, the crown prince, 'Abbas Mirza, created a standing army of 12,000. Though it continued to exist on paper after the Perso-Russian wars, it disintegrated and attempts to revitalize it in the 1830s and 1840s had no lasting effect.[59] It was not until 1879 that the so-called Cossack regiments were set up under Russian officers. These were the only effective military force at the disposal of central government other than the king's bodyguard. By 1890, there were two Cossack regiments of 560 privates, 13 noncommissioned and 25 commissioned officers each. These constituted what came to be known as the Cossack Brigade, changed to the Cossack Division in 1916.[60] It is true that the Shah's son Zill al-Soltan, the governor of the provinces of western and southern Iran, had an efficient private standing army that numbered 15,800 men in 1886.[61] However, this very fact made Naser al-Din Shah so apprehensive about the potential for rebellion that he ordered most units of Zill al-Soltan's army to be disbanded and deprived him of all his governorships except that of Isfahan in February 1888.

Having surveyed the fiscal and military bases on which it rested, we can now turn to the nature and goals of government in Qajar Iran. The government was unmistakably patrimonial in Weber's classic typology. The kingdom was an extension of the household of the ruler and subject to his personal authority. The ruler's authority was delegated to officials who represented the royal person in administration. There was a natural tendency to delegate administrative authority to the members of the royal family and to personal servants (*nokar*) and stewards of the royal household.[62] Qajar princes were appointed to provincial governorships, and if they were minors, a tutor was put in charge of their instruction and government transactions. A *vazir* was also appointed to assist the prince and look after the interests of the central government. Each prince set up a provincial court, which was a replica of the Shah's court in Tehran. This practice points to a striking feature of Qajar patrimonialism, or more precisely, of the political economy of royal polygamy. Aqa Mohammad was a eunuch but his successor, Fath 'Ali, made up for his infertility. The humorous historian, Rostam al-Hokama, having enumerated Fath 'Ali Shah's one hundred and two sons and given each a real or fictitious title, proceeds to remark: "Let it not be hidden to scholars that many of these princes each have ten, twenty, thirty, forty, fifty, sixty, seventy and eighty children and offsprings in the name of God. Furthermore there are as many female children [to the king] as male, may God preserve them from affliction."[63] The English traveler Fraser, commenting on the expensive mode of providing for the royal progeny, points out that "every chief town and district, nay, every petty *ballook* and considerable village, comes to be assigned to one or other of these royal scions. . . . Nor is this all—each of these princes, taking an example from his father or grandfather, must have a huge harem filled with women of all sorts—a perfect hotbed of profligacy. . . ." Fraser proceeds to deplore the financially exhaustive and morally corrupting consequences of "the sensuality and vice of a race of royal drones, the most profligate and depraved, and the most noxious to their country, that perhaps any land and age have ever produced."[64]

Under Qajar patrimonialism, the kingdom was quite literally an extension of the royal household and the harem. This is clearly demonstrated in the revealing diary kept by the courtier E'temad al-Saltaneh in the 1880s and 1890s. Amineh Aqdas, the Shah's favorite wife in his later years, frequently intervened in administration of the state on behalf of influential individuals.[65] To reclaim the regiment assigned to a cousin of his, the general Shehab al-Dawleh took sanctuary (*bast*) [*sic!*] at the doorstep of the same Amineh Aqdas, incidentally, a Kurdish peasant who had entered the harem as a servant. He got his regiment back.[66] Other individuals approached the harem favorites, male and female, or the influential midget eunuch, Agha Mohammad, to gain favors from the Shah.[67] In 1887, Naser al-Din Shah's favorite, Malijak II, a boy of eight, received the commissions of brigadier and then general. In the subsequent year, he was given charge of a cavalry regiment of 500 while his younger brother was given the rank of brigadier.[68] A figure like the blind Shaykh Asadollah, who was the Koran cantor (*qari*) and instructor of the ladies of the harem (also a multiple secret agent who reported to the Shah on the public opinion, to prime minister Amin al-Soltan about the Shah and the harem, and vice versa) must be counted among the most influential men of the realm.[69]

Delegation of authority on the basis of personal and family ties did not stop at the royal level and pervaded all central and provincial administration. The prime minister also appointed sons, relatives, kinsmen, and stewards to the most important offices. The personal nature of authority entailed the prevalence of widespread nepotism and created a strong hereditary tendency. Some of the more notorious instances of inheritance of office are to be found in the latter part of the century when minors occupied such important offices as that of the minister of finance and *imam jom'eh* of Tehran, and Naser al-Din's last prime minister, Amin al-Soltan, inherited forty offices from his father, the head of the royal kitchen.[70] Furthermore, each important official maintained a host of retainers.[71] The personal nature of authority and utter insecurity of tenure also fostered the development of patron-client networks. There was a tendency for the weak to attach themselves to the strong, and the strong needed to secure enormous revenues to attract clients by their openhandedness and hospitality.[72] The personal nature of authority and the ubiquity of clientism explains the curious fact that the high officials were at once very wealthy and at the same time heavily in debt in order to maintain their influence by lavish entertaining, gifts, and large entourages.[73] Clientism, intrigue, and insecurity of office also go a long way toward explaining the inability of Qajar government to pursue consistent policy and the tremendous influence of the British and Russian missions among the notables, many of whom sought to put themselves under the protection of Britain and Russia.[74]

As in other patrimonial governments, office was not distinct from property. Offices were often sold and farmed, and office holders treated them as property. Furthermore, the property of the official, as a rule accumulated as a result of tenure of office, was not immune from confiscation and mulcting upon dismissal.[75]

In theory, the purpose of the Qajar king, as the Shadow of God on Earth, was to rule with justice, provide internal security, defend the country against external enemies, and promote the prosperity of the subjects. In practice, the twin goals of Qajar government can be said to have been the maintenance of the royal household and taxation and its distribution to notables, government officials, Shi'ite clerics, and others as benefices for past and present services to the ruling House. This assertion can be sustained by the figure for government expenditure given in Table 5 (see Appendix). It should also be pointed out that distribution of benefices was fully sanctioned by the ethos of patrimonialism. By bestowing such benefices, the ruler became the benefactor (*vali-ne'mat,* literally "the lord of benefit") and the recipient became bound to him for having "tasted his salt," obligated to be loyal to the ruler's house and to pray for its prosperity and everlasting rule (*du'a-gu'i*).[76]

Not only did the Qajar state enable the official class and their retinue to enrich themselves from tax collection, tax farming, subdelegation of offices, and grants of land as *tuyul,* it also distributed about one-quarter of its annual revenue in the royal household and among the notables as pensions and allowances. As can be seen in Table 5 (see Appendix), pensions (*vazifeh*), or annuities (*mostamarri*), as distinct from salary (*mavajeb*), constituted an important component of government expenditures, accounting for some 15 percent. These carried with them no office service as obligation, were usually granted in perpetuity, and were subject to a scramble among the heirs, relatives, and competitors upon the decease of the recipient.[77] Furthermore, the enormous expenditure on the army included a large amount of revenue for officials and the notables. This very substantial distributive component of the military budget consisted of the difference between the cost of maintaining the nominal strength of the army in the government accounts and that of the actual number of men serving and paid. During the last years of Naser al-Din's reign, his son Kamran Mirza, the Regent, who was also the governor of Tehran and minister of war, received money to maintain a nominal army of over ninety thousand when the actual number under arms was between fifteen and thirty thousand.[78]

In sharp contrast to the distribution of government revenue to the royal household and the class of officials and notables stands the miniscule expenditure on national education. Colleges (*madrasas*) were financed from the existing religious endowments or new endowments by princes, officials, and notables as individuals. The state spent virtually nothing on public works, hospitals, and social services.[79] What there was in the way of these services was supported by charitable endowments by the individual princes and notables, by the Shah, and by other individuals such as wealthy merchants.

Emergence of the Modern Idea of the State and Failure of Centralizing Reforms

Against this background let us examine the incipient attempts at the reform and modernization of the state. Some comparisons with Western Europe are helpful at the start. In his concise and magisterial essay on state formation, Barker points out that, in the seventeenth century, the European state was still not distinct from "family, property and society."[80] From about 1660 onward for some two centuries, there followed the "disengaging of the idea of the state as a service-rendering organization." The institution of absolutism and the proclamation of national sovereignty were the two hallmarks of this gradual process of disengagement. Fiscal factors were of considerable importance in the institution of absolutism. The *intendants,* the instruments of absolutism in seventeenth century France, began as temporary tax commissioners but gradually became permanent administrators controlling a whole range of economic, but also judiciary, matters. An analogous development took place with the Prussian Councillors of Taxes (*Steuerräthe*), who were transformed into administrators of towns.[81] No parallel development can be found in Iran. There, the functions of the tax commissioners (*mohassels*) remained restricted to *ad hoc* exaction of taxes in the eighteenth century. They became noticeably less conspicuous in the nineteenth century.

Crucial to the process of disengagement of the state from dynastic, proprietary, and social interlinkages was the development of the *idea* of the state. The Hohenzollern rulers of Prussia, who assumed the title of King in 1701, built their absolutism around the idea of impersonal devotion to duty of state. In Prussian theory, the King was the "carrier of State power," and "the first servant of the state." The contribution of the French Revolution to the process was the establishment of the doctrine of national sovereignty as the foundation of the state. With Napoleon's repeated endorsement by national plebiscites, the ruler of the modern nation-state also became "the first representative of the nation."[82] These ideas, central to the disengagement of the idea of the state from dynastic rule and patrimonial government, began their impact on Iran's political ethos early in the nineteenth century.

The sociopolitical consequences of the incorporation of Iran into the world economy was negligible before the very end of the nineteenth century. By contrast, the consequences of the incorporation of Iran into the modern international system of sovereign states were considerable. The most important of these was the reception of the modern idea of the state. This is not to say that internal developments were irrelevant. The collapse of the Safavid dynasty after two and a half centuries created a serious and unresolved crisis of legitimacy that continued until the very end of the eighteenth century. This prolonged crisis of legitimacy fostered the growth of the abstract idea of "the Iranian State" (*dawlat-e Iran*) and helped disengage it from the concept of dynastic rule as a God-ordained "turn" (*dawlat*) in power.[83] Nader, having served as a lieutenant of the Safavid princes (and nominal kings), Tahmasp II and 'Abbas III, had himself "elected" king in 1736 by a nation-wide gathering modeled after the Mongolian tribal pattern. He also sought to revive the legitimacy of the rule of the Turkman tribe. Despite these efforts, his dynasty lacked legitimacy and his one descendant who did rule in Khorasan until the last decade of the century was forced to lay heavy emphasis on his maternal descent from the Safavids. Even the Qajars at first claimed to be of Safavid descent and gave up this pretention only when Fath 'Ali Shah was fully assured of an alternative legitimation by a close alliance with the Shi'ite hierocracy.[84] Meanwhile, the rulers of Iran (most notably Karim Khan Zand, who never assumed the title of Shah) occasionally sought to legitimatize their rule as "servants of the state (*dawlat*) of Iran." This abstract idea of the state developed hand in hand with the growing notions of Iranian nationality. Nader's formula for disengaging his rule from God-ordained Safavid dynastic sovereignty is highly significant. The inscription on his seal is as follows: "The Seal of State and Religion having been displaced [an allusion to the collapse of the Safavid dynasty], God has given order to Iran in the Name of Nader."[85] The formula, though somewhat confused, indicates a shift of emphasis from religiously legitimized dynastic rule of the Safavids as the propagators and protectors of Shi'ism to the destiny[86] of Iran. In the Zand period the shift of emphasis became much stronger and the implicit notion of Iran's destiny became linked to that of serving "the Iranian State." Quite possibly with a touch of anachronism, 'Ali-Mardan Khan, one of the triumvirate of Bakhtiyari-Zand coalition of 1750, which ruled in the name of a Safavid minor, is reported to have said, "We do not claim sovereignty. We are the servants of the Iranian State (*khidmatgozar-e dawlat-e iran*)."[87]

Karim Khan Zand, who assumed the titles of Representative of the State (*vakil al-dawleh*) and Representative of the Subjects (*vakil al-ra'aya*) in preference to king, was more emphatic in his references to the Iranian State he claimed to represent, and in one instance is reported to have specified "service to the Iranian State" as the sole criterion for the receipt of salaries and pensions from the government.[88]

These internal developments were undoubtedly important. Nevertheless, it should be noted that most of the instances of references to the Iranian State and to the ruler as its representative are in the diplomatic context, in remarks to the British emissary or those transmitted to the Ottoman Sultan. In the nineteenth century, diplomatic representation of the powers in Iran became permanent, and Iran was incorporated into the international system of sovereign states. This incorporation greatly enhanced the notions of the state and service to the state. The modern conception of the state as an impersonal organization in charge of the commonweal of the nation became more current and more clearly understood.

Qa'em-Maqam, the elder, a Zand bureaucrat, entered the service of the Qajars, became the minister of the crown prince, 'Abbas Mirza, and conducted the ultimately abortive military reforms of the first decades of the nineteenth century in Azerbaijan. His son, Mirza Abo'l-Qasem Qa'em-Maqam, who succeeded him as the crown prince's minister and became the prime minister of Mohammad Shah in 1834, was an opponent of absolute arbitrary authority of the ruler and a strong advocate of service to the state. On the occasion of what he considered excessive royal largesse, he opposed the king and reprimanded him, saying: "We both have authority in the service of the Iranian State and are not entitled to more than one hundred thousand *tumans* of the subject's money. You are greater than me in the service of the state. If you wish to be in charge of the hospitality of the nation of Iran, appoint yourself to that task and take eighty thousand *tumans,* leaving twenty thousand to me. Otherwise, I will undertake this function and you should graciously make do with twenty thousand *tumans.*"[89] When cutting many of the pensions and benefices, Qa'em-Maqam ridiculed a basic tenet of the ethos of patrimonialism that justified the distribution of benefices by reportedly saying: "the state needs soldiers, not *du'a-gu*s (those praying for the perpetuity of the dynasty in exchange for their benefices)."[90] The new political ethos was unpalatable to the king, who ordered his prime minister to be executed after some eighteen months in office. Qa'em-Maqam's successor, Haji Mirza Aqasi is, according to *Sadr al-Tavarikh,* the first grand vazir to be designated "The First Person of Iran" (*shakhs-e avval-e iran,* that is, the first servant of the state)."[91] Significantly, it was the representatives of foreign states (*doval-e kharejeh*) who first designated him thus. A decree of Naser al-Din in November 1872, establishing some ministries and a cabinet, designated the prime minister as the "First Person of the State" responsible to the king.[92] By the end of the century, the use of the title First Person for the prime minister became routine.[93] The decree of August 5, 1906, which granted Iran a parliament, referred to the prime minister as the First Person of the State. Later in the twentieth century, this designation is modified to the "First Person of the Nation" (*mamlekat*) and applied to the king.

The incorporation of Iran into the international system of sovereign states induced attempts at modernizing the state, which in Iran as in the Ottoman empire began with efforts to modernize military organization and create a standing army. To underscore the importance of the international influence, it is worth pointing out that the Ministry of Foreign Affairs (set apart from all others in Table 4 on government expenditure) was the first to become a hierarchically organized and highly differentiated department on the basis of functional division of labor.[94] Furthermore, beginning with Amir Nezam in 1848, the impetus to the modernization of the state and administrative rationalization of other departments came from men in the diplomatic service.[95] The international influence also softened the political ethos of patrimonialism and put an end to brutal practices necessitated by the excessive personalism of the system of delegated authority. Although previous prime ministers such as Haji Ebrahim, Qa'em-Maqam, and Amir Nezam had been executed upon dismissal, usually with their relatives and dependents, by 1873 Naser al-Din was concerned that Iran might look "uncivilized and barbarian" in European eyes.[96] When the king dismissed the reforming prime minister, Moshir al-Dawleh, in September, he remained alive; henceforth, Iranian prime ministers could count on that after dismissal from high office.

Even though the idea of the modern state gained currency and became the ideal of the reformers, repeated attempts to realize it by reorganizing and modernizing Qajar government ran into insurmountable obstacles and ended in failure. Inspired by the abortive military reforms of the Ottoman Sultan Selim III, the first reforms to modernize the Iranian state began in Azerbaijan by the crown prince and regent, 'Abbas Mirza, and his minister, Mirza Bozorg Qa'em-Maqam, during the first decade of the nineteenth century. The defensive character of this modernization is evident from the complete primacy of military organization, from their historical context—the Perso-Russian wars—and from their geographical concentration on Azerbaijan, the region on the border of Czarist Russia. The measures to reorganize the military system were referred to as *nezam-e jadid* (the New Order). It consisted in the creation of a new standing army of 12,000, organized and trained according to modern military methods, first by fugitive Russian officers and then by French, British, and other European advisors. This standing army was supported by 24,000 tribal horsemen and by irregular volunteers in wartime. 'Abbas Mirza developed military industries and sent students to England to be trained as military engineers and technicians. He also built Western-style fortresses under the supervision of a French engineer.[97] To sustain his military reforms, the crown prince and his minister undertook a number of centralizing measures along the traditional lines. Lands belonging to the *khaleseh* (royal domains) were identified, surveyed, and registered, and repair of the irrigation network of connected wells (*qanat*) was undertaken to increase government revenue from water dues, to be charged after an initial period of grace.[98] New judges were appointed and an attempt was made to centralize the judiciary system through a *divan-khaneh* (the ruler's bureau of justice); tax assessments were to be submitted for approval to the same organ.[99] To improve communication, hostels and carriage houses were set up on major roads.[100]

'Abbas Mirza's reforms ended in failure after Iran's devastating defeat by Russia in 1827. Nevertheless, their importance is attested by the permanent imprint on the Persian language: the term *nezam* (order) and its derivative, *nezami,* have come to mean "army" and "military," and the designation he introduced for the new infantry, *sarbaz* (he who is ready to lose his head), has become the common word for "soldier" in modern Persian.[101]

The most serious attempt to reform and centralize the state was undertaken by Naser al-Din Shah's first prime minister, Mirza Taqi Khan Amir Nezam from 1848 to 1951. As was the case with 'Abbas Mirza and Qa'em-

Maqams, the cornerstone of Amir Nezam's modernization program was military reorganization.

Amir Nezam sought to impose the obligation to supply soldiers—or the equivalent of their wages—by geographical area instead of granting land (*tuyul*) in exchange for the provision of contingents. This measure was aimed at assuring the loyalty of soldiers to the central government, rather than to the landlords, and at rationalizing recruitment. He also upgraded the fortifications, built guardhouses, forbad billeting and the customary direct exactions by soldiers and, to some extent, purged the army accounting books of pay for fictitious troops.[102] The nominal strength of the army—that is, the potential mobilizing number—appears to have been increased from under 93,-000 in January 1849 to over 137,000 in January 1852 (94,750 infantry, 23,419 cavalry). The effective strength of the army must also have been increased considerably because Sheil estimates it at about 70,000.[103] Military reforms were as usual supported by increased centralization. Amir Nezam fixed the young Shah's personal allowance at 2000 *tumans* a month, cut pensions, undertook reassessment of land taxes, established tax auditors directly responsible to him, and collected overdue taxes.[104] The *divan-khaneh* was given increasing prominence in the judiciary system and attempts were made to control the religious courts indirectly.[105] Last but not least, Amir Nezam set up an extensive network of secret agents in order to increase the control of central government.[106]

Amir Nezam's military reforms were effective immediately and had a marked effect on improved internal security. Their effect, however, was not lasting and gradually eroded. By 1890 the nominal strength of the army was just over 90,000, but Curzon estimated the actual number serving to be one-third of this figure.[107] Nor was soldier recruitment rationalized: Landlords, especially tribal landlords, continued to be relied on for supplying contingents. Cities did not supply soldiers for the army. Soldiers were haphazardly conscripted for life, badly and irregularly paid, and dismissed for part or most of the year. The cavalry usually sold the fodder allotment for their horses and let their beasts graze in the meadows.

Real gains were made in strengthening central authority by Amir Nezam's reforms. Furthermore, improved communications, especially the creation of telegraph lines in the second half of the century, put provincial governors within instant reach of central government and greatly enhanced its authority. Reforms were halted for the best part of a decade with the dismissal of Amir Nezam, but Naser al-Din then carried out a number of successive administrative reorganizations in 1858, 1864, and 1866. These set up six centralized ministries, eliminating the position of prime minister and putting the ministers under the direct control of the Shah, but they proved unsatisfactory and Naser al-Din reverted to appointing Mirza Hosayn Khan Moshir al-Dawleh as prime minister in 1871. His reforming ministry lasted from November 1871 until September 1873. Reforms in the Ministry of Justice came first; they had in fact already begun in March 1871. Building on two earlier decrees issued in 1858 and 1862, Moshir al-Dawleh continued the effort to centralize the judiciary system by strengthening the central *divan-khaneh*. A court of appeals, an executive court, a legislative court to draw up regulations, and three specialized courts for dealing with criminal, commercial, and property cases were established in the Ministry of Justice. Other measures sought to regularize and rationalize the judiciary process.[108] No radical reorganization of the army was attempted, and the reforms were confined to preventing peculation and other abuses along traditional lines. On the other hand, an attempt was made to distinguish salary (*mavajeb*) from pension (*mastamarri*) sharply. Salary was to be paid to the office and not, as had been the practice, to the person, and an administrative order by the prime minister early in 1872 stated that the crown had no special relationship to sons, kin, and government.[109] In 1876, the Shah abolished the *qarasuran* corps, which were controlled by provincial governors and local notables, thereby limiting the governors' power and increasing central control of the armed forces.[110]

Given the familial nature of authority under patrimonialism, it is not surprising that, when Moshir al-Dawleh denied pensions to families of officials after their death, he was called immoral by his critics among the nobility. Furthermore, contrary to Moshir's reformist proclamations and intent, offices continued to be sold, especially to provincial governors who drew no salary. Nor did the crown cease to have a special relationship to sons, kin, and government.[111] Naser al-Din also set up a Council of State, modeled after Napoleon's, which functioned intermittently from 1871 to 1891. This consultative body often met with no agenda and it was difficult to stop story telling and irrelevant discussions. The Shah also sought to promote functional division of labor among the ministries, but government continued to be plagued persistently by interference of one powerful official in the departmental domain of another.[112] Frustrated by these difficulties, Naser al-Din became increasingly indifference to administrative reform by the 1880s.

The one lasting result of attempts at reorganizing government administration was expansion of the central bureaucracy, which in the latter part of the century dwarfed that of the Safavids'. In the absence of increased sources of revenue, this set the stage for the bankruptcy of the Iranian state in the 1890s and 1900s. After 1890, there was rapid deterioration of state finances and fragmentation of power. To secure cash through *pishkesh* (gift), the Shah increased the turnover of governors, who tended to grant tax remissions and new pensions, especially to the 'ulama, in order to forestall disturbances in the provinces to which they were newly appointed.[113] The Shah appointed a number of children with sonorous titles to high office. In addition to provincial governorships, departments of state were farmed out. There were flagrant court scandals and widespread demoralization. The Shah's harem grew to comprise some two hundred wives, and the court spending became reckless, while the treasury was run as a private bank by the prime minister and his brother. Tax collectors became more oppressive and there was a general breakdown of order caused by the *lutis* (local toughs), urban riots, and brigandage in the countryside.[114] Thus, the gains made in strengthening the authority of central government during the early years of Naser al-Din Shah's reign were eroded in his last years. The conditions deteriorated further under his successor, Mozaffar al-Din Shah (1896–1907), especially after the premiership of Amin al-Dawleh from 1897 to 1898 and the failure of his proposed reforms,[115] and Iran entered the twentieth century with a disintegrating central authority and a financially bankrupt central government.

The imposing presence of the imperialist powers, Russia and Britain, in the closing decades of the nineteenth century offered the crisis-ridden Iranian state alternatives for warding off insolvency other than increased centralization and more efficient taxation: granting concessions to foreign companies in exchange for royalties and foreign loans. From 1890 onward, these alternatives, especially loans, were used to prevent bankruptcy. Consequently, by the eve of the Constitutional Revolution in 1906, the Iranian state had borrowed £800,000 and 32,500,000 rubles (about £3,250,000)—or three times its annual revenue—from Britain and Russia.[116] Foreign debt hurt national interest by greatly increasing Iran's dependence on the imperial powers and concessions were injurious to the economic interests of the Iranian mercantile class. Once the Qajar government resorted to these measures, it cemented the alliance between the Shi'ite hierocracy and the urban mercantile class, who as champions of national interest both opposed the state.

It can therefore be said that in the nineteenth century no real progress had been made in disengaging the state from dynastic and patrimonial encumbrances, and transforming it into a service-rendering organization under national sovereignty. The task awaited the twentieth century.

2

The Constitutional
Revolution: 1905-1911

As we have seen, from the last decade of Naser al-Din Shah's reign the Iranian state decayed and its armed forces became dilapidated. We now turn to the significant growth of the civil society, its political awakening, and to the expression of that awakening, the Constitutional Revolution.

During the nineteenth century, Iran became more integrated into the world economy, although to a lesser extent than Egypt and the Ottoman Empire. From 1800 to 1913, the volume of Iran's foreign trade increased twelvefold in real terms, as compared to a fifty- or sixtyfold increase in Egypt, a fifteen- or twentyfold increase for Turkey, and a fiftyfold increase in world trade.[1] The penetration of Iran by foreign capital was also much smaller than Egypt, the Ottoman Empire, and India. Nevertheless, the increasing incorporation into the world system had a significant impact on Iran's economy, notably the decline of domestic handicraft, growth of cash crops, and after 1870, production of carpets for exportation.[2] The impact of this process on social stratification and class formation remained negligible until the last decades of the nineteenth century, but by the 1880s and 1890s, the configuration of Iran's urban society was somewhat modified by the rise of wealthy entrepreneurial merchants and, less significantly, a small working class consisting overwhelmingly of migrant workers employed in the neighboring regions of Russia. Thus, "to the traditional classes of landlords, military, officials, merchants, craftsmen and peasants, the country's economic development added the nucleus of two new ones: an industrial, commercial and financial bourgeoisie and an industrial working class."[3]

Until the closing decades of the nineteenth century, the urban strata—craftsmen, shopkeepers, and merchants—did not constitute a nationwide political force. This changed under the impact of a set of factors that broke through the self-sufficiency of urban communities and increased national integration in the last quarter of the century: introduction of new telegraph lines and a postal service, improved roads, and internal security under Naser al-Din Shah; increased international trade; and the appearance of newspapers.[4] The merchant communities in Baku, Istanbul, India, and elsewhere played an important role in the political awakening of Iran's civil society, as did the newspapers published in these foreign centers.[5] Furthermore, education of individuals with private means in Europe and the educational reforms of the late Qajar period, feeble though they were, produced a small but vocal intelligentsia that advocated reform and modernization of the state, with a significant number of its members seeing parliamentary democracy as the essential prerequisite for modernization. This group included a number of enlightened bureaucrats whose crucial role in constitutional reforms of the state will be discussed at length.

The social background of the intelligentsia at the turn of the century was undoubtedly diverse and included clerical, bureaucratic, landowning, and mercantile elements.[6] But this diversity of social background did not prevent their unification on the basis of a single ideology comprised of the philosophy of the Enlightenment, the Victorian conception of progress, and the political ideas of nationalism and of parliamentary democracy. Nor did it prevent the intelligentsia from acting as the agent of mobilization and political enfranchisement of the growing civil society on the basis of that same ideology.

Thus, the Constitutional Revolution of 1906 to 1911 was both a nationalist revolution and a democratic revolution and has commonly been recognized as such. This characterization, however, does not do justice to the teleology of the Constitutional Revolution in that it leaves out a primary goal—*the* primary goal for many of the participants—of that revolution: the reform of government, creating a strong state capable of overcoming Iran's backwardness. The Constitutional Revolution also represented the nationwide, concerted political action of the urban mercantile strata, acting as a class for itself, and has accordingly been interpreted as a bourgeois revolution. However, this does not do full justice to the dynamic of the Constitutional Revolution any more than the previous interpretation. It should be clear from our picture of Iran's social structure in Chapter 1 that, in the early decades of the twentieth century, civil society, though growing in economic importance, was nevertheless quite small and weak. The mercantile bourgeoisie could not act effectively without seeking support from the hierocracy, and the urban alliance of the mosque and the bazaar could not fail to draw the military might of the tribal periphery into the political arena. In fact, civil society was swamped before long by the forces in the existing structure of hierocratic and political power that reacted to its revolution, and with which it had to compromise ideologically and practically.

A new interpretation is offered in this chapter to do justice to the neglected aspects of the teleology of the Constitutional Revolution. This interpretation highlights two distinct elements in the teleology of the Constitutional Revolution and the inchoate, disparate reactionary movement that these elements set in motion. These components mark the three overlapping phases of the Constitutional Revolution into which this chapter is divided.

Revolution and Parliamentary Democracy: April 1905–June 1908

Parliamentary representation and constitutional government constitute the first important element of the teleology of this twentieth century political revolution in Asia. This element, which properly informs the revolution's designation as constitutional, originated in the discussions of the political societies formed among the intelligentsia during the preceding years. Except for the earlier sporadic activities of the free masons, the first appearance of semi-secret groups dates from the end of Naser al-Din Shah's reign. These associations became known as "national societies" (*anjomanha-ye melli*) and persisted into the twentieth century. About 1903, a Secret Center was organized among the members of a group associated with the Public Library and the periodical *Ganjineh-ye Fonun* (the Treasure of Crafts) in Tabriz. In Tehran, a secret society was formed around the National Library in May 1904, subsequently referred to as the Revolutionary Committee; and in February 1905, another noteworthy association, the Secret Society, was formed in Tehran. The programs of the two Tehran societies variously emphasized the evils of tyranny and the benefits of the rule of law, the desirability of the form of government found in the progressive nations and the necessity of the reform of the state, the government, taxation, the army, and the judiciary system.[7] The activities of these societies played an important part in canalizing growing discontent in Iranian society and in giving direction to the popular protest movement that later broke out in reaction to chronic misgovernment.

We know that revolutions can occur as a result of a heightened sense of injustice and relative deprivation when things are changing for the better. This insight, which we owe to Tocqueville,[8] should however be interpreted and applied carefully. Economic life in Iran at the turn of the century was getting better, but there was no improvement whatsoever in political life, which may in fact have been getting worse. Misgovernment, tyranny, and injustice may have been as chronic in earlier periods in the nineteenth century but they could *not* have been worse. We may therefore assume that the level of absolute deprivation in the political sphere remained constant. What the accompanying economic improvement, increasing international contacts, and political awareness did was to make the unchanging dismal political life intolerable and thus created a revolutionary situation.

Against a background of recurrent bread riots, fresh instances of misgovernment in the provinces, and despotic rule of the new governor of Tehran,[9] public discontent over the Shah's wasteful impending trip to Europe and the arrogant demeanor of the Belgian director of the Customs Administration, Monsieur Naus, set in motion a wave of agitation by merchants that activated an alliance between the hierocracy and civil society. On April 26, 1905, a group of bazaar merchants, grocers, and money changers took sanctuary in the shrine of 'Abd al-'Azim near Tehran.[10] The conflict between the merchants and the governor of Tehran and the Director of Customs remained recurrent and unresolved throughout the summer and fall. In mid-December 1905, a much larger group, including the leading *mojtaheds* of Tehran and their clerical retinue and students, took sanctuary in 'Abd al-'Azim and demanded the dismissal of Monsieur Naus and the creation of a governmental house of justice (*'adalat-khaneh-ye dawlati*).

Inclusion of the House of Justice among the demands represented a victory for the advocates of the rule of law and constitutional government. The manner in which this inclusion was brought about is instructive: it shows how a small group of intellectuals, which included only one *mojtahed,* influenced the general direction of the protest movement over and above the particularistic goals of different groups of its supporters. The original list of seven demands, to be submitted by the clerical leaders of the protest to the Shah through the good offices of the Ottoman ambassador, were concrete and highly specific. Two items, the dismissal of Naus and of the governor of Tehran, can be taken as the specific goals of the mercantile class. Three items reflected the particularistic interests of the clerical estate only, and two other items were outright trivial. The Constitutionalist educator, Yahya Dawlat-abadi, who was in charge of transmitting the message to the Ottoman ambassador, has recorded his astonishment at the short-sightedness and lack of concern for any general reform expressed by these demands. He took the initiative in adding a "general proposition," which eventually took the form of the demand for a house of justice.[11] This demand was accepted by the Shah in January 1906, and it became the focus of subsequent agitation that led to another gathering of merchants and clerics in July, this time taking sanctuary at the British legation. Meanwhile, the *mojtaheds* of Tehran and their entourage had left the capital for Qom in protest against the expulsion of a preacher and the killing of a seminarian. In this instance, another prominent Constitutionalist intervened to orient the direction of protest movement. When the protestors had gathered at the British legation, "deputations were sent secretly to Sani-ed-Dawleh, who formulated for them the idea which they had only vaguely formed of a Constitution with the object of reform."[12]

These two interventions were decisive in determining the teleology of the Constitutional Revolution. On August 5, 1906, the Shah issued a *farman* (decree) to the new prime minister, ordering him to set up a national consultative assembly (*majles-e shura-ye melli*):

> we do enact that an Assembly of delegates elected by the Princes, the Doctors of Divinity (*'ulama'*), the Qajar family, the nobles and notables, the landowners, the merchants and the guilds shall be formed and constituted, by election of the classes above mentioned, in the capital Tehran; which Assembly shall carry out the requisite deliberations and investigations on all necessary subjects connected with important affairs of the State and Empire and the public interests.[13]

For the teleology of the Constitutional Revolution to unfold without hinderance, the Constitutionalists had to transcend not only the particularistic goals of the clerical estate and the mercantile class, but also the constant preoccupation of the masses with bread and meat.

When the news of the granting of a Majles by the Shah reached the city of Tabriz, which was governed by the absolutist crown prince, in September 1906, the Constitutionalists in Tabriz overcame their trepidations and organized public rallies in favor of the constitution. In one such rally, according to the diary of the Constitutionalist Shaykhi cleric Theqat al-Eslam,

The people were asked: "What do you want, cheap bread and meat? Or do you have another objective?" The mass which was privy to no knowledge said yes, we want cheap bread and meat. It was suggested once more to them: "Do you want constitutional government (*mashruteh*)?" This time they said yes three times in a loud voice.[14]

In the following days, Theqat al-Eslam reported intense debates on the meaning of constitutional government, and a cleric's suggestion that such government be termed *mashruteh mashru'eh* (from Sacred Law *shar'*, to emphasize and assure its conformity with the Sacred Law).[15]

The elections began and the Majles was opened in October 1906, without even waiting for the provincial deputies to arrive. Delegates began work immediately on the Fundamental Law of the constitution, which was completed and ratified by Mozaffar al-Din Shah on his death bed on December 30, 1906. Iran had been granted a constitution, and the National Consultative Assembly set up with very little bloodshed.[16]

The ratification of the Fundamental Law on December 30, 1906, did not mean the secure achievement of the first goal of the Constitutional Revolution, namely the enlargement of the political society through representation under the principles of rule of law and national sovereignty. In fact, the subsequent eighteen months were a period of protracted struggle between the Majles, which championed the principle of national sovereignty, and an increasingly diffident new monarch. We have already mentioned the secret political societies that had come into being in the few years prior to 1906. After the grant of a parliament in August 1906, there was no longer any need for secrecy, and a large number of popular political associations (*anjoman*s) came into being. In addition to these popular *anjoman*s, the Constitutionalists considered representative participation in official *anjoman*s in the form of provincial assemblies (*anjomanha-ye ayalati va velayati*) and municipal councils (*anjomanha-ye baladi*) an essential aspect of democratic government. Laws regarding these assemblies were among the earliest enacted by the First Majles in April 1907, and the essence of the law on provincial assemblies was included in a special section (Articles 90–93) in the Supplementary Fundamental Law of October 7, 1907.[17]

The spontaneous growth of the popular political associations alarmed the new monarch, Mohammad 'Ali Shah, who unsuccessfully sought to suppress them.[18] The Shah's hostility and the fear that he may revoke the Constitution in turn resulted in the mushrooming of popular associations, intensification of their activity, and their progressive radicalization between May and October of 1907, by which time some 130 of them were active in Tehran alone.[19] Meanwhile,

the authority of the Central Government had been reduced to almost nothing and could only be exercised through the provincial *anjoman*s. . . . The moral authority of the old regime in the provinces had been destroyed by the provincial *anjoman*s; and the framework of such elementary administration as had once existed had almost disappeared.[20]

The chaos created by the absence of clear demarcation between the functions and authority of the executive, the Majles, and the *anjomans*[21] also alarmed some of the Constitutionalists who began to perceive the goal of increased popular political participation as being in contradiction to the other main goal of the Constitutional Revolution, namely the reform of the state.[22] Nevertheless, popular political societies and provincial assemblies must be considered an important component of the constitutionalist movement and the expression of one of its primary goals. After open hostilities between the Shah and the Majles in mid-December 1907, political societies began to recruit and arm nationalist militia. Before long they had some 2000 men, and there were plans for the creation of a national guard of 200 men to protect the Majles.[23] The extent to which power had passed to the *anjomans* by the end of April 1908 is demonstrated by their success in releasing the persons arrested by the police for attempting to assassinate the Shah on February 28, 1908.[24]

When the Shah bombarded the Majles and restored autocracy in June 1908, political societies and provincial assemblies were banned throughout the country. The provincial assembly of Tabriz, however, defied the monarch and carried on the torch of constitutionalism during the following months. (Its resistance was finally broken by the entry of Russian troops on April 29, 1909.) As Mohammad 'Ali Shah's restored autocracy gradually foundered from January to July of 1909, provincial assemblies were revived one after another in Isfahan, Hamadan, Tonekabon, Rasht, Shiraz, Mashhad, Kermanshah, and Bushehr.[25] However, after the deposition of Mohammad 'Ali Shah and the restoration of constitutional government in July 1909, the activities of popular societies paradoxically declined.[26] This was due primarily to the increasing prominence of state building as a Constitutionalist goal in the second phase of the Constitutional Revolution. It was also in part due to the attempt by Naser al-Molk, the new regent, to channel institutionalized political participation into political parties. In March 1911, he put considerable pressure on the Majles deputies to form two political parties, and on the prime minister to present the "program" of the majority party to the Majles for formal approval.[27] In less than a month, a two-party system was instituted in the Majles, and the parties—the Moderates (*E'tedaliyyun*) and the Democrats— drew away the intellectuals who could supply the guiding spirit of popular political societies.[28]

In the long run, the political parties declined, as had the political societies and provincial assemblies, and it was the Majles alone that, as the organ of national sovereignty, represented the realization of the democratic goal of the Constitutional Revolution. Furthermore, certain important corollaries of the principle of national sovereignty became permanently established by the First Majles. Articles 18, 22, 23, 24, and 25 of the Fundamental Law asserted the authority of the Majles, as representatives of the nation, over the natural resources of the country, concessions, foreign loans and economic treaties, and government expenditure. Articles 94 and 96 of the Supplementary Fundamental Law established the control of the Majles over taxation.

Acceptance of the principle of ministerial responsibility to the parliament was gradually forced on a begrudging government during the first months of 1907 and successfully put to test at the end of April when the Majles brought down the cabinet of the acting prime minister, Vazir Afkham.[29] The principle was then written into the Supplementary Fundamental Law (Articles 60–67).

Two early goals that the First Majles failed to achieve are also worth mentioning, as they, too, were implicitly conceived as economic corollaries of the principle of national sovereignty: the establishment of a national bank and the freeing of financial administration from control of foreign personnel.[30] The Constitutionalists' persistent demand for the dismissal of Naus were finally met in February 1907, but before the year was out they had by necessity proposed the hiring of a French financial advisor. As we shall see, a much greater role had to be assigned to foreign financial advisors after the restoration of constitutional government, and the government had to resort to Swedish officers in the military reorganization planned in 1911.

Constitutional Reforms of the State: 1907–December 1911

This brings us to the second element of the teleology of the Constitutional Revolution, which is as crucial as it is neglected in the historiography of the revolution: the goal of reform and centralization of the state. From the very beginning, the Constitutionalists included a number of important advocates of reform and centralization.[31] These enlightened bureaucrats were joined by other sympathetic Qajar bureaucrats once constitutional government became established. The most notable members of this group were Sani' al-Dawleh and his brother, Mokhber al-Saltaneh, Naser al-Molk, and Vothuq al-Dawleh in the younger generation. The attempts by this group to reform the state ended in failure. However, failure to build a strong state has obscured unduly the significance and consequences of the early reforms of the Constitutionalists. The result has been a one-sided and, therefore, deficient understanding of the teleology of the Constitutional Revolution and of the counterrevolutionary trends set in motion from 1907 onward. To rectify the picture, let us survey the early attempts at reform of government and the obstacles they encountered.

It is clear from the debates of the first months of the Majles that financial reform was the constant preoccupation of the Constitutionalists. It was generally considered the most important objective of the Majles after the fundamental laws of the constitution.[32] The Constitutionalists would have concurred with W. Morgan Shuster, the American financial advisor they hired five years later, that "Persia's sole chance for self-redemption lay with the reform of her broken finances."[33] The task, needless to say, was a formidable one. As far as sources of government revenue, there was little practical value in the theoretical right of the state over the land assigned as *tuyul,* which seemed irretrievably alienated, and the *khaleseh* land, as we have seen, had diminished as a result of repeated sales. The perennial problems of tax collection were aggravated by the disintegration of central authority as a result of the confrontation between the Shah and the Majles. On the expenditure side, the wasting of foreign loans for nonproductive purposes during the preceding decade, such as the monarch's extravagant trips to Europe, meant that the government now had to service an enormous debt. By 1910, servicing this debt was absorbing four-fifths of the revenue from Customs, the most secure source of government revenue.[34] The army accounts were as inflated as ever, and constituted a huge demand on government revenue. In 1911, Shuster estimated the cost of maintaining an efficient army of 15,000 men at two million *tumans*. "Yet the annual amount demanded by the Ministry of War, which could not muster 5,000 rugged and underfed troops in the entire Empire, was 7,000,000 tumans!"[35] The payment of "pensions" constituted the other perennial financial problem. It was slightly aggravated by the fact that the Majles had adopted the vice of the old patrimonial system and awarded pensions to its own servants and friends.[36]

> One of the most remarkable examples of Persia's peculiar financial chaos was this system of "pensions." According to the loosely kept records of the different Ministries the Government was expected to pay out each year to nearly 100,000 different people throughout the Empire the sum of about 3,000,000 tumans, in money and gain.
>
> The greater part of this strange burden had been inherited by the Constitutional Government from the regime of the former Shahs. Some pensions had, however, been decreed by the Medjlis, to priests and others who had served the Nationalist movement, and to the relatives of men who had been killed while fighting for the Constitution.[37]

However, this dismal picture should not obscure the fact that the objective conditions for increasing the fiscal resources of the state had become more favorable from the 1880s onward as a result of expansion of foreign trade and growth of the urban economy. The development of trade creates the possibility of levying indirect taxes, to quote Ardant, "Various kinds of entry taxes, duties on internal trade, and tariffs were important in the fiscal system of the nineteenth century: the ease in levying this kind of relatively painless tax was one of the bases for the growing power of states."[38] Expanding trade can, therefore, remove a powerful constraint on the growth of centralized states, namely the minimal tax potential of agrarian societies. In 1722 (the year of the fall of the Safavid empire), the British government had a much easier task of collecting taxes than the Iranian one because only 25 percent of its revenue came from land tax; during the second quarter of the nineteenth century, land and property tax consistently accounted for only 10 percent of its revenue while customs and excise accounted for three-quarters of it.[39] Turning to Iran, for most of the nineteenth century land taxes and revenue from the royal domains appear to have accounted for nearly 90 percent of the state revenue and customs for almost all of the rest. After about 1880, the picture began to change. As Table 6 in the Appendix shows, the state's dependence on land tax was greatly reduced in the twentieth century (to under 25 percent), and more easily collectible revenue from customs and other sources came to account for the bulk of its revenue. A variety of customs farming practices were terminated and the administration of customs was given to Belgian functionaries in 1898. Customs Administration became the government's most efficient source of revenue.

It took fifteen years for Iran's state builders to be able to take advantage of these more favorable conditions to increase the fiscal resources of the state. All attempts by the Constitutionalists between 1907 and 1911 ended in failure because of three main factors: the difficulty of forming a strong government under an unsympathetic monarch; the opposition of imperialist powers, especially Russia; and the vested interests of the grandees of the old regime who had accepted the constitutional government.

It is useful to distinguish three phases in the fiscal reforms. The first phase, extending from the fall of 1906 to the middle of December 1907, is dominated by Naser-al-Molk, minister of finance until September 1907 and prime minister and minister of finance from October 25 until the crisis of mid-December 1907, when he was detained by the Shah, who reportedly spared his life only after the intercession of the British. In this phase, the Finance Committee of the Majles, chaired by its first vice president, Vothuq al-Dawleh, played a vigorous role. The second phase, interrupted by the restoration of autocracy and civil war, is dominated by Sani' al-Dawleh, minister of finance from February 29 to May 31, 1908, and from October 30, 1910, until his assassination on February 6, 1911. The third phase consists of the reforms of Morgan Shuster as the treasurer-general of Iran from May to December 1911.

The search for a strong personality to carry out the state-building effort first led the prominent Constitutionalists to Amin al-Soltan, Atabak-e A'zam, who had served both Naser al-Din Shah and Mozaffar al-Din Shah as prime minister. Amin al-Soltan was unpopular among the radicals and political societies because of his career under the ancien régime. However, he was strongly supported by leading figures of the mercantile, clerical, and bureaucratic factions of the Constitutionalist movement,[40] who managed to rally the overwhelming majority of the Majles deputies behind him.[41] Amin al-Soltan claimed his journey to Japan had converted him to the cause of constitutional government and sought to present himself as the only man capable of building a strong state. In April 1907, a few days before his arrival from Europe, his supporters published a plan for administrative reform, "written and sealed by Amin al-Soltan twenty years ago."[42] He had, in fact, proposed reforms during his last term as prime minister in 1903—including proposals to abolish some pensions, to put provincial governors on regular salaries, and to deflate the army payroll by cutting the nominal strength of the army by one half.[43] Amin al-Soltan was appointed prime minister early in May 1907. However, Mohammad 'Ali Shah feared Amin al-Soltan,[44] and the radicals hated him. He was trapped between the Scylla of a distrustful monarch and the Charybdis of hostile radical political societies and was assassinated on August 31, 1907.[45] Meanwhile, significant measures for financial reform were being taken by the Finance Committee of the Majles and by the finance minister, Naser al-Molk.

These measures were continued with greater vigor when Naser al-Molk himself became prime minister in October 1907.

Naser al-Molk had begun his career as a reformer as the minister of finance under Amin al-Dawleh in 1897–98. He had centralized the collection of all revenues—from the land tax, Customs, the post, and so forth—and disbursement of all salaries in the Treasury, and had issued new directives for tax collection. He was also responsible for putting the administration of Customs under Belgians,[46] under whom "the principle of salary against work was established."[47] Naser al-Molk's new method of tax collection, however, was discontinued after the fall of the reforming cabinet of Amin al-Dawleh in 1898. Naser al-Molk returned to the Ministry of Finance in September 1903 and had resumed his reform measures by the onset of the Constitutional Revolution.[48]

Once the Majles was instituted in the fall of 1906, financial reform assumed a high priority on its agenda. In November 1906, the Majles asked the government to submit an annual budget for approval. In December, a Finance Committee was set up and given the task of preparing a budget and general plans for financial reform. It was chaired by Vothuq al-Dawleh, who had been a tax superintendent (*mostawfi*), and Taqizadeh, a radical deputy from Tabriz, who later became one of the Majles's most active members.[49] The Finance Committee was responsible for a number of important measures to increase the revenue of the central government. Its chairman reported frequently to the Majles from March 1907 onward and submitted a bill on the *tuyul* and taxation in early May.[50] It abolished the practice of *tas'ir* (conversion of taxes in kind to cash [on the basis of antiquated rates]) and made *tafavot-e amal* ("difference of the operation," the amount collected over the entries in the tax books) illegal, while proposing to incorporate it into new official tax figures. However, it is interesting to note that the Majles did not dare assert the theoretical right of the state to all *tuyul* by simply recalling all land assigned as such. Instead, it abolished *tuyul* as an institution, compensating the *tuyul*-holders by an annual stipend.[51]

Naser al-Molk became prime minister in late October 1907 with strong support from the Majles. The Finance Committee completed its assignment during his premiership. In November 1907, the first budget in Iranian history was submitted to the Majles and overwhelmingly approved. It incorporated the last set of financial measures the Finance Committee had been working on. The committee had examined tens of thousands of pensions individually. In the November budget, some of the more spectacular cuts—in the pensions of the princes—were announced. More spectacular still, the annual expenditure of the Shah's court was reduced by over a third to 500,000 *tumans*.[52]

The crisis provoked by these financial reforms will be mentioned in the next section. It led to the departure of Naser al-Molk from Iran but did not put an end to the constitutional reforms of the state. The next cabinet was formed on December 21, 1907, by the octogenarian Nezam al-Saltaneh whose function was to reconcile the Shah and the Majles. The dominant figures of the reformist cabinet were the Hedayat brothers, Mokhber al-Saltaneh, the minister of justice who initiated the reform of the judiciary system, and, especially, Sani' al-Dawleh, who held the portfolios for the ministries of education and public welfare and became minister of finance while retaining public welfare after the cabinet reshuffle of February 29, 1908. Sani' al-Dawleh differed from the preceding reformer, Naser al-Molk, in personality and education. Mokhber al-Saltaneh said of Naser al-Molk, with characteristic terseness: "Because he is educated at Oxford, he cannot make decisions."[53] Sani' al-Dawleh, by contrast, was decisive. He had been educated in Germany and was familiar with the German writings on public finance and the welfare goals of the state. Once work on the fundamental laws of constitutional government was complete, Sani' al-Dawleh had resigned as the president of the Majles (September 6, 1907) to devote himself to the task of the reform of the state. On November 1, 1907, he submitted a tract on the aims and fiscal characteristics of the modern state to serve as the basis for the reform of constitutional government and the modernization of the state. In this short but brilliant tract, published under the title of *The Road to Salvation* (*Rah-e Nejat*), he argued that the promotion of public welfare and the economic transformation of Iran required a fiscally strong state. He related Iran's backwardness and, conversely, the progress of the Western nations and even the Ottoman Empire to their respective levels of per capita taxation. Increased taxation would give the state the resources necessary for the promotion of the public good: for the effective maintenance of law and order, provision of internal security, construction and maintenance of roads and channels of communication and commerical intercourse, and the provision of the education for its citizens in general and technical schools in particular:

> Therefore, for the inhabitants of any country to live with ease and felicity, four things are necessary: First, military and security forces to protect the property and life of people without intermediaries. Second, a judiciary system to protect the inhabitants from wrongdoing to one another through fraud. Third, the establishment of an educational system to teach the inhabitants gainful occupations. Fourth, roads for the provision of necessities from near and far.[54]

Sani' al-Dawleh put much emphasis on government investment in the infrastructure, and gave the highest priority to the construction of a trans-Iranian railway.[55] He promised to propose concrete means for financing it to the Majles.

He had the chance to do so when he became the minister of finance in 1908. He proposed a tax on sugar and tea to finance the railway construction and to build roads. He also submitted a second bill proposing new taxes on cultivated land and on real estate in cities. One half of the amount collected in urban taxes was to be spent on the creation of primary schools and on municipal services.[56] The tax on sugar and tea was opposed by the imperialist powers, Russia and Britain, who found it in contravention of the Treaty of Golestan, concluded with the Russians in 1813 after Iran's defeat. Furthermore, during the last day of May, while the Law of Pensions was being passed, Sani' al-Dawleh came under the personal attack of Taqizadeh and some of the radicals, despite broad support for his proposals.[57] At this juncture, the aged prime minister, Nezam al-Saltaneh, resigned. Sani' al-Dawleh attempted to form a cabinet. The fate of his ambitious scheme for fiscal reform was tied to this attempt. The attempt failed in the first days of June 1908;[58] and the tide changed. Three weeks later, the Shah bombarded the Majles.

Sani' al-Dawleh had a second chance when he became finance minister in the reform cabinet of Mostawfi al-Mamalek at the end of October 1910. He drew up a plan for the consolidation of Iran's foreign debt, which was implemented after his death,[59] and revived the idea of the tax on sugar and tea. Sani' al-Dawleh eventually had to settle for a tax on salt, a domestically produced item which was not covered by the Treaty of Golestan but whose tax was far less lucrative and more difficult to collect.[60] Furthermore, he entered negotiations with the Germans for the construction of railways.[61] On February 6, 1911, he was assassinated by two Russian subjects who subsequently left Iran under Russian protection.[62]

Despite their pronounced nationalism, it became clear to the Constitutionalists that fiscal reform could not be achieved without the help of Western financial advisors. They retained the Belgians other than Naus for the administration of Customs and chose a Frenchman, Monsieur Bizot, as financial advisor. Bizot was proposed to the Majles in October 1907 and approved in January 1908. He became the financial advisor to the government of Iran with special responsibility for the creation of a national bank. Bizot distinguished himself by inactivity with regard both to his general and special responsibilities. He did, however, draw up a plan for the reorganization of state finances.[63] After the restoration of constitutional government, the leading advocates of reform and reorganization of the state were elected to the Second Majles; on November 30, 1909, the government proposed a program of reform that included Bizot's plan.[64] The plan, however, made little headway. In July 1910, the Democrats came to power for the first time under the premiership of Mostawfi al-Mamalek and, after the assertion of the authority of central government over the nationalist militia in a bloody clash in Tehran, pursued a much more ambitious reform program. As we have seen, they were joined by Sani' al-Dawleh, first as minister of public welfare and then as the minister of finance. Meanwhile, the foreign minister, Navvab, had been particularly active in the search for foreign financial advisors since August and eventually proposed an American team headed by W. Morgan Shuster. The Majles approved the employment of the Americans a few days before Sani' al-Dawleh's assassination early in February 1911.

Shuster arrived in Iran on May 12, 1911. He avoided contact with the Russian and British legations and associated himself closely with the Majles. The Majles in turn demonstrated its very strong support for Shuster by speedy approval of his proposals, most notably, the Law of June 13, 1911, which established the Office of the Treasurer-General of Iran as the central organization in charge of collection and disbursement of all government revenue.[65]

Shuster prepared proposals for the redemption of pensions, which as we have seen, the Majles had reduced but not dared to abolish, and for railway construction.[66] He also made a firm attempt to eliminate bogus military expenditures.[67] But his most significant measures concerned the collection of taxes. For this purpose, he set up a Treasury gendarmerie, which was trained by four American officers and whose number, according to Shuster, reached 800 by November 1911 and 1100 by the end of that year.[68]

Shuster's success in collecting taxes and curbing bogus army pay enabled him to finance the campaigns of the Constitutionalists against the deposed Mohammad 'Ali Shah, who returned to Iran in July 1911. However, it also displeased the Russians, who demanded his dismissal and occupied northern Iran in November 1911. Furthermore, his vigor in tax collection had offended Mohammad Vali Khan, Sepahdar-e A'zam, the first prime minister under whom he served; the grandees, many of whom enjoyed the protection of the foreign powers; the provincial tax collectors, who had a vested interest in the weakness of central government; and even some of the reformers, such as Vothuq al-Dawleh, whose father was the tax collector of Tabriz.[69] Finally, his resistance to making payments for inflated military expenditures offended a few influential figures including the subsequent prime minister, Samsam al-Saltaneh, a grandee in his own right and a tribal chief who constantly demanded money for his Bakhtiyari troops.[70] He and his foreign minister, Vothuq al-Dawleh accepted the Russian ultimatum. The Bakhtiyari leaders staged a coup with their tribal horsemen and the newly created gendarmes; the Regent, Naser al-Molk, dissolved the Second Majles on December 24, 1911. Shuster was dismissed the next day.[71] Constitutional reforms of the state came to an end, and the effort to build a strong state was given up.

More clearly than in the previous years, the events of 1911 show Russian imperialism along with British acquiescence as a major obstacle to the reform and centralization of government. They also bring out the basic contradiction in early attempts to modernize the state: The Iranian civil society could attain the goals of its revolution only through the power structure of the ancien régime. To overthrow the restored autocracy of Mohammad 'Ali Shah in 1909, the Constitutionalists had to enlist the support of his chief marshal, Sepahdar-e A'zam,[72] and to draw in the military force of the Bakhtiyaris from the tribal periphery.[73] They themselves could only muster some 800 militiamen, who were put under Sepahdar's command, as compared to the 2000 Bakhtiyari horsemen.[74] The new ethos of constitutional government was not pleasing to the Sepahdar[75] and other grandees of the ancien régime who had accepted constitutional government. And the Bakhtiyari leaders, whose tribal troops sustained the constitutional government, wanted the spoils of office and clashed with Shuster when he resisted their "wholesale attempts at looting the treasury."[76]

Given the failure of the financial reforms to secure the necessary resources for the central government, it is not surprising that little was done by way of military reorganization. Military reorganization had been at the center of the modernization efforts in the nineteenth century and was naturally included in the aims of the Constitutionalists. The first session of the Majles was, significantly, inaugurated at the military academy in Tehran.[77] During his premiership, Naser al-Molk brought up the issue of military reform and complained of the insubordination of the Cossack Brigade. The efforts of his government to subordinate the Cossacks to the ministry of war, however, were unsuccessful.[78] The most important obstacle to their attainment was undoubtedly Russian control of the Cossacks, which continued even after the collapse of Mohammad 'Ali Shah's autocracy in 1909. In a gathering on January 13, 1911, twenty-three Iranian officers of the brigade[79] swore loyalty to the Majles. They were severely reprimanded by their Russian commander whose initial decision had been to discharge them.[80]

After the restoration of constitutional government in 1909, the most important military reform was the creation of the gendarmerie. Owing to the crucial importance of the Bakhtiyari troops for the constitutional government, there was also some discussion of the role of the tribal forces, and a bill to regularize recruitment of cavalry from the tribes was submitted to the Majles by Minister of War Sardar As'ad Bakhtiyari in February 1911.[81] But such compromise measures ran counter to the requirements of military modernization and the creation of a standing army, and they were doomed to failure. The plan for the creation of the gendarmerie, however, was implemented. Negotiations for the recruitment of several Swedish officers to set up and train an Iranian gendarmerie were completed in May 1911 and approved by the Majles. The Swedish officers arrived in the middle of August 1911 and began to organize the Iranian force effectively despite the opposition of Russia.[82] After Shuster's dismissal, his treasury gendarmerie was taken over by the Swedish officers and became the national gendarmerie and the only disciplined force other than the Cossack Brigade.

Judiciary reform began with considerable vigor during the first two months of 1908 by Minister of Justice Mokhber al-Saltaneh. He laid the foundation for a centralized hierarchical judiciary system comprising both the secular and the religious courts. The vigor of these measures alarmed the *'ulama,* who perceived the threat to their total control over the religious courts,[83] and alarmed the Shah after he heard a suit had been filed against his steward in Zanjan.[84] Pressure was put on Mokhber al-Saltaneh to relinquish the Ministry of Justice in the cabinet reshuffle at the end of February 1908.[85] The one significant legal reform after 1909 was the enactment of the Law of Registration of Deeds, passed by the Majles on May 11, 1911.[86] It sought to nationalize the haphazard registration of deeds with the *'ulama,* thereby setting in motion a trend that was to deprive the latter of an important function. More importantly, coming after the abolition of *tuyul,* it gave legal sanction to unconditionally held private property in land and, thus, constituted the legal basis for emergence of a new class of landlords who from now on would hold their titles independent of the state.

Finally, mention should be made of the constant preoccupation of the First Majles with bread and meat.[87] This preoccupation has often been ridiculed by the observers as a reflection of confusion regarding the respective functions of the executive and the legislative. Yet, together with the establishment of a national educational system envisioned in Article 19 of the Supplementary Fundamental Law and such measures as sending students to Europe by the Second Majles in 1911,[88] it formed a rudimentary idea of the welfare state, which was to develop later.

At the practical level, the efforts of the Constitutionalists did not change the character of the state, which remained geared to distribution with little concern for welfare or development. This is reflected in the expenditure estimates for 1911. While the army absorbed over a quarter of government expenditures, with the same proportion for pensions and the court, only some 2 percent of it was spent on education and welfare; development expenditure was truly miniscule. (See Table 7, Appendix.)

Its practical failures notwithstanding, the Constitutional Revolution radically altered the conception and purposes of taxation and, with this, of the purposes of the state itself. The principles of consent to taxation through the parliamentary mechanisms and of citizen participation in the taxation process became established. More importantly, provision of public services became the sole legitimate purpose of taxation. The state's accountability to the nation became irrevocably established. This last step completed the incorporation of the idea of the modern state into the political ethos of Iran.

Traditionalist Counterrevolution and Tribal Civil War: 1908–March 1912

The counterrevolution that set in from 1907 onward lacked coherence and consisted of a number of divergent trends that Mohammad 'Ali Shah tried to exploit in a fairly haphazard fashion. To make intelligible the heterogeneity of the forces of reaction and the complete lack of long-term coordination among its components, we need to distinguish between two phases of the counterrevolution. The counterrevolution's first phase began in 1907 and was briefly triumphant from June 1908 until July 1909 when it collapsed. It had a pronounced ideological character and its direction was shaped by an emergent ideology of Islamic traditionalism. It can be understood as a reaction to three factors: (1) direct attack on traditional authority by Constitutionalist militants, (2) challenge to traditional authority posed by the creation of the legal framework of constitutional government in the form of the Fundamental Law and its Supplement, and (3) the financial reforms of the Constitutionalists. The reaction to the first two factors set in motion the development of a rival traditionalist ideology by some of the leading members of the Shi'ite hierocracy. This ideology of Islamic traditionalism provided a platform, or at least a pretext, for all those who were alienated by the financial reforms of the Constitutionalists, including the Shah and the courtiers.

It was clear to Mohammad 'Ali Shah as he ascended the throne that constitutionalism could not be fought by a royalist ideology reviving the ethos of patrimonialism and Persian theories of kingship. His only possibility for mobilizing popular support lay in an appeal to Islam and to the hierocracy.[89] In this phase, the Shah controlled the Cossacks through the Russian officers, and the tribal forces of the periphery only played a secondary part.[90]

The second phase of the counterrevolution was completely nonideological. It was set in motion by the return of the now deposed Mohammad 'Ali Shah and the royalist princes in July 1911 and relied exclusively on the tribal forces that had remained loyal to the Shah. The movement was militarily defeated in September 1911, but its definitive end can be dated on March 10, 1912, when the ex-Shah finally left the Iranian soil for Russia.

Of these two phases, the traditionalist counterrevolution is by far the more significant, both because of its immediate impact on Iranian society and because it foreshadows the teleology of the Islamic revolution of 1979.

Implanting the teleology of the Constitutional Revolution in particularistic group interests had been fairly easy within the mercantile class, which quickly became convinced that law and order, security of property, and immunity from arbitrary power could all be achieved by importing parliamentary democracy from the civilized nations of Europe. With the clerical estate, the *'ulama,* it was much more difficult. They had to be persuaded that constitutional government would strengthen the hierocracy vis-à-vis the monarchy or, at the very least, that it would not affect hierocratic power adversely. Alternatively, they had to be pressured into joining the movement from fear of loosing their popular backing as a result of the competition of the Constitutionalist pamphleteers and publicists.

Both these mechanisms worked and hierocratic power was effectively exploited to establish the Majles in 1905 to 1906. However, with the ratification of the Fundamental Law on December 30, 1906, the Constitutional Revolution entered a new phase in which the functions and jurisdiction of the Majles were progressively defined. With this progressive definition of parliamentarianism, the hierocracy faced the necessity of reassessing its stand and clarifying its position. Thus, it was only after the signing of the Fundamental Law and during the debates on its Supplement that the crystallization of the attitudes within the hierocracy toward constitutionalism took place.

In July and August of 1906, while the *'ulama* were in Qom, the secular Constitutionalists had won a very important point: the initial wording, "Islamic Consultative Assembly" (*majles-e shura-ye eslami*), had been changed to "National Consultative Assembly" (*majles-e shura-ye melli*) in the final August 5 version of the royal decree that granted Iran a constitution.[91] This, however, had not passed entirely unnoticed. Before the return of the *'ulama* was secured, a confirmation of the decree had to be issued that referred to the "Islamic Consultative Assembly" whose constitution was to be in conformity "with the laws of the holy *Shari'a* [Sacred Law]."[92]

Signs within the hierocracy of dissatisfaction with the Constitution and the turn of events made their appearance on the wake of the ratification of the Fundamental Law. Already at the end of January 1907, a number of clerics were obstructing the elections of the deputies in the provinces.[93] In February, as the Majles began discussions of the Supplementary Fundamental Law, Shaykh Fazlollah Nuri, the most popular of the three *mojtaheds* of Tehran, emerged as the leading figure in opposition to the Constitutionalists. He set out to build a traditionalist constituency by capitalizing on the issue of the differential rights of the Muslims and the religious minorities, while at the same time seeking a *rapprochement* with the Shah.[94] Mohammad 'Ali Shah responded by demanding from the Majles that "the law be written in accordance with the Sacred Law of Mohammad (*Shari 'at-e mohammadi*)."[95] The opportunity offered by the ceremonies of Moharram was used by another cleric to preach against the Constitution.[96]

The *'ulama*'s dissatisfaction was as yet unfocused and their opposition inchoate. In March and April, peasants arrived at the Majles, complaining of oppression and extortion by two of the landowning clerics,[97] and very serious clashes broke out between the Constitutionalists and the prominent *mojtaheds* of Rasht[98] and Tabriz. The expulsion of the latter, together with a number of other *'ulama,* made a particularly deep impression on the clerical estate.[99]

It took until some time in May for serious clerical agitation under the leadership of Shaykh Fazlollah Nuri to begin. He proposed his "principle"— subjecting all parliamentary legislation to the ratification of a committee of five *mojtaheds* of the highest rank—while his supporters continuously gathered outside the Majles to demonstrate and to intimidate the unsympathetic deputies. In addition, Nuri formed a political association (*anjoman*) with the exiled *mojtaheds* of Tabriz and Rasht. Other clerics formed similar associations in provincial cities. Nuri's principle was passed as Article 2 of the Supplementary Fundamental Law by the Majles in the second week of July 1907; not considering this parliamentary victory sufficient, he took sanctuary in the shrine of 'Abd al-'Azim.[100]

From the very beginning, there had been an element of reaction against centralization in the popular movement of 1905–6. The hierocracy was annoyed by one of the measures taken by Naser al-Molk in his financial pre-Revolution reforms: the imposition of a stamp duty on the pensions paid to the clerics by the state. Sayyed 'Abdollah Behbehani and the other protesting religious leaders demanded the abolition of this stamp duty in December 1905.[101] More serious were the repercussions of financial reforms of the Majles in 1907, especially the abolition of *tuyul,* conversion of taxation in kind, and the examination and reduction of pensions. The abolition of *tuyul* in May 1907 alienated the *tuyul*-holding landlords from the Majles.[102] In Nuri's Islamic traditionalism, these dispossessed landholders found a convenient ideology for opposing Constitutionalism. Early in June, the cleric Naqib al-Sadat Shirazi organized a series of religious meetings (*rawzeh-khani*) as the platform of opposition to the Majles by the *tuyul*-holders in coordination with Nuri's activities.[103]

It was during the three months of sanctuary at the shrine of 'Abd al-'Azim, from mid-June to mid-September, that Nuri coherently formulated the clerical objections to parliamentarianism, thus transforming their common though unspecified orientation into a consistent traditionalist ideology. He did so in a series of open letters published and distributed from 'Abd al-'Azim that became known as the *Ruz-nameh* (*Journal*) of Shaykh Fazlollah.

Though the opening of schools for women and proposals to use funds allotted to religious ceremonies for the purpose of building factories and European (*farangi*) industries were not spared from attack, Nuri's chief principled objections to parliamentarianism and to the Constitutionalists centered around the following themes: "the inauguration of the customs and practices of the realms of infidelity; the intention to tamper with the Sacred Law, which is said to belong to 1300 years ago and not to be in accordance with the requirements of the modern age; the ridiculing of the Muslims and insults directed at the *'ulama;* the equal rights of nationalities and religions; spread of prostitution; and the freedom of the press which is 'contrary to our Sacred Law'."[104]

The primary aim of the campaign was "the protection of the citadel of Islam against the deviations willed by the heretics and the apostates."[105] The *'ulama* wished to awaken the religious brethen and rectify their misconceptions and errors so as to protect them from the heretical Babis and the materialists who were engaged in beguiling "the masses who are more benighted than cattle" (*'avamm, azall min al-an'am*).[106] We shall not tolerate, Nuri declared, the weakening of Islam and the distortion of the commandments of the Sacred Law.[107]

In marked contrast to the modernist stratagem of presenting Western political concepts and practices as embodiments of the true spirit of Islam, these proponents of Islamic traditionalism highlighted their imported and alien quality, stressing the Europeanness of the parliament and of the Fundamental Law.[108] "Fireworks, receptions of the ambassadors, those foreign habits, the crying of hurrah, all those inscriptions of Long Live, Long Live! (*zendeh bad*): Long Live Equality, Fraternity. Why not write on one of them: Long Live the Sacred Law, Long Live the Qur'an, Long Live Islam?"[109]

No one denied the desirability of a majles. What had been originally demanded from the state in 1906 had been a "majles of justice (*majles-e ma'delat*)" "so as to spread justice and equity and enforce the Sacred Law; no one had heard the National Consultative (*shura-ye melli*) or Constitutional (*mashruta*)." The Majles should not be contrary to Islam and should enjoin the good, forbid the evil, and protect the citadel of Islam.[110] But the Constitutionalists "want to make Iran's Consultative Assembly the Parliament of Paris. . . . We see today that in the *Majles-e Shura* they have brought the legal books of the European parliament(s) and have deemed it necessary to expand the law . . . whereas we the people of Islam have a heavenly and eternal Sacred Law."[111] Seeking to coin a concept for the system of government he envisages, Nuri proposed that the qualification *mashru'a* (conforming to the Sacred Law) be added to the term constitutional (*mashruta*). The notion of *mashruta-ye mashru'a,* which can loosely be translated as "Islamic Constitutionalism," is thus given currency.[112]

Nuri's propaganda caused quite a stir. It articulated an ideology for the opponents of parliamentary government whose interests were threatened by the proposed judiciary and financial reforms, *and* it turned against parliamentarianism a considerable number of those who had piously withdrawn from politics, or were marginally interested in politics. Parliamentarianism was now presented as a threat to the Islamic tradition. The impact of Nuri's *Journal* was immediate. It affected the religious craftsmen and shopkeepers, and many of the Majles deputies.[113] But above all, it gave definitive form to the orientation of the clerical estate toward parliamentarianism. It caused a split within the hierocracy. A few of the prominent *mojtaheds* who were firmly convinced of the virtues of parliamentary democracy supported the Constitutionalist cause unflinchingly. Others lent their support when convenient. But in all these cases, the sympathetic clerical support was henceforth implicitly conditional, and they too insisted on the supervisory veto power of religious authority, the restriction of freedom of the press, and the disavowal of any reforms entailing the secularization of the judiciary and educational system.

One of the most important consequences of Nuri's propaganda was the mobilization of the piously apolitical *'ulama* against parliamentarianism. This brings us to Nuri's resounding success in Najaf, an interesting and neglected fact in the history of the constitutional revolution. The news of the events of 1905 to 1906 had an electrifying effect on the torpid atmosphere in the *madrasas* (religious colleges) of the holy cities in Iraq. Journals and political literature poured in from India, Egypt, Lebanon, and Iran. The *tollab* (seminarians) were politicized overnight. Even the withdrawn and pious scholars who called themselves "the Army of God" (*jond Allah*) joined the Constitutionalists. Two of the foremost Shi'ite dignitaries, Akhund-e Khorasani and Hajj Mirza Hosayn, supported by the less eminent Shaykh 'Abdollah Mazandarani, responded affirmatively to the general enthusiasm and, benefitting from the independent telegraph line of the East India Company, turned Najaf into an important center for the transmission of pro-Constitutional injunctions. The constitutionalist seminarians soon gathered around Akhund-e Khorasani, an eminent scholar whose reputation in jurisprudence was not matched by his small following and scanty financial resources as a *marja'-e taqlid*. The other religious authorities, most notably Sayyed Mohammad Kazem Yazdi, remained neutral and refused to intervene in politics.[114]

By the time Nuri's son was in Najaf to enlist support, Yazdi had come under constant pressure from his politicized students to intervene in politics on behalf of the Constitutionalists. He persistently refused to do so, and the students' pressure turned into intimidation and threats of assassination. At this point, Yazdi responded to Shaykh Fazlollah's call, dispatched his son to the shrine of 'Abd al-'Azim, and spoke out against Constitutionalism. He was supported by his Arab followers, the Ottoman governor of the province, and "the Army of God." Clashes between his supporters and those of Khorasani ended in the former's complete victory. Constitutionalist seminarians were molested, especially by Arab tribesmen who were followers of Yazdi. Yazdi emerged as the undisputed master of the holy cities. "The religious party pulled back from the Constitutionalists. The local Arab supporters of autocracy and the uneducated Iranians forced Sayyed Mohammad Kazem Yazdi out of neutrality and made him the bearer of the banner of autocracy."[115] According to Kasravi, a few thousand Iranians and Arabs would pray behind Yazdi every day, whereas no more than thirty were to be found following Khorasani at the daily public prayer.[116]

Meanwhile, clerical support for Nuri was growing in Iran itself. The alliance of the mosque and the bazaar, forged in common opposition to the state in 1905, was falling apart. Many clerics had assumed important, and often leading, positions in the provincial and municipal assemblies outside of the capital. As the power shifted from the central government to these assemblies during the summer and fall of 1907, divergences of opinion between the clerics on the one hand, and the merchants and tradesmen on the other, came into the open and caused frequent clashes.[117] Consequently, many of the provincial clerics were becoming increasingly distant from the Constitutionalists. It is therefore not surprising that an increasing number of prominent *'ulama* were pledging their support to Nuri. Furthermore, 'the court-connected clerics were joining forces with him, linking his party to the Shah and the absolutist nobility.[118] Let us note, in passing, that in the clerical anti-parliamentarian party of 1907 and 1908, we encounter not only staunch reactionaries who had even opposed the demand for a House of Justice, but also the clerics whose humiliation and banishments were among the precipitating causes of the protest movement of 1905–1906.[119]

Nuri's success in organizing concerted political action with the Shah and the other anti-Majles factions and groups in 1907 and early 1908 was much less impressive than his feat of formulating the ideology of Islamic traditionalism. Mohammad 'Ali Shah and the nobility made their peace with the Constitutionalists immediately after Amin al-Soltan's assassination on August 31, 1907, leaving the clerical party out of the deal. The deal, however, did not work. As we have seen, Naser al-Molk's budget of November 1907 cut considerably the stipends of the princes and the amount allotted to the Shah's court. Furthermore, the Shah was coming under the vituperative attack of the Constitutionalist press and began to look again for active support from the clerical party. Many had rallied to the Shah in reaction to the abolition of *tuyul* and conversion of in-kind taxes[120] and to the measures regarding pensions that probably affected as many as 100,000 persons and their families. In the middle of December 1907, when the Shah arrested Naser al-Molk, servants, grooms, and muleteers of the court, protesting their pay cuts, staged disorderly demonstrations with Shaykh Fazollah's followers against the Majles. Such an array of distinguished *mojtaheds* amidst the arak-drinking ruffians of Tehran during the demonstrations in *Tup-khaneh* Square, does not give one the impression of a well-planned and well-organized political action. In any event, it is clear that Mohammad 'Ali Shah changed his mind about the planned coup d'etat at the last minute, without prior consultation with his clerical allies, and even without the slightest concern for enabling them to save face.[121] Nuri's part in the event and his association with the unsavory hirelings of the Shah earned him the *takfir* (excommunication) of Khorasani and his colleagues in Najaf and brought him into temporary disrepute.

The anti-Constitutionalist clerics had to wait six more months until the monarch would once again decide to confront the Majles. Meanwhile, Mirza Hasan Mojtahed and his retinue returned to Tabriz, swayed the bulk of the body of *'ulama* to the absolutist side, and formed the Islamic Society (*anjoman-e eslamiyyeh*), whose manifestos proclaimed the incumbency of holy war (*jihad*) on the grounds that the heretics intended to destroy Islam in the name of Constitutionalism.[122] A Society of the House of Mohammad (*anjoman-e al-e Mohammad*) with the slogan *"Al-e Mohammad mashruteh nemikhahad"* (the House of Mohammad does not want a Constitution) was formed by a son of the Constitutionalist *mojtahed,* Sayyed Mohammad Tabataba'i. Nuri pledged his support.[123] Mohammad 'Ali Shah asked his clerical confidante, Mirza Abu-Taleb Zanjani, to determine whether or not he should destroy the Majles through bibliomancy (*estekhareh*). Zanjani consulted the Koran, and here is his oracular reply: "Action must be taken in this matter. Victory is certain, even though there may be initial troubles."[124] The Majles was accordingly bombarded on June 23, 1908.

Early in July 1908, flanked by the *imam jum'ehs* of Tehran and Khoy, Mohammad 'Ali Shah proclaimed that he had closed the Majles to protect Islam and would reopen it in due course. On July 11, 1908, the Shah sent his carriage to bring Shaykh Fazlollah to the royal palace. Having been profusely honored by the monarch, Nuri spoke of the desirability of "Islamic Constitutionalism" (*mashruta-ye mashru'a*) and a limited *Majles,* but not chaos." On the following day, the triumphant Nuri issued an injunction excommunicating all journalists.[125] In mid-August, he proceeded to excommunicate the Constitutionalist *mojtahed*s of Najaf.[126]

In November 1908 came Nuri's chance to define the precise institutional translation of his Islamic (*mashru'a*) government. A week later, the date he had previously set for the convocation of the Majles, Mohammad 'Ali Shah reneged on his promise on the basis of Nuri's statement that constitutionalism was contrary to Islam and the election of a Majles would result in "the destruction of religion and the annihilation of the Islamic norms (*navamis*)."[127] Nuri, seconded by the state-appointed *imam jum'eh* of Tehran and on occasion other clerics, then sent a series of telegrams in late November to early December 1908 to the provinces announcing the inauguration of an alternative assembly: a Greater Consultative Assembly (*majles-e dar al-shura-ye kobra*) with fifty handpicked members.[128]

Nuri's supremacy made uneasy a number of the *mojtahed*s at the highest echelon of the amorphous Shi'ite hierocracy, while the renewed Constitutionalist activities in Istanbul and Najaf under the impetus of the Young Turks' Revolution spelled fresh trouble. In December 1908, some of the *'ulama* of Tehran, responding to renewed signals from Najaf and Istanbul, took sanctuary in the Ottoman Embassy. Another mollah, indignant that the Shah had not paid his debts in recognition of his part in the demonstrations of *Tup-khaneh,* took sanctuary in the shrine of 'Abd al-'Azim.[129]

Nuri reacted promptly to stem the tide of opposition to his Islamic regime and vowed to prevent the restoration of constitutional government as long as he lived. He acted vigorously to maintain the momentum of the anti-Constitutionalist movement and, as regards the manipulation of the clerical estates in Iran, with good chances of success. With Nuri's reaction, "the situation changed drastically."[130] But not for long, owing to factors outside of Nuri's control: Central government failed to assert its authority and rule effectively in the provinces, or to do anything about its insolvency. The economic conditions of the country deteriorated sharply,[131] and law and order broke down outside the capital.

In January 1909, Nuri was seriously injured as a result of an assassination attempt. To the news of the disturbances in Rasht and Mashhad were added those of the fall of Isfahan to the pro-Constitutionalist Bakhtiyaris. The Constitutionalist *mojtahed*s of Najaf issued an injunction forbidding the believers to pay any taxes to the state.[132] Out of prudence, the *'ulama* of Tehran shifted one by one to the Constitutionalist camp around the beginning of March 1909.[133] In April, 1909, the *imam jum'eh* of Tehran, like Mohammad 'Ali Shah himself, reminded his constituents of his well-attested sympathies for constitutionalism! A royal decree banishing Shaykh Fazlollah from Tehran was issued by the Shah but later revoked. Only the men of firm conviction and the fools among the mollahs stood firm and continued to condemn constitutionalism.[134]

In May 1909, Mohammad 'Ali Shah belatedly capitulated to the Constitutionalists, earnestly promising them—doubtless to Nuri's sorrow—a "parliament" (rather than the equivalent *Majles-e Shura*).[135] In July of that year, Mohammad 'Ali was overthrown. As for the anti-parliamentarian *'ulama,* the prudent escaped unharmed, withdrawing to their estates or making themselves otherwise inconspicuous, or being given refuge by Constitutionalist colleagues and friends.[136] A number of the imprudent diehards were killed by the Constitutionalist militiamen. Shaykh Fazlollah Nuri, having comported himself with great dignity during his captivity and trial, was hanged on July 31, 1909.[137]

The second phase of the counterrevolution—the rebellion of the tribes against the constitutional government in support of the deposed Shah in 1911—can be treated more briefly. If the Constitutionalists had drawn hierocratic power into their political struggle with the king, so, too, had Mohammad 'Ali Shah drawn the military force of the tribal periphery into his struggle against civil society and constitutionalism. Furthermore, just as the hierocracy was politically split, the naturally fractious tribes of the periphery did not uniformly support the king. In fact, one important tribe, the Bakhtiyaris, thanks to the Constitutionalist convictions of the educated chief 'Ali-qoli Khan, Sardar As'ad, made an alliance with the Constitutionalists and, as we have seen, played the critical military role in the defeat of Mohammad 'Ali Shah and the restoration of constitutional government in July 1909.

The tribes, however, had no particular love for constitutionalism. The Shahsevan under the leadership of Rahim Khan, who were fighting the Constitutionalists of Tabriz on behalf of Mohammad 'Ali Shah, never acknowledged the authority of the restored constitutional government and continued their depredations until Rahim Khan was finally defeated in January 1910.[138] Furthermore, the predominance of the Bakhtiyaris in Tehran[139] alarmed the rival tribes of Fars and Lorestan who vowed to fight what they perceived as an attempt "to change the dynasty and establish the absolute and unrivalled rule of the Bakhtiyaris."[140] In February 1910, the Qashqa'i chief, Sawlat al-Dawleh, the Lor chief and hereditary governor Posht-e Kuh, Sardar Ashraf, and the hereditary governor of Khuzestan, Shaykh Khaz'al, concluded a pact to check the growth of Bakhtiyari power. However, the participation of the energetic governor of the Fars province, Nezam al-Saltaneh, meant that the parties in the pact paid lip service to constitutionalism and pledged "to protect the Holy Majles."[141] Furthermore, the Kalhor Kurds in the west, the Turkman tribes in the northeast and the Baluch in the southeast were showing clear signs of open insubordination.[142]

Relying on tribal support in the west and the northeast, and preceded by his brother and his agents by some two months, the deposed Mohammad 'Ali landed at the Caspian port of Gumish Tappeh in the region of the Yamut Turkman tribe on July 17, 1911.[143] The news of his arrival alarmed the Constitutionalists, who turned to the Bakhtiyaris in Tehran. On July 26, the Bakhtiyari chief, Samsam al-Saltaneh became prime minister and minister of war, inaugurating the period of "Bakhtiyari domination," which lasted until his resignation in January 1913.[144] On the other hand, the Kalhor Kurds, the Lors and the Shahsevan in the west, and the Turkmans in the northeast declared their support for the former Shah. These tribes supplied his brother, Salar al-Dawleh, with an army of 6000 in the west, and Mohammad 'Ali himself with a smaller number of horsemen in the northeast, including a main army of 1000 Turkman horsemen.[145] The royalist forces thus clearly outnumbered the forces of the constitutional government, which consisted of the same 2000 Bakhtiyari horsemen and a small but very significant group of seasoned Armenian freedom fighters under the chief of Tehran police, Yefrem Khan.

There is no evidence whatsoever that Mohammad 'Ali thought of reviving Nuri's traditionalist platform, which he had cynically exploited before, nor any evidence that he approached the hierocracy to mobilize popular support. Given his popularity outside of Tehran[146] and the widespread disenchantment with the weak constitutional government, such an attempt might have made all the difference. Be that as it may, the traditionalist *'ulama* remained out of the picture and disillusioned about both sides. The outcome of the struggle between the former Shah and the constitutional government was decided swiftly in a series of battles in August and September 1911. Mohammad 'Ali had brought two machine guns with him across Russia, but they remained unpacked as he had no one to operate them. The Bakhtiyari and Armenian contingent of the constitutional government, by contrast, were supported by a machine gun operated by a German artillery instructor.[147] The decisive defeat of the former Shah's forces came on September 8, that of the royalist army of the west on September 27. "In both cases, the machine gun employed by the outnumbered government forces panicked the tribesmen and decimated their fleeing ranks."[148] Mohammad 'Ali remained in Iran for six more months. He began a new three-pronged offensive on the capital in February 1912, only to be told he would not be recognized by the Russians who occupied northern Iran and were now satisfied with the Bakhtiyaris in power. He left Iran for Baku on March 10, 1912.[149]

Compromises in the Constitutional Revolution

The traditionalist counterrevolutionary trend of 1907 to mid-1909 demonstrated the demise of the alliance between the bourgeois strata and the hierocracy, which had set in motion the protest movement of 1905–1906 against the state—a movement financed by the merchants[150] and led by the hierocracy. As we have seen, the hierocracy was split before long, and the majority of its members supported Nuri's Islamic traditionalism. What needs to be stressed now is that the attitudes of the *'ulama* who sided with civil society, too, became affected by the Islamic traditionalism of Nuri's counterrevolution. Though the uncompromising anti-parliamentarianism of Nuri during the restoration of autocracy was doomed by Mohammad 'Ali Shah's eventual failure, an earlier variant of his position, the advocacy of Islamic constitutionalism (*mashruta-ye mashru'eh*), was in effect accepted by the vast majority of the Constitutionalist *'ulama* as well. Full cognizance of every major point made by Nuri was taken by the Constitutionalist *'ulama,* in theory and in practice.

The earliest clerical pro-Constitutionalist tract known to me was written fresh from the influence of Nuri's propaganda in November or December 1907, and it took great pains to stress the restriction of parliamentary legislation to matters secular, to insist on the unimpaired judiciary authority of the Shi'ite hierocracy—restricting the jurisdiction of the Ministry of Justice to customary law (*'orfiyyat*)—and to underline the necessity of the clerical supervision of the press in the interest of religious orthodoxy.[151] The subsequent Constitutionalist tracts written by religious scholars do likewise.[152]

In practice, too, the political action of the Constitutionalist *'ulama* was premised on the observance of the principles specifically formulated by Nuri. Khorasani and other Constitutionalist *mojtaheds* fully supported Nuri's "principle," which was ratified as Article 2 of the Supplementary Fundamental Law. Khorasani also supported Nuri's unsuccessful demand for the inclusion of an article against heresy.[153] More vigorous still was the opposition of the Constitutionalist *'ulama* to judiciary reform. In September 1907, Sayyed 'Abdollah Behbehani was finally induced to accept Article 27 of the Supplementary Fundamental Law relating to the judiciary only after being threatened at gunpoint.[154] When the bill for reform of the judiciary came up for discussion during the last sessions of the First Majles in early 1908, the proponents of judiciary reform Hajj Yahya Dawlatabadi and Mokhber al-Saltaneh Hedayat incurred excommunication (*takfir*) by Behbehani and by another consistently Constitutionalist *mojtahed.*[155] Nuri's objections to the principle of equality of Muslims and the religious minorities also found some resonance among the clerical members of the First Majles.

In fact, after the execution of Nuri, the mantle of the traditionalist opposition to parliamentarianism, and to the secular state, fell upon the shoulders of the Constitutionalist *mojtahed* of Tehran, Behbehani. After the conquest of Tehran in July 1909, Sepahdar went out of his way to let it be known that the *'ulama* would henceforth be excluded from parliamentary government, as did other Constitutionalist leaders.[156] The Constitutionalist *'ulama* were quickly disillusioned with the restored parliamentary regime. It is worth emphasizing that at the time of his assassination on July 16, 1910, Behbehani intended to embark on a campaign of opposition against the irreligious Westernizers with the backing of the Constitutionalist *mojtaheds* of Najaf.[157]

Similarly, civil society had to compromise with the forces of the tribal periphery that it had drawn into the struggle against restored autocracy. The predominant Bakhtiyari presence in the capital, oppressively felt ever since its reconquest in July 1909, became the outright Bakhtiyari domination of July 1911 to January 1913. By 1912, the Bakhtiyari khans were not only completely dominant in the cabinet but also held seven important provincial governorships.[158] Nevertheless, as was the case with the concessions to the traditionalism of the Shi'ite hierocracy, constitutionalism ultimately prevailed. There was widespread apprehension that the Bakhtiyari coup d'etat of December 24, 1911, was the prelude to the establishment of a new tribal dynasty in place of the Qajars.[159] In the same fateful day of December 24, Sardar As'ad made the following disclaimer: "I desire to give every assurance that myself and my family do not in the least think of acquiring the Regency or of establishing a Bakhtiyari dynasty."[160]

The Bakhtiyaris were the last tribe to have the chance of establishing a new dynasty along the lines of the pattern discovered by the great medieval historian, Ibn Khaldun. That they did forego it demonstrated the triumph of constitutionalism. The Constitutionalist Sardar As'ad and other Bakhtiyari chiefs could not pursue the state-building reforms at the expense of the basis of their power—their own tribe—with any vigor. But a decade later, Reza Khan, a soldier who was to found the first and last important nontribal dynasty in over eight hundred years, could and did.

3

Formation of the
Modern Bureaucratic State
in the Twentieth Century

The Constitutional Revolution of 1906 did not succeed in setting up a strong modern state. On the contrary, as we have seen, it was followed by the restoration of autocracy and civil war and, finally, the Russian occupation of northern Iran in the last months of 1911. In sharp contrast to the French revolutionaries of 1789, the Russian revolutionaries of 1917, and the Iranian revolutionaries of 1979, the Constitutionalists of the first decade of the twentieth century did not inherit a centralized state. This fact goes a long way toward explaining the fifteen years of anarchy and disintegration that followed the Constitutional Revolution.

Central authority disintegrated during World War I, with one government and an adolescent king in Tehran; a pro-German provisional government in Kermanshah from July 1916 to March 1917, followed by Ottoman rule by proxy in Azerbaijan; the city of Kashan in the hands of the bandit Nayeb Hosayn and Isfahan similarly dominated by outlaws; and British troops and the British-organized South Persian Rifles in control of southern Iran after 1916. After the war, the government of Vothuq al-Dawleh took significant steps to eliminate the bandits and reassert the authority of government,[1] but central authority did not fully recover. In 1920, central government faced the rebellions of Shaykh Mohammad Khiyabani in Tabriz, of the Za'feranlu tribe in Quchan, and of Kuchek Khan in Gilan. The last rebellion was the most serious. It created the Soviet Republic of Gilan and was not put down until the autumn of 1921, after the assumption of power by Reza Khan. In the same year, the autonomist rebellion of Colonel Mohammad Taqi Khan Pesyan in Khorasan foundered.[2]

Modernization of the State Under Reza Shah

Iran's international relations were radically altered after World War I. Russian imperialism was removed from the picture by the October Revolution, and Britain was favorably disposed toward the creation of a strong nationalist state to withstand the threat of bolshevism. Vothuq al-Dawleh sought to entrust the entire enterprise of state building to the British. The treaty of August 1919, signed by him but never ratified by the Majles, would have made Iran a virtual British protectorate. Not surprisingly, it offended all nationalists and aroused the instant fury of a significant number of Constitutionalist leaders and of the Shi'ite religious leaders. Vothuq al-Dawleh was forced to declare martial law in Tehran and banish five prominent Constitutionalists to Kashan but failed to have a pro-treaty Majles elected.[3] He fell from power in July 1920 but after a British general and a British high financial advisor had arrived in Tehran and had begun their functions according to the provisions of the 1919 agreement. A group of Anglophile political activists led by Sayyed Zia al-Din Tabataba'i, a young journalist with strong British connections,[4] plotted a coup d'etat that would preserve the monarchy but would aim at "the formation of a strong government to prevent the penetration of the bolshevik creed which was rapidly gaining in popularity."[5] The projected coup had three eminently modern features. Its goals included the prevention of the spread of revolutionary bolshevism, it was planned by an intellectual—a journalist about to discard his traditional clerical garb for modern civilian attire, and it was to be carried out by a modern military officer.

In the preceding fifteen years, a military school for training Iranian army officers for the Cossack Brigade had been established in Tehran, and a small number of Iranians had received military training in European military academies. The white Russian officers of the Cossack Brigade had lost the support of their government after the October 1917 Revolution, but had retained its command. The first Iranian commander of the brigade, Sardar Homayun who was a graduate of St. Cyr, was appointed no sooner than the end of October 1920. Sayyed Zia approached him first, but failed to win him over. He then turned to the next officer in rank, Colonel Reza Khan of Mazandaran, a man of humble background who lacked the prestige and military education of his superior.

Reza Khan carried out the coup d'etat on February 21, 1921, and held the military power as the commander of the Cossack Brigade. The Shah appointed Sayyed Zia prime minister. In a proclamation issued on February 26, 1921, Sayyed Zia deplored the frustration of the aims of the Constitutional Revolution during the past fifteen years. He declared the treaty of August 1919 null and void, and promised military and administrative reforms. To these were added extensive social reforms and land distribution in order to forestall the thunder of bolshevism.[6] His favorite reform, and the only one he had the chance to make headway with during his brief tenure, was that of the municipalities.[7] Late in May 1921, after less than a hundred days in office, Sayyed Zia was ousted by Reza Khan, to whom he had initially denied the war ministry but who had nevertheless emerged as his chief rival. Once appointed minister of war, over the prime minister's head, on April 25, 1921, Reza Khan's ascent was apparent to all and met with little effective opposition.[8]

The rise of Reza Khan and the creation of the modern army are closely interwoven. He began to create a standing army as soon as he became the minister of war. The sundry armed corps of some 23,000 men were dissolved and then reorganized as the National Army in February 1922. Reza Khan's reorganization of the army encountered resistance from the gendarmerie, which resented the dominance of the Cossack commander. The gendarmerie had been dissolved with the disintegration of central authority in 1915 but reconstituted after the war. Its officers were popular and enjoyed a reputation for nationalism. It is significant that Sayyed Zia's first minister of war whom Reza Khan ousted, Major Mas'ud Khan Keyhan, was not only a graduate of St. Cyr but also a gendarme. Colonel Mohammad Taqi Khan Pesyan, whose insubordination to central government shortly after this ouster has already been mentioned, was the commander of Khorasan gendarmerie and the brother of Major 'Ali-qoli Khan, who had commanded the gendarmerie of Shiraz and defied the order for its dissolution with nationalist support for a short time in 1915.[9] On February 1, 1922, reacting to the combination of a three-months pay arrearage and imminent loss of autonomy resulting from the proposed incorporation into the Cossack-dominated unified army, the gendarmerie division of Azerbaijan rebelled under the leadership of Major Lahuti in Tabriz. The uprising was put down by government forces a week later.[10]

Minister of War Reza Khan was not deterred. He appropriated the revenue from the royal domains and indirect taxes[11] and pursued his military reorganization with great vigor. The reorganized National Army of 1922 consisted of 1,944 officers and 27,725 soldiers. An Iranian Air Force was also created, which had three German and four Russian airplanes in 1924, and twenty combat aircraft by 1941. In 1923, a group of forty-seven officers was sent to France for training; in the following years, the Air Force sent a group of students to Russia to be trained as pilots. Early in 1922, various military schools were amalgamated to form the Military Academy, which produced 2,095 officers between 1923 and 1941. In addition, some five thousand officers, mostly of lower ranks, were trained in schools for lieutenants, reserve officers, and military technicians. Finally, compulsory conscription was introduced in June 1925.[12] (The navy did not come into being until the early 1930s.)

As he was building the new army, Reza Khan was also consolidating his political power. Despite the outcry from the press, he successfully used the device of martial law, appointing some officers as martial law administrators and provisional governors, and other officers in civilian garb as commissars to represent him in departments of government, thus spreading a military net over Iran.[13] In April 1922, Reza Khan put considerable pressure on Prime Minister Moshir al-Dawleh to muffle the press opposed to the military and forced his resignation in May. Qavam al-Saltaneh then became prime minister, but real power remained with Reza Khan. In October, Reza Khan resigned in protest to continued criticism of martial law and his control over government finances while carefully orchestrating clamorous support from the military, from some newspapers and from the preachers. One editorial declared that the departure of the minister of war was the end of Iran.[14] The minister was persuaded to stay on, but insisted on the control of the police and all security forces. A year later, having foiled a plot to assassinate him that involved former premier Qavam al-Saltaneh, Reza Khan became prime minister. The talk of a republic with Reza Khan as its president began to spread immediately.[15]

In 1924, after some flirtation with republicanism,[16] Reza Khan decided in favor of monarchy in response to the anxiety of his clerical supporters caused by Atatürk's abolition of the Caliphate in Turkey. In February 1925, by overwhelming majority, the Majles made Reza Khan commander-in-chief of the armed forces, a position reserved for the monarch in Article 50 of the Supplementary Fundamental Law.[17] At the end of October 1925, the Majles abolished the Qajar dynasty and ordered the election of a Constitutent Assembly to change the Fundamental Laws. Reza Khan, who had recently assumed the surname of Pahlavi with a view toward reviving the imperial glory of pre-Islamic Iran in his forthcoming reign, was voted Shah by the Constituent Assembly in December 1925. The Qajar dynasty was thus replaced by the Pahlavi dynasty, with a monarch almost as new to the name chosen for his dynasty as to the throne itself.[18]

Reza Khan used the title of *Sardar-e Sepah* (Marshal of the Army) until he became Shah in 1925. Through the end of his reign and beyond, Pahlavi rule remained closely associated with the rule of the new elite of army officers—with the uniform, the military boot, and the officers' clubs. In 1941, Reza Shah left behind an army of 183,863 men. It is worth emphasizing the importance of the army in Reza Shah's rule. This importance is reflected in Table 8, which contrasts the numerical strength of the army under Reza Shah with the earlier and later periods (see Appendix). It should be noted that only in the 1970s did the absolute size of the army exceed that of the last year of Reza Shah's reign and, more importantly, that the military participation ratio never reached the peak of 1941 again.

However, it would be completely wrong to assume that he rose to power on the basis of coercion or without broad political support. The Constitutional Revolution had not drawn the peasants into national politics and they remained on the periphery of political society, and Reza Shah's regime was neither of nor for the tribes. But in urban political society, he enjoyed wide political support and used it adroitly for his rise to supremacy. The old Constitutional elite preponderantly supported him, as did the younger generation of politicians associated with the reformist and radical political parties of the Fourth and the Fifth Majles.[19] So did the Shi'ite hierocracy.[20] Even the four deputies of the Fifth Majles who opposed the change of dynasty were in favor of Reza Khan's centralization policies. The most famous of the four, Dr. Mohammad Khan, Mosaddeq al-Saltaneh, had in fact used his legal expertise and reputation for supplying an elaborate rationale for the transfer of the supreme command of the armed forces from the monarch to the prime minister, Reza Khan, in February 1925.[21]

Reza Khan's political astuteness in manipulating political forces and personalities is undoubtedly an important factor in explaining his ascent.[22] But far more important was his success in realizing, at last, the second goal of the Constitutional Revolution: modernizing the state. The broad early support for him was actually support for the creation of a strong centralized state. Reza Khan was no less the child of the Iranian Constitutional Revolution than Napoleon was of the French Revolution. Not only his military reforms but all his important early measures had been long contemplated by the Constitutionalists and were widely popular in the Majles: the monopoly on sugar and tea (May 1925), the introduction of national military service (by the law of June 1925, which the Majles passed unanimously); the construction of a Trans-Iranian Railway (proposed in November 1925, approved in February 1927);[23] the building of major roads; and the creation of the National Bank of Iran (law of May 1927, inaugurated in October 1928).

Even while the army was being reorganized and modernized, Reza Khan and his officers began to subdue the tribes. As Max Weber emphasized, the modern state is characterized by its monopoly of the legitimate use of coercion and by the concentration of the forces of coercion under its control. The counterpart of the creation of a standing army at the center of the state was the disarming of the tribes at the periphery. The disarming of the tribes began in the spring of 1922 and continued throughout Reza Khan's reign as Shah. The first resounding victory of the government forces was the capture of Chehriq fortress from the Shakkak Kurds, whose leader, Esma'il Aqa, fled Iran in August 1922. It was celebrated in all major cities, and praise was showered on the minister of war. In the following year, the government captured Maku and dislodged the other Shakkak chief, Eqbal al-Saltaneh; his resplendent treasury was directly appropriated by the Ministry of War.[24] The Shahsevan, the Bakhtiyari, the Kurds, the Lors, the Boyr Ahmad, and the Turkmans were defeated and disarmed in a series of campaigns from 1922 to 1925. Conscription was applied to the tribes, and the drafted young tribesmen were scattered in garrison towns and outposts throughout the country away from their own tribes.[25] There was renewed tribal unrest in the northeast and in the south in 1928 and 1929:[26] The Qashqa'is rebelled in the summer of 1929 and were joined by other tribes of Fars. The Baluch were pacified after the arrest and execution of their chief, Dust Mohammad Khan, early in 1929.[27] The Mamasanis and the Boyr Ahmad in Fars were defeated, with considerable army casualties in 1930.[28] The Qashqa'is rebelled for the last time in 1932,[29] and the tribes of the more remote areas of Lorestan were subdued in the early 1930s with brutal severity.[30] Tribal unrest in Kurdestan continued until the late 1930s.[31]

Reza Shah also adopted the tribal policy of the pre-Pahlavi centralizers of power. Like Fath 'Ali Shah a century earlier, Reza Shah forced the tribal leaders and other tribal hostages to reside in Tehran, and imprisoned and killed them if necessary.[32] During the last decade of his reign, he had most of the tribal leaders executed. Sawlat al-Dawleh, "the Crownless King" of the Qashqa'is, was in forced residence in Tehran with his eldest son, Naser Khan, from 1926 and died in prison in 1933.[33] Ja'far-qoli Khan Bakhtiyari, Sardar As'ad, while still minister of war, was suddenly arrested in November 1933 and

died in prison shortly thereafter.[34] Less important tribal chiefs and their followers were also double-crossed, arrested, and killed despite guarantees of amnesty sworn and often written on the Koran.[35]

But Reza Shah went beyond the earlier policy toward the tribal leaders
and also sought to destroy the very social and economic organization of the
nomadic tribes. In 1933, a policy of forced sedentarization was enforced,
with devastating social consequences. Military governors were appointed for
tribes, and the tribal elders and headmen were forced to discard their traditional garb for Westernized uniform clothes.[36] The electoral law of October
1934 abolished all tribal constituencies.[37] In August–September 1941, after the
occupation of Iran by the Allies and the abdication of Reza Shah, many of
the Khans returned to their tribesmen, who left the settlements to resume nomadic pastoralism, and avenged themselves by killing some of the officers in
their charge.[38]

It is true that one of the first and most important uses Reza Khan put to
his new army and newly acquired German airplanes was the subjugation and
destruction of the tribes. However, there is more to the story. Reza Shah took
full advantage of the factious character of the tribal periphery and often used
the help of one tribe in fighting and disarming another.[39] While engaged
against the Shakkak Kurds, Reza Khan accepted the apology of the Bakhtiyari chiefs for the massacre of an army camp in the summer of 1922 and left
the punishment of the culprits to them.[40] In June 1927, he shrewdly made
Sardar As'ad the minister of war to assure the loyalty of Bakhtiyaris while
he was tackling the rival Qashqa'i, Khamseh, and Lor tribes of the southwest
and the south. That he needed to exploit the contentious nature of the tribes
to divide, disarm, and rule them was demonstrated by the serious consequences of the failure of this policy in one instance: the widespread rebellion
of the tribal confederations of Fars in the summer of 1929, which even spread
to the Bakhtiyaris. This rebellion showed that the army could not easily
prevail against a unified tribal uprising; the government only brought the rebellion to an end with a mixture of concessions, bluffs, and pressure on the
tribal chiefs in forced residence in Tehran.[41] What decisively changed the
balance in favor of the army was the completion in the 1930s of major roads
along the migration roots of the nomadic tribes. These obviated the need for
the army to penetrate the mountainous tribal territory and enabled the mechanized army, with the aid of a few observation planes, to confront the tribes
at most advantageous strategic points during the tribes' inevitable migration
from the winter to the summer quarters.[42]

From about 1930 onward, Reza Shah's rule became increasingly dictatorial. He used the police to strengthen his personal rule and was blatant in
the political misuse of the newly organized judiciary. The Majles, whose
deputies he handpicked, became a rubber stamp for the Shah's measures. Far
from remaining the first servant of the state, he became a neo-patrimonial
ruler who chose his appointed ministers on the basis of personal loyalty to
him and treated and abused them as his servants. Nor did he draw a sharp
line between the management of his own property and that of the affairs of

the state: Karim Khan Buzarjomehri, one of Reza Shah's closest military associates, retained an important military command while serving as the mayor of Tehran and was at the same time in charge of the management of the newly acquired Pahlavi estate in the north, frequently using army officers for this purpose.[43] Reza Shah even reintroduced the patrimonial practice of killing disgraced ministers and political opponents in jail, though now without legal or cultural sanction.

As we have seen, attempts to reorganize and rationalize Iran's fiscal system through the use of American and French financial experts had been underway since the restoration of constitutional government in 1909. These attempts were taken up and seriously pursued by Dr. Arthur Millspaugh from 1922 until 1927, when he was forced to resign because he resisted Reza Shah's demands for state funds, especially for the army and for his incessant travels to the four corners of Iran.[44] Nevertheless, the deficit in government spending was gradually eliminated, and an annual budget was published from 1924 onward. The budget figures for the Reza Shah period reflect both the increased fiscal strength of the Pahlavi state, as compared to the Qajar state, and its different goals. Over the years, they also reflect the considerable growth of the centralized bureaucratic state.

The Pahlavi state inherited the relatively more favorable fiscal structure of the late Qajar and Constitutional period. (See Appendix, Table 6.) Now less than one-fifth of the government revenue came from taxation on land. A somewhat smaller proportion came from oil royalties and a little less from indirect taxes. A greater proportion of government receipts—about one-third—came from customs duties. An income tax was introduced in 1930 that henceforth supplied about one-tenth of government revenues. Government monopolies, notably on tobacco, wheat, and cotton, were established in addition to those on sugar and tea, resulting in a considerable increase in government revenue. By 1937, revenue from monopolies accounted for as much as one-third of the total government revenue.[45] In this connection it is worth noting the similarity of Reza Shah's attempts to increase the fiscal resources of his state and those of other state builders. Frederick William, the Great Elector (1640–88), for instance, had introduced an urban excise and a monopoly of salt to support his state building and establish a standing army. Similarly, Napoleon established a tobacco monopoly and levied indirect taxes.[46] Reza Shah's financial and fiscal reorganization, too, followed the *étatiste* line. The Ministry of Finance was completely centralized. The National Bank was founded on the basis of the law of May 5, 1927; it was given the right to issue notes and began to do so in July 1930.[47] A state monopoly on foreign trade was set up in March 1931. Finally, from 1939 onward, with the National Bank in place, Reza Shah inaugurated the modern fiscal technique of deficit financing with government expenditure exceeding its revenue by one-tenth or more.

Reza Shah used these fiscal resources to build his army and bureaucratic state. Government revenue and expenditure increased more than tenfold in real terms during his reign.[48] This increase covered the creation of a centralized bureaucratic state. The first law regulating the civil service was enacted in December 1922, and special classes to train civil servants were set up by the Ministry of the Interior in 1932. Successive laws in November 1936 and January 1938, following the French pattern, finally organized the administration of the country into a highly centralized hierarchical structure controlled from the capital. The country was divided into ten departments (*ostan*) and subdivided into urban administrative units (*shahrestan*); these were in turn subdivided into districts (*bakhsh*). Provincial governors, mayors, police officers, and other municipal officials were all appointed by the Ministry of the Interior in Tehran.[49]

Hand in hand with the centralization of administration, a parallel reorganization and centralization of the judiciary took place under the determined direction of 'Ali Akbar Davar from 1927 until 1940, when, presumably in anticipation of the Shah's wrath, he committed suicide. The judiciary system under the Qajars had been markedly dualistic. The state had run the customary tribunals while the religious courts were under clerical control, independent of the state. The duality of the traditional legal system had been recognized and endorsed in the Supplementary Fundamental Law of 1907 and the first Civil Code of 1911, and the religious courts indeed remained the dominant element of the judiciary system well into the 1920s. However in 1922, state courts were given partial appellate jurisdiction over the religious courts. From 1926 to 1940, Reza Shah's judiciary reforms gradually reduced the purview of the religious courts until the Civil and Penal Codes of 1939 and 1940 finally omitted all reference to the Sacred Law and to religious courts. The entire judiciary system was secularized and incorporated into the centralized state as the Ministry of Justice. A hierarchy of courts corresponding to the administrative divisions of the country was set up with a supreme court in Tehran at its apex.[50] For the first time in Iranian history, the division between sacred and customary law was obliterated and law became codified. The substance of the code was largely borrowed from the continental European legal material but also included many of the provisions of the Shi'ite Sacred Law, especially on family law.[51] An important step in the complete secularization of the judiciary was the law of registration of documents and property of March 1932, which followed upon regulations already in effect since 1928. It required legal documents to be registered by official state notaries, thereby depriving the religious courts of one of their most lucrative functions. Clerics who did cling to that function were before long forced to remove their turbans and assume civilian garb. In addition, the principle of national sovereignty was given expression in the extension of the jurisdiction of the new judiciary over foreign nationals and abolition of the consular jurisdiction over the latter category (which had been termed "capitulation") in May 1927.

The increased fiscal resources of the state, the creation of the new army, and the demise of the tribes thus made possible the rise of a centralized bureaucratic state. In contrast to the Qajar state, the new state was conceived of from the beginning as a service-rendering organization. Reza Shah had no inhibition in using his position and the coercive arm of the state to acquire a huge landed estate in the north.[52] Nevertheless, in theory if not always in practice, the goals of the new state and those of the ruler became clearly differentiated. Needless to say, the state automatically served as the instrument of maintaining the new House of Pahlavi, the new military, and to a lesser extent, the bureaucratic elite of the regime. But it also sought to develop the country by building roads, railways, and factories and to promote the welfare of the population by building schools and hospitals.

The goals of the new state are reflected in its pattern of expenditures. Defense spending accounted for the same proportion as in the 1880s, about 40 percent, but this time the money was spent on a real and not a fictitious army. It no longer included a preponderant element of sinecures and hypothetical soldiers. Other sinecures also disappeared. Millspaugh had already reduced the pensions to practically zero, freeing some 13 percent of the state revenue to be utilized for public works. The size of this category increased progressively. Opening a new session of the Majles in December 1930, Reza Shah declared the beginning of the era of economic reforms, and by the end of Reza Shah's reign, roughly 40 percent of government expenditures could be classed as development expenditures.[53] This category indicates the primary nonmilitary goal of the Pahlavi state: the development of the country. The distributive state of the Qajar period thus gave way to a modern state functioning as a service-rendering organization in charge of development. The new state did enable the ruler and certain members of the elite to enrich themselves indirectly or illegally, but the use of taxation for distribution of resources (from the poor to the rich) ceased to be its explicit function. Under Millspaugh's financial administration, irrigation schemes and light mechanization on state land (royal domains) had begun, and 200 trucks were purchased for transportation of grain to areas hit by the famine of 1925. Reza Shah set up textile factories and other light industry, built roads, and above all, built the Trans-Iranian Railway between 1927 and 1938. Urban development, too, absorbed Reza Shah's energy. Many of the tree-lined boulevards in Iranian cities date from his reign.[54]

One item in Reza Shah's developmental program requires special attention because of its political consequences. In 1900, there were no more than 800 miles of road in the entire country. Between 1923 and 1938, Reza Shah built some 14,000 miles of new roads, over 3000 miles of which were classified first class roads (highways).[55] Between 1920 and 1933, the cost of transportation of merchandise was reduced threefold and the time spent in transportation tenfold.[56] This economic aspect, however, does not exhaust the significance of the new roads. Good roads became the infrastructure of Reza Shah's centralized power as they had been the infrastructure of the Roman empire. The Pahlavis were able to move their troops from Tehran and the provincial centers into peripheral areas many times faster than the Qajar kings could. It is no accident that Reza Shah's minister of roads, General Tahmasbi, was killed on a road being built in March–April 1928 to connect the tribal areas of Lorestan to the national networks.[57]

Education was given lower priority, receiving about 2 percent of the budget at the beginning of Reza Shah's reign and about 5 percent of it at the end. Nevertheless, education was a lot cheaper than railway construction and the 5 percent went a long way. A secular system of national primary and secondary schools along Western lines was set up by the Ministry of Education, which in a radical departure from tradition, also included girls. By 1941, well over a quarter of a million children were enrolled in the new primary schools (as compared to some 37,000 still educated at the traditional Koranic schools run by the mollahs), and nearly 30,000 in secondary schools. A number of students were sent abroad for higher education, and the University of Tehran was inaugurated in 1935. By 1940, 500 graduates had returned from abroad and nearly as many were completing their studies.[58] Reza Shah's educational reform was his best investment in the development and welfare of the country, though neither he nor his son reaped its benefits. Furthermore, by creating a new middle class, it had a momentous impact on Iranian society.

Social welfare assumed even lower priority, but efforts were made in the area of public health. The first modern hospital—significantly, for the army—was inaugurated in Tehran in 1933, and a number of municipal hospitals were built subsequently. In 1940, the Bureau of Public Health in the Ministry of Interior was upgraded into the Ministry of Health.

With the rise of the new state, there was an enormous increase in the bureaucratic redtape; an elaborate system of registration and a prolific machinery for repairing and issuing permits and licenses came into being. More importantly, there was the propagation of the official ideology of *étatiste* nationalism. Reza Shah's assumption of the surname of Pahlavi was indicative of the forthcoming attempts to capitalize on the memory of the imperial glory of pre-Islamic Iran. The old Persian names of the months were revived to replace the Arabic and Turkish ones, and the Arabic lunar calendar was replaced by the Iranian solar calendar, in 1925.[59] The modernization program of the Pahlavi state also entailed significant secularization of culture. New branches of learning, the history of pre-Islamic Iran, Ferdawsi's *Epic of the Kings,* and the secular nationalist ideology of the Pahlavi state were propagated by the new system of national education in the 1930s. This ideology bypassed constitutionalism and was emphatically monarchist, as best illustrated by the order of the three words in the motto it inscribed in the minds of the whole generation of its school children: God, the King, the Fatherland.[60] Perhaps the most spectacular aspect of the state-promoted secularization of culture was the unveiling of women in 1935, a forced but nevertheless courageous break with the Islamic tradition.

An end was put to Reza Shah's reign when the Allies, apprehensive about the pro-German sympathy of the Shah and some of his army officers and wanting to open a new corridor to Russia, invaded Iran in August 1941. In September, Reza Shah was forced to abdicate in favor of his son Mohammad Reza and left Iran on a British ship.

Impact of the State on Iranian Society

The Pahlavi state survived its founder and continued to operate sluggishly from 1941 to 1962. After the Allied invasion, massive desertion reduced the army to some 65,000 men, although it did not disintegrate and gradually grew to some 80,000 men by mid-1943.[61] Millspaugh returned as Iran's financial czar in 1942. The tribes resumed nomadic pastoralism. The Soviets occupied the north until the end of 1946 and set up the short-lived Soviet republics of Azerbaijan and Kurdestan. With this and unruly tribes in the southwest, political turmoil surrounding the nationalization of oil, and an unassertive young monarch on the throne, the presence of the state on the political scene was less conspicuous. Nevertheless, it was present by the end of the 1940s; and the primacy of the promotion of economic development was underlined by the creation of the Plan Organization in 1949, and by the subsequent implementation of a series of seven-year and five-year developmental plans. The first two plans showed an increase in expenditure on education and health (about 8 percent and 5 percent of actual expenditure respectively), though their level remained quite low by international comparisons.

However, economic development and some growth of welfare services do not by any means represent the most important component of the social change brought about by the new state. Two other consequences of the rise of the centralized bureaucratic state are sociologically of far greater significance. The modern state radically altered the social structure of Iran, greatly weakening its tribal component, consolidating landholders into a class, and creating a new urban middle class; *and* it linked the elements of the social structure more closely, that is, it acted as the major agency of national integration.

The major steps taken toward the demise of the tribes were taken by Reza Shah. What was not completed by the major thrust of state formation and centralization under him was brought about indirectly as a result of national integration, economic development, voluntary migration and sedentarization, and city-oriented economic policies of the state after World War II.[62] Table 9 depicts the demise of tribalism in the twentieth century (see Appendix).

The tribal population dropped from about 25 percent of the total in the period from 1900 to 1920, before the rise of the Pahlavi state, to around 1 percent on the eve of the Islamic revolution. Even allowing for the bias of census statistics from 1966 and 1976, the general picture is unmistakably clear.[63] A change from 25 percent at the beginning of the century to about 8 percent in the 1930s represents the impact of the Pahlavi state building. As has been pointed out, the trend reversed after the fall of Reza Shah in 1941, and many forcibly settled tribesmen and their families resumed pastoralism. The change from 12 to 13 percent in the 1940s to about 1 percent in 1976 represents the effects of national integration and economic development as well as the continued hostility of the state and its adverse policies.

During Reza Shah's rule, the urban proportion of the population remained constant at just over one-fifth. Its composition, however, underwent considerable change, a change which continued after his fall. The many officers and the state employees themselves embody the most significant component of this change. We have already considered the growth of the army and can now turn to the civilian employees of the bureaucratic state: government functionaries and school teachers. Table 10 represents the growth of the mammoth bureaucratic state (see Appendix). It shows that the Pahlavi state, built in the midst of economic backwardness and poverty, survived to flourish and grow in the rosier years of the 1960s and 1970s. On the eve of the revolution about a fifth of all civilian households depended directly on the state for their livelihood.

Indirectly, too, the modern state facilitated the growth of a professional middle class and an entrepreneurial bourgeoisie of government contractors. Lawyers came to figure prominently in the former groups; road builders, civil engineers, and architects among the latter. As a consequence of secularization of the juriciary system, a constant demand for legal services was created. The laws of 1935 and 1937 radically revised the earlier regulations of the Constitutional period, setting rules governing the professional conduct of lawyers and recognizing their bar association. They soon came to constitute a small but significant segment of Iran's civil society, and played a prominent role in the politics of the nationalist era under Mosaddeq. (In fact, the bill assuring the independence of the Bar Association was prepared by Mosaddeq early in 1952 but did not become law until after his fall.)[64] Unlike the government contractors who developed close personal and financial ties with the functionaries of the Pahlavi state and, increasingly, with the members of the royal family, the lawyers generally maintained their distance from the regime and constituted the backbone of the liberal-nationalist opposition to it.

A very important sociological consequence of the reforms of the legal system was the enhancement of private property in land and the creation of a homogeneous class of landowners. Building on the Law of May 1913, the land and property registration laws and regulations of 1922, 1928, 1932, and 1939 converted a variety of conditional, de facto and tribal—often communal—holdings into unconditionally held private property. It thus created a landowning class, a class with an independent economic and sociolegal base that was subsequently referred to as the "Thousand Families." Reza Shah himself became before long the biggest landlord in the country. Furthermore, the Democrats, when in power in 1911, had abolished property qualifications for voters and had instituted universal male suffrage.[65] An unintended consequence of this ideologically motivated populist measure was in fact to strengthen the landlords in the Majles, as they had the rural vote in their pockets. By the end of Reza Shah's reign, the landlords occupied over a quarter of the Majles seats (as compared to under a tenth in the First

Majles) and constituted an important bloc.[66] The 1935 law concerning the village headmen and other regulations of the relationship between peasant and landlord tended to enhance the position of the latter.[67] These laws also went a considerable way toward converting the surviving tribal chiefs into big landlords.[68] In the longer run, the legal framework of the Pahlavi state, and especially the laws governing agrarian private property, facilitated the hegemony of the Thousand Families which lasted until Mohammad Reza Shah's land reform in the early 1960s.

In addition to these important changes in the social structure resulting from its formation, the Pahlavi state also acted as the primary agency of national integration. The introduction of conscription in 1925 meant not only the penetration of the rural and tribal periphery by the central state but also a considerable increase in national integration. The centralization of administration and establishment of a national educational system had a similar effect. So did the building of roads and railways and the setting up in 1940 of a state radio in Tehran. Road buildings continued under Mohammad Reza Shah, and the expenditure on communications and telecommunications accounted for about a third of the total expenditure in the developmental plans of the 1950s and 1960s.[69] The stage was set for the penetration of the villages by television sets and telephone lines in the 1970s.

The development of transportation and communication, compulsory universal male military service, and a uniform system of national education turned the regionally and tribally segmented Qajar society into a reasonably homogeneous nation. Nothing can attest better to the success of the nation building by the Constitutionalist and Pahlavi governments than the abysmal failure of the Soviet-sponsored attempt to set up an independent republic among Turkish-speaking Azerbaijanis (the largest ethnic group in the country) after World War II. It is also worth noting that the measures contributing to nation building and national integration built on the essential precondition of religious uniformity due to the efforts of Shi'ite hierocracy from the seventeenth century onward. The absence of separatism among the Azerbaijanis is no doubt in large measure due to their being Shi'ite. Conversely, it is amongst the Sunni ethnic groups—the Kurds, the Baluchis, and the Turkmans, who constitute less than 10 percent of the population— that refractory resistance to integration into the Iranian nation-state has persisted.

Mohammed Reza Shah's "White Revolution" and Its Consequences

From late 1944 onward, Iranian politics became increasingly dominated by the issue of nationalization of oil. Reza Shah's old foes, Sayyed Zia al-Din Tabataba'i, Qavam al-Saltaneh (now Ahmad Qavam) and Mosaddeq al-Saltaneh (now Dr. Mohammad Mosaddeq) had returned to the political scene, and the Majles had again assumed the importance of the early constitutional period. Another important change was the advent of the Tudeh on the Iranian political scene—a well-organized and disciplined Communist Party with an effective machinery for mass mobilization. With the predominance of the Majles, political instability and short-lived cabinets also became typical once more until Mosaddeq was appointed prime minister in May 1951. Mosaddeq championed Iranian nationalism and had broad support from the bazaar and from the new middle class of teachers, civil servants, and lawyers. He had spoken against the employment of foreign financial advisors and for full control of the finances of the state by Iranians.[70] But above all, since his introduction of a bill forbidding foreign concessions, which passed the Majles in December 1944, he had become the advocate of the nationalization of the Anglo-Iranian Oil Company. The nationalization of oil was the constant preoccupation and main achievement of his premiership. It was the final application of the principle of national sovereignty to Iran's most important natural resource, oil, and thus completed that aspect of the teleology of the Constitutional Revolution.

Furthermore, Mosaddeq also sought to restrict the neo-patrimonial powers of the Shah and to reduce him to a constitutional monarch and a ceremonial figurehead. To achieve this second Constitutionalist goal, he forced a showdown with the Shah in July 1952 and won. But this alienated the Shah and the army, who became increasingly apprehensive as Mosaddeq's power grew. On August 19, 1953, with the support of the U.S. Central Intelligence Agency and the backing of the influential clerics of Tehran, General Fazollah Zahedi and a number of other officers carried out a *coup d'état* against Mosaddeq. The Shah, who had lost his nerve and fled the country a couple of days earlier, returned to the throne.[71]

With the political instability of the first dozen years of Mohammad Reza Shah's reign, the capacity of the centralized bureaucratic state to act upon society was not put to any consistent use except for the Seven- and Five-Year Development Plans. The one man likely to resuscitate the reforming state, General Hajj-'Ali Razmara, the best product of the Pahlavi military-administrative complex who had been among the first group of officers sent to St. Cyr by Reza Khan in 1923—was not trusted by the young Shah, lacked popularity, and was in any event assassinated after a brief premiership in March 1951. Nor did the Shah resolve to do so in the remaining years of the 1950s. In 1961, however, under considerable pressure from the Kennedy administration, the Shah asked 'Ali Amini to form a cabinet with a reform program. Amini's minister of agriculture, Hasan Arsanjani, energetically initiated and pursued the land reform. By early 1963, apprehensive of the popularity of the sponsors of reform, Mohammad Reza Shah overcame the unassertive reserve of the earlier years of his reign and took full responsibility and credit for the reform program. In January 1963, he held a national referendum that approved the six points of his reform program: (1) the land reform, (2) sale of some state-owned factories to finance the land reform, (3) the enfranchisement of women, (4) nationalization of forests and pastures, (5) formation of a literacy corps, and (6) institution of profit-sharing schemes for workers in industry.[72] This reform program was officially referred to as the "White Revolution" and as the "Revolution of the Shah and the People" after 1967. For what slogans are worth, the White Revolution of the 1960s did set in motion a revolution of the Shah and the people. However, there was a sharp disjunction in the latter revolution. The Shah did his revolution first, completing it by 1978. Then the people began theirs. The revolution of the people is the main subject of this book, but let us take a brief look at the consequences of the Shah's revolution, which set it in motion.

By far the most important of the six points was land reform. Although its redistributive effects fell short of what could be expected, it liquidated the big landlords as a class—and a nationally dominant class at that. During the first, and only genuine stage of the land reform in 1962 and 1963, the landowning Thousand Families, including the tribal chiefs, lost their socio-legal base and were thus liquidated as a class. Though many of its members retained large holdings of land and became mechanized commercial farm-ers, joining the petrobourgeoisie in the prosperity of the 1970s, and many even remained in the Pahlavi political elite, there can be do doubt that the traditional peasant-landlord relationship, which was the power basis of the landowning class and accounted for its prominence in the Majles, was de-stroyed.[73] Furthermore, by failing to give any or enough land to the ma-jority of the peasants, the land reform accelerated the massive migration from the rural areas into the cities.

Next in significance was the formation of a literacy corps. This program was pursued with some vigor; it sent conscripted men with high school di-plomas to small towns and villages in fulfillment of their compulsory mili-tary service. The establishment of the literacy corps furthered the process of national integration and can be presumed to have brought about some political awakening in the torpid rural periphery and small towns. Also in-strumental in increasing national integration was the establishment of the Health Corps by a royal decree in 1963. By the end of 1966, some 1000 doctors and paramedics had already served in it.

The profit-sharing schemes for workers in industry, together with close supervision and favorable labor legislation, did succeed in winning the loyalty of the relatively prosperous industrial workers for the Pahlavi regime.

While the reform program was in operation, the importance of economic development as a goal of the modernized state was reaffirmed by the Third (1963–67) and Fourth (1968–72) Development Plans. These plans shared an emphasis on infrastructural development and on state incentives to indus-trialization in the private sector, and can be presumed to have contributed to the impressive performance of the industrial sector in this period.[74]

As was the case with his father, Mohammad Reza Shah's reform and modernization programs of the 1960s and 1970s went hand in hand with the strengthening of his personal rule and the establishment of neo-patri-monial dictatorship. The Majles once more became the rubber stamp for royal dictates. The secret police, the SAVAK, set up in 1953 with the CIA advice to extirpate the Tudeh underground cells and prevent the return of

Communism, became increasingly omnipresent and increasingly hated. The army was expanded to an efficient force of 400,000 men equipped with technologically advanced U.S. weapons.

In contrast to Reza Shah's reign, the period 1953 to 1978 witnessed two new types of social change: urbanization and the much accelerated expansion of higher education. As Tables 11 and 12 show, between 1956 and 1976, the urban population of Iran increased from 31 percent to 47 percent of the total population (from 6 to 16 million). Rural-urban migration accounts for a substantial proportion of this growth in urban population, over one-third for the decade 1966 to 1976 (see Appendix), the proportion being much higher for Tehran. The decade 1966 to 1976 also witnessed an unprecedented expansion of higher education. The number of persons with higher education quadrupled to about 300,000 and the enrollment in universities and professional schools in Iran trebled to about 150,000 (see Table 13, Appendix). About as many were studying in the institutions of higher learning abroad. The indirect relation between these changes and the Islamic Revolution of 1979 will become clear in subsequent chapters.

The revolution of the Shah was by no means coextensive with the impact of his reform program. In fact, one may argue that it was in larger part the indirect result of the consequences of the rise of the OPEC oil prices in 1972, in which the Shah played a leading role. But the economic chaos caused by massive injection of oil money in the mid-1970s also belongs to a later chapter that deals with the revolution of the people.

4

Shi'ism Versus Statism

Sources of Conflict Between the State and Hierocracy

The state and the hierocracy are two institutions of legitimate authority. As such, a certain degree of rivalry between them for the loyalty of the subjects is inevitable. As Weber stated,

> Whenever hierocratic charisma is stronger than political authority it seeks to degrade it, if it does not appropriate it outright. Since political power claims a competing charisma of its own, it may be made to appear as the work of Satan.[1]

The precondition for the strength of hierocratic charisma is its autonomy, an autonomy that the world religions of salvation grant in principle but that does not always find adequate institutional translation. The most important feature distinguishing Shi'ite from Sunni Islam since the end of the eighteenth century is the separation of political and religious authority, and the corresponding autonomy of the religious institution from the state. This separation could, under a number of circumstances, induce a negative evaluation of political charisma on the part of the men of religion. This negative evaluation of political power by religious authorities could, in theory, manifest itself in at least two distinctive manners: (1) pious *withdrawal* from the political sphere and indifference toward politics; and (2) active endeavor to *subjugate* political power and impose ethico-hierocratic regulation upon the political sphere.

Comparative history teaches us that once the separation of political and hierocratic domination has been established, given the indisputable superiority of God over earthly powers, theocratic monism is but a further logical step that could theoretically be taken at any time.[2] In Western Christianity, the separation of hierocratic and political authority on the basis of "freedom of the Church" was established under Gregory VII (d. 1085). Gregory's arguments for the freedom of the Church went hand in hand with the assertion of the superiority of hierocracy over monarchy, thus implying theocratic monism:

> The Son of God had given Peter and his successors the power to bind and loose souls, a power, that is to say, which is spiritual and heavenly;

how much more, then, can Peter dispose of what is purely earthly and secular?[3]

However, other conditions had to be fulfilled before theocratic monism could take the form of the legal theory of papal monarchy. In particular, over a century passed before the Crusades would show "to the alert and sensitive canonistic mind the immense possibilities that existed for the expansion of papal power" in the form of the papalist claims to world monarchy.[4]

An argument analogous to Gregory VII's—that limitation of spiritual power to the soul alone is unreasonable—could be constructed all the more easily in Shi'ite Islam, where the dividing line between matters spiritual and temporal is less sharply drawn and where the Imam, the original source of hierocratic authority, held both temporal and spiritual authority. But it would only be made under specific circumstances. What alerted the mind of Khomeini and a number of other Shi'ite jurists to the immense possibilities for the expansion of the Shi'ite hierocratic power was (1) the opportunity to lead a crusade against foreign, imperialist domination that had presented itself recurrently since the nineteenth century; and (2) a desperate struggle for the very survival of the Shi'ite religious institutions against the onslaught of the modernizing Pahlavi state, beginning in the 1960s and becoming more intense in the mid-1970s.

To approach the problem from the opposite angle, from the viewpoint of the state, let us turn to a pre-Islamic—a Sasanian—tract on rulership and statecraft that has come down to us in an eighth century Arabic translation, the testament attributed to Ardashir, son of Babak, the King of Kings to his successors among the Persian kings:

> Know that kingship and religion are twin brothers; there is no strength for one of them except through its companion, because religion is the foundation of kingship, and kingship the protector of religion. Kingship needs its foundation and religion its protector, as whatever lacks a protector perishes and whatever lacks a foundation is destroyed. What I fear most for you is the assault of the populace. Be attentive to the teaching of religion, and to its interpretation and understanding. You will be carried by the glory of kingship to [display] disdain towards religion, its teaching, interpretation and comprehension. Then there will arise within religion leaders lying hidden among the lowly from the populace and the subjects and the bulk of the masses—those whom you have wronged, tyrannized, deprived and humiliated. And know that a clandestine leader in religion and an official leader in kingship can never coexist within a single kingdom, except that the leader in religion expropriates what is in the hands of the leader in kingship. [This is so] because religion is the foundation and kingship the pillar, and the lord of the foundation has prior potency over the entire edifice as against the lord of the pillar.[5]

This passage, which very probably dates from the sixth century after the mass heretical religious uprising of Mazdak, is remarkable in retrospect as a foretelling of the Islamic Revolution of 1979. It sees the roots of popular revolutionary uprisings in the disturbance in state-hierocracy relations. From the nineteenth century onward, the most important source of such disturbance—and the source that made it inevitable—was the modernization of the state.

I have elsewhere presented the establishment of a dual structure of authority in nineteenth century Iran as the institutional unfolding of the logic of Twelver Shi'ism after the disappearance of the last caesaropapist remnants of the era of millenarian extremism. By 1800, Fath 'Ali Shah had recognized the power and legitimatory influence of the hierocracy and given up his predecessor's futile claim to descent from the caesaropapist Safavids. He came to terms with the *mojtaheds* some of whom expounded on the normative foundations of kingship. The prominent *Osuli* jurist and the most important *mojtahed* resident in Iran, Mirza Abo'l-Qasem Qomi, wrote on the necessity of concord between religion and kingship along the lines of Ardashir's testament.[6] However, at the very time the dual division of authority was being established in Qajar Iran, the seeds of discord and inevitable antagonism in hierocracy-state relations were being sown by the modernizing reforms of 'Abbas Mirza. It is true that, at first, 'Abbas Mirza and his minister went out of their way to cultivate good relations with the *'ulama* and to justify the military reforms in Islamic terms. They also succeeded in enlisting the support of all the major *mojtaheds* for mobilizing the masses for the 1812 war with Russia by declaring *jihad*. However, the prince came to regret this policy bitterly and, in 1827, ordered Qa'em-Maqam, the younger, to cease all association with the *'ulama* who were no more useful "than overfed horses who have forgotten their function of running," and to "shake himself of their dust."[7]

The immediate causes of 'Abbas Mirza's disillusionment with the policy of enlisting the support of the Shi'ite hierocracy for the state were that the *mojtaheds* had pushed him into a second, totally disastrous war with Russia when he did not want it and that a cleric, Mir Fattah, had arranged behind his back the surrender of Tabriz to the Russians. Yet, the conflict between the state and the hierocracy had a more fundamental cause: Modernization and the expansion of the power of the state was, sooner or later, bound to entail restricton of hierocratic power.

With the beginning of 'Abbas Mirza's military reforms during the first decade of the nineteenth century, we also witness a keener interest in the pre-Islamic imperial past of Iran, an interest that was to reach its culmination under the Pahlavis. 'Abbas Mirza chose the names of most of his sons from Ferdawsi's *Epic of the Kings* and raised his sons and grandsons with secular education in Persian literary and historical culture. The more gifted in the new generation of princes wrote histories of ancient Iran, the first ones after centuries of relative neglect. 'Abbas Mirza's son, Mohammad Shah, received the first Persian translation of the inscriptions on Bisotun from Sir Henry Rawlinson. Yet after Fath 'Ali Shah, while the nineteenth and twentieth century monarchs' attention was increasingly drawn to the glories of pre-Islamic imperial Iran, with one possible exception, they neglected the fundamental precept of pre-Islamic statecraft contained in the testament attributed to Ardashir, founder of the Sasanian Empire. This time, the eventual result was not a heretical mass uprising but a massive revolt led by the beleaguered guardians of Shi'ite orthodoxy. During the reign of Mohammad Shah (1834–48), the tension caused by the centralization policies, first of Qa'em-Maqam the younger and then of Haji Mirza Aqasi, aggravated by the strong Sufi inclination of the monarch and Aqasi, produced a rift between the state and the hierocracy. Naser al-Din Shah's (1848–96) first prime minister, Amir Nezam, had no doubts about the necessity of breaking the hierocratic power for the success of his centralizing reforms.[8] Nevertheless, the Babi millenarian uprisings, whose aim was the overthrow both of the Qajar state and the Shi'ite hierocracy,[9] drew the *'ulama* closer to the state. The rift between the hierocracy and the state was largely repaired after the fall of Amir Nezam in 1851.[10] The theory of the two powers, and of the interdependence of kingship and religion, was reiterated during the reign of Naser al-Din Shah, as they were in the first decade of the present century by Shaykh Fazlollah Nuri.

The rift between the hierocracy and the state could not become critical so long as the attempts at centralization and modernization of the state remained feeble and largely ineffective, as they did in the nineteenth century. It did become critical and irreparable in the twentieth century when effective measures towards centralization and modernization of the state were taken under the two Pahlavis.

Confrontations with the State During the Constitutional Revolution

Despite the endemic possibility of conflict, the dual structure of religious and political power prevailed throughout the Qajar period. The existence of this dualism in the power structure offered possibilities that would not be neglected by leaders of political movements in the modern period. Ironically, it was the late nineteenth century advocates of reforms, as well as the large merchants, who first thought of exploiting the influence of the leading figures in the Shi'ite hierocracy, and the latter's independence from the state, for the purpose of putting pressure on the ruler to carry out badly needed reforms and to preserve the national economic interests against imperialist encroachment. The idea worked brilliantly. A handful of intellectual activists, with the strong support of the important merchants, were able to uncover the tremendous political potential of the use of clerical domination over the masses for the purpose of mass mobilization. A nationwide embargo on the use of tobacco could thus be successfully orchestrated in 1891 and early 1892, and it led to the repeal of a monopolistic tobacco concession to a British company.[11] A decade and a half later the endemic rivalry within the hierocracy and the clashes between the hierocracy and the state were exploited by the advocates of constitutional government to generate a national movement and to obtain the grant of a constitution from the monarch in 1906.

Both the above movements were at the time viewed not in terms of interests of any estate, class, or civil society, but as confrontations between the nation (*mellat*) and the state (*dawlat*), with the Shi'ite hierocracy assuming the position of "leaders (*ru'asa*) of the nation." Although the structure of authority in the Qajar period was dual, the separation of religious and temporal authority was *not* based on a secular conception of society or nation. Even though his authority over the subject did not, in principle,[12] extend beyond matters temporal, the ruler was yet "the King of Islam" and of the Shi'ite nation. This meant that the protection of Islam and of the Shi'ite nation against the enroachments of the infidels was one of his prime responsibilities. If the ruler failed to carry out his foremost duty of the protection of Islam and of the Shi'ite nation or community, the hierocracy would consider it incumbent upon them to carry out this obligation by assuming the leadership of the nation conceived of as no other than the Shi'ite community.[13]

Against this background, the leading role of the *'ulama* in protests against the government's sale of Iran to the infidels during the first years of this century[14] and in the popular movement of 1905–1906 becomes easily understandable once they are viewed as the contemporaries viewed them: a continuous struggle between the nation (*mellat*) and the government (*dawlat*), which was not only tyrannical but was also engaged in selling the country to foreign imperialists. The autonomy of the Shi'ite hierocracy from the Qajar state made the conflict between the two institutions possible, indeed endemic. The administration of bastinado to the *mojtahed* of Kerman and harsh treatment of other clerics actualized the potential for conflict between the hierocracy and the Qajar state and were thus among the chief causes precipitating the crisis of 1905.[15] In this conflict between the Shi'ite nation and the government, the famous orator Sayyed Jamal would find it natural to appeal, in December 1905, to the *'ulama* to discharge their responsibility of leadership as "the leaders (*ru'asa*) of Islam and the deputies of the Imams."[16] It would be equally natural to speak of the letter of the Constitutionalist *mojtahed*, Tabataba'i, to Mozaffar al-Din Shah as the letter of "the leader of the nation (*ra'is-e mellat*) to the First Person of the State (*shakhs-e avval-e dawlat*);"[17] and for Tabataba'i to sign his telegram to the Emperor of Japan as "the leader of the Islamic nation" (*ra'is-e mellat-e Islam*).[18]

The accounts of the first year of the Constitutional Revolution leave us in no doubt that the prominent *mojtaheds* of Tehran were under constant pressure, not infrequently in the form of assasination threats, to assume their responsibility for the leadership of the nation, to act as spokesmen for the people, and not to relent before obtaining their demands in full.[19] Early in 1906, two *mojtaheds* of Tehran, Sayyed Mohammad Tabataba'i and Sayyed 'Abdollah Behbehani, responded affirmatively and took up the cause of the Shi'ite nation, being supported by a number of less important clerics. In mid-July 1906, the expulsion of a popular preacher and the assassination of a young cleric exacerbated the hierocracy-state conflict, forcing the *'ulama* to act not only as the spokesmen of the people but also in defense of the honor of their own clerical estate. The *'ulama* migrated to Qom. Even the eminently worldly and staunchly conservative *mojtahed* of Isfahan, Aqa Najafi, who could not be suspected of harboring any sympathy for the popular demands, would feel obliged to demonstrate his solidarity with his colleagues against the state by sending an envoy to Qom.[20]

If the conflict was conceived as a conflict between the nation and the state (*mellat va dawlat*), so was its resolution universally hailed as the "union of the nation and the state." The last demand of the *'ulama* before leaving Qom had been the foundation of a "gathering and assembly of justice" (*majma' va majles-e 'adalat*) in order to attain "the important Islamic purpose": "the strengthening of Islamic kingship." When what was intended to be such a *majles* was inaugurated in the presence of seventeen *mojtahed*s as the leaders of the Shi'ite nation in October 1906, due stress was laid on the "union of the nation and the state" in Mozaffar al-Din Shah's inaugural speech.[21]

However, as we have seen in Chapter 2, the situation changed rapidly as the functions and jurisdictions of the Majles were gradually defined. From then on, eminent Shi'ite *'ulama* began to realize one by one that, despite their indispensable role in mobilizing the masses and their continued ceremonial prominence, the directing spirit was that of the Westernized intellectuals, a spirit that posed a serious threat to their vested cultural and material interests. The majority of the religious dignitaries followed Shaykh Fazlollah Nuri in supporting the restoration of autocracy in 1908, but judiciously withdrew from the royalist camp once the failure of the attempt became evident in 1909. By contrast, the eminent *mojtahed* of Najaf, Akhund-e Khorasani, seconded by Shaykh 'Abdollah Mazandarani, persisted in championing the cause of constitutionalism and thus assuming the "leadership of the nation" in confronting the restored autocracy.

With the restoration of constitutional government came a serious menace to hierocratic power in the form of political radicalism which, by 1910, had claimed the lives of two eminent *mojtahed*s, Nuri and Behbehani,[22] and a number of lesser clerics. By 1911, the Shi'ite hierocracy had become disillusioned with constitutionalism and were predominantly hostile to it.

The Era of Pahlavi Modernization

It is interesting to note that it was Teymurtash, Reza Shah's powerful minister of court, who commissioned the translation of the *Testament of Ardashir* into Persian.[23] Yet it is the complete neglect of the testament of Ardashir by the Pahlavi rulers—and this would necessarily be true of any modernizing autocrat from Peter the Great onward—rather than any tendency within Shi'ism that aggravated the rift between the Shi'ite hierocracy and the centralized and modernizing Pahlavi state. Distrustful of the secular intelligentsia in general, and of the leftist radicals in particular, the Shi'ite hierocracy twice turned to the Pahlavis, twice giving them crucial support in critical periods. During his rise to supreme power (1921–25), Reza Khan feigned ostentatious displays of religiosity and successfully courted the hierocracy, exploiting their fears of anarchy and of Westernism and especially their alarm at Atatürk's abolition of the Caliphate in Turkey early in March 1924, which they saw as the inevitable result of republicanism.[24] Late in March 1924, Reza Khan went to Qom to visit the Shi'ite religious leaders about to return to the holy cities of Iraq after their expulsion by the British. In May, he ordered religious ceremonies to be held by the army and the police. In June, he received, with military pomp and ceremony, a portrait of Imam 'Ali, the Commander of the Faithful, sent to him from Iraq by the returning *mojtahed*s.[25] In December, he visited those *mojtahed*s in Najaf during a pilgrimage to the holy shrines obviously undertaken for publicity. The leading Shi'ite religious dignitaries supported Reza Khan and helped him to oust the Qajars and ascend the throne.[26] At least two of them had publicly branded those who opposed Pahlavi's rule as enemies of Islam. In return, Reza Khan had stated that he discarded the idea of a republic and opted for monarchy in deference to their views.[27] Reza Shah did not fail to express his gratitude in the speech he delivered at his coronation on April 24, 1926:

> My special attention has been and will continue to be given to the preservation of the principles of religion and the strengthening of its foundations because I consider the complete reinforcing of religion one of the most effective means of achieving national unity and strengthening the spirit of the Iranian society.[28]

Reza's son, Mohammad Reza Shah, was also given crucial clerical support at the most critical moment of his reign. Though commentators have shown astounding forgetfulness of the fact, the support of the Ayatollahs Kashani and Behbehani—son of the Constitutionalist *mojtahed*—was as important as that of the CIA in staging the return of Mohammad Reza Shah after his flight to Rome in 1953.[29]

If the hierocracy had expected anything in return for their important support, they were to be rudely disappointed. In both instances, the cordiality between the hierocracy and the Pahlavis lasted only as long as it suited the latter: for some half decade after 1921 and a slightly longer period after 1953. Once securely ensconced on the throne, both Pahlavi monarchs felt they could dispense with the support of the hierocracy. Reza Shah carried out a vigorous program of modernization and centralization in the 1920s and 1930s, Mohammad Reza Shah in the 1960s and 1970s. Pahlavi centralization and modernization of the state destroyed the Qajar division of authority in the polity between the hierocracy and the state and devastated the institutional foundations of clerical power and cultural influence. The Shi'ite hierocracy became irreconcilably alienated from the Pahlavi state.

In 1927, there were very clear signs that the honeymoon between Reza Shah and the Shi'ite hierocracy was over. In the early months of 1927, Davar took over the Ministry of Justice with an ambitious reform program. With the abolition of the special consular jurisdiction of foreign powers in May 1927, new procedural regulations for the religious courts were laid down. The hitherto independent religious judges and other members of the Shi'ite hierocracy found these disagreeable. A more serious violation of the traditional autonomy of the hierocracy was the conscription of seminarians (*tollab*) and young clerics during the implementation of the compulsory national military service law. And because cities did not supply soldiers in pre-modern Iran, conscription was also hard for the urban families to accept; those whose sons were being taken away from them for the army turned to the hierocracy to voice their protest against the state. Serious disturbances occurred in the cities, notably Isfahan, and a number of prominent *'ulama* assembled in Qom in protest against the conscription of the young clerics. Minister of Court Teymurtash favored the use of force, but Prime Minister Mokhber al-Saltaneh, a devout man with little taste for the secular culture and its libertinism, went to Qom in person and managed to appease the hierocracy. But the truce was brief. Mokhber al-Saltaneh's advice on the necessity of maintaining a measure of concord with the hierocracy was ignored. Late in 1928, the state declared war on the hierocracy by the issuance of rigid restrictions on wearing clerical garb and the forcible removal of turbans as part of its campaign for enforcing the laws for adoption of European clothes and hats and by licensing liquor stores in the holy city of Qom.[30] Reza Shah himself traveled to the holy city and publicly humiliated Ayatollah Bafqi in the shrine of Qom by beating him and having him dragged by the beard. Internal exile began to be the fate of an increasing number of disgruntled clerics.[31]

The hierocracy-state relations deteriorated much further in the 1930s. What Reza Shah brought back from his visit to Atatürk in Turkey in June 1934 was explicit anti-clericalism. In a speech in Turkey he vowed to eradicate religious superstitions during his reign. The teaching of the Koran and religious instructions in schools were prohibited. In March 1935, from the first day of the Persian new year, the police were ordered to remove the women's veil in the streets by force.[32] The unveiling of women in particular outraged the hierocracy as the most violent rape of Islam. It provoked a serious confrontation, which ended with the bloody suppression of a clerically organized anti-government gathering in the mosque of Gawharshad in Mashhad in the summer of 1935.[33]

The erosion of clerical control over education had begun even before the Constitutional Revolution. It culminated in the creation of a secular, national educational system with the implementation of Reza Shah's educational reforms. Control over education was the least defensible of clerical prerogatives as it was a contingent fact, lacking any doctrinal basis. But more defensible clerical citadels also fell under the attack of the centralizing state. The 1930s witnessed the major defeat of the hierocracy in the judicial sphere, a sphere where clerical domination rested on a firm doctrinal basis. Religious courts were replaced by the secular courts of the Ministry of Justice, and the function of authentication and registration of deeds was transferred to the notaries of the state by the regulations of 1928 and the law of 1932. Finally, the Endowments Law of 1934 established a centralized control over religious endowments throughout Iran, which had largely been under direct or delegated control of the hierocracy.[34]

These developments seriously weakened the hierocracy. They also had another important consequence: *The differentiation and the separation of religious and political powers became virtually complete.* The embeddedness of the hierocracy in the Iranian polity was further undermined by Mohammed Reza Shah's land reform of the 1960s, which resulted in the redistribution of land owned by mosques and seminaries. The religious institutions became totally independent of the state. This independence was sustained by the one source of income inevitably immune from state encroachment: the voluntary payment of religious taxes to the chief Shi'ite *mojtahed*s as the vicegerents of the Hidden Imam. With the economic prosperity of the 1960s and 1970s, this revenue increased considerably.

The loss of judicial and educational functions on the one hand, and the loss of control of the religious endowments and of land ownership on the other, meant that the Shi'ite hierocracy became by and large "disembedded" from the Pahlavi regime. They became, in the words of an observer, a *declassé* stratum.[35] This economic and political disengagement of the hierocracy was strongly complemented by their social "disembeddedness": the upper echelons of the hierocracy formed a highly endogamous quasi caste, the entry into which by bright young men was almost invariably accompanied by marriage to daughters of their teacher.[36] What is even more important is the greatly reduced intermarriage and affiliation between the *'ulama* and the political elite in the Pahlavi era. Thus, by 1970s, the *'ulama* retained very few of the economic, social, and political ties that had bound them to the court, the state, and the landlords and *tuyul*-holders in 1907. At the same time, they had become more homogeneous and much more distinct from the secular intellectuals and professionals produced by the modern educational system. In his recent work on revolutions, Eisenstadt emphasizes the relevance of the degree of autonomy and disembedment of a leading social stratum—an elite—to the generation of revolutionary social change. Such autonomy facilitates development of coalitions with broader groups, and tends to result in far-reaching restructuring of social institutions.[37] The disengagement of the Shi'ite hierocracy from the Pahlavi regime, the increased homogeneity of the *'ulama* as an estate (status group), and the sharpened distinctness of their identity from secular intelligentsia go a long way toward explaining how they came to lead the first successful traditionalist revolution in modern history.

The disengagement of the hierocracy from the Pahlavi polity further disposed the religious institution toward perpetuating the alliance with the bazaar. On the side of the bazaar were the merchants and the guildsmen, who constituted the traditional sector of Iran's urban economy in the 1970s, and who correctly perceived the threat to their long-term economic interests posed by Mohammad Reza Shah's policy of industrialization and sought to cement their bonds with the hierocracy.[38]

Furthermore, the hierocracy, progressively disengaged from the state, increasingly reaffirmed their engagement with the people. In the preceding paragraphs, we discussed the historical roots of unrivaled clerical domination over the masses. This domination entailed a considerable measure of populism in the orientation of the hierocracy. Not unlike the Spanish clergy from the seventeenth century to the first decade of the nineteenth century, the Iranian *'ulama* frequently took up the cause of the oppressed against the arbitrary excesses of temporal authorities. More important, from the 1870s onward the hierocracy emerged as the champion of the Islamic nation against the economic penetration and cultural influence of foreign powers. Not unlike the Spanish priests and monks who led the masses in the war of independence against Napoleon and "atheistic France" and whose wrath was subsequently turned against the "atheistic liberals,[39] the Shi'ite hierocracy led the Muslim nation of Iran against the economic domination *and* the cultural influence of the imperialist infidels. To their traditional pattern of activity in defense of Islam—the persecution and killing of heretics, such as Sufis and Babis (most of whom subsequently became Baha'is)—was added combating foreign cultural influences and the violation of traditional cultural and religious norms by an increasingly Westernized political elite.

However, a serious emphasis on social justice was lacking in the teachings of the Shi'ite religious institution. This was due in part to the extensive participation of the *'ulama* in the Qajar polity and also to the fact that prior to the advent of modern (party) politics the loyalty of the masses could be taken for granted once heresy was suppressed. The situation changed drastically from 1962 to 1978 when the religious institutions came under relentless attack by the Pahlavi state and had to court the masses more assiduously in order to mobilize them in their defense. Its populism became markedly more pronounced and an emphasis on social justice began to enter the writings of the clerical pamphleteers.[40]

During the 1928 to 1941 period, the hierocracy were perhaps too surprised and stunned to react effectively. In any event, the foremost religious leader of the time, the Grand Ayatollah 'Abdo'l-Karim Ha'iri, opted for political quietism and building up of a center of religious learning in Qom. (This apolitical action was continued in the period after World War II by the Grand Ayatollah Hajj Hosayn Borujerdi, who led the hierocracy until his death in 1961.) After the fall of Reza Shah, a nationwide agitation of the hierocracy was led by the Grand Ayatollah Tabataba'i Qomi in 1944. The hierocracy demanded a more strict observance of the provisions of the Shi'ite Sacred Law on morality and succeeded in removing the prohibition on wearing the veil and clerical garb.[41] Imitating cultural patterns of the Western infidels came under heavy attack. A collaborator of Qomi and an early supporter of Reza Shah, Ayatollah Abo'l-Qasem Kashani, remained active in politics and became a dominant figure on the political scene until 1953. Emerging from the rigors of Reza Shah's dictatorship, the hierocracy showed an appreciation for the constitution, which subjected the power of the monarch to very considerable restraints. Kashani's platform, therefore, combined the elements of opposition to foreign domination over the Islamic people (the Anglo-Iranian Oil Company in Iran and Israel in the Middle East) and appeal to the Sacred Law, with a somewhat novel stress on the Constitution as the source of legitimacy.[42]

As we have pointed out, by 1911 the Shi'ite hierocracy was predominantly hostile to Constitutionalism. Nevertheless, because of their amorphous organization, the hierocracy did not, and could not, act as one body. The religious leaders had played a prominent role in the initial phase of the Constitutional Revolution and a few religious dignitaries, most notably Sayyed Hasan Modarres, had remained active in the nationalist provisional government in Kermanshah and in the parliamentary politics of the early 1920s. It was therefore possible for the religious leaders to appeal to the Constitution plausibly and effectively from the 1940s onward in order to protest against the arbitrariness of the state. At the same time, most of the *'ulama,* including Grand Ayatollah Borujerdi, were becoming increasingly alarmed by the growth of the Tudeh party and drew closer to the monarch. Even Ayatollah Kashani abandoned Mosaddeq for the Shah. In 1925, the leaders of the Shi'ite hierocracy had supported the foundation of the Pahlavi Dynasty as a bulwark against republicanism; in 1953, they supported the preservation of monarchy and the return of the Shah as a safeguard against the spread of communism.

Thus, most religious leaders had forgotten their old grievances against the first Pahlavi by the 1950s and were ready for an accommodation with the young Shah, who was more than conciliatory while his rule remained precarious[43]—most but not all, and certainly not Ayatollah Ruhollah Khomeini, who saw the reassertion of royal power and the initiation of a new reform program by Mohammad Reza Shah in the 1960s as replete with motifs already encountered during the dreadful reign of the first Pahlavi. In 1962 Khomeini emerged as one of the leading figures in an anti-government protest that the Islamic revolutionary leaders of the 1980s are unanimous in regarding as the beginning of their movement.[44] Early in October 1962, the government publicized a bill for the election of town councils that eliminated the profession of Islam as a condition for the electors and the candidates, used the term [oath by] "the Heavenly-Book" instead of the Koran, and enfranchised women. Khomenini vigorously reacted against all these propositions. He denounced the bill as the first step toward the abolition of Islam and the delivery of Iran to the Baha'is, the presumed agents of Zionism and Imperialism who were implicitly enfranchised by the bill alongside women. The enfranchisement of women was vigorously denounced as a ploy to destroy family life and spread prostitution. Clerical agitation continued and was intensified after the Shah proposed, in January 1963, a national referendum on six principles of his reform program, subsequently to be called the "White Revolution." Khomeini denounced the referendum vehemently. The Shah's suddenly increasing social activism and mobilization drive, in the form of a widely publicized Peasants' Congress to celebrate the land reform in January 1963, must have alarmed Khomeini and roused his apprehension. In March 1963, holding a copy of the Koran in one hand and a copy of the Constitution in the other, Khomeini publicly accused the Shah of violating his oath to defend Islam and the Constitution.[45] The authoritarian rule of the Shah was denounced as a violation of the Constitution, and he was attacked for the maintenance of relations with

Israel. Massive demonstrations by Khomeini's followers were brutally suppressed in June 1963. Khomeini himself, having been imprisoned during the violence, was exiled to Turkey in November 1964 and settled in Najaf in Iraq the following year.

The Shah scored a victory by presenting Khomeini's opposition as "black reaction" to his reform program. During the subsequent fifteen years of relentless opposition from exile, Khomeini did, on occasion, dismiss the Shah's reform program as a fraud. The brunt of his attack, however, was against the following: (1) the Shah's autocratic rule, culminating in the violent denunciation of the celebration in 1971 of the twenty-five hundredth anniversary of the founding of the Persian Empire; (2) denunciation of close ties with and subservience to the United States; and (3) the disregard of Islamic morals and government-sponsored "spread of prostitution" to corrupt the nation and perpetuate the imperialist cultural domination.

Meanwhile, the Shah had initiated a ruthless attack on religious institutions. In the 1960s and 1970s, he took a series of severely repressive measures, which included assaults on the main theological college of Qom (1963 and 1975) and the destruction of most of the theological seminaries of the holy city of Mashhad in 1975 under the pretext of the creation of a greenspace around the shrine of the eighth Imam. Furthermore, the Shah replaced the Islamic calendar with a fictitious imperial one, and embarked on an attempt to invade the religious sphere proper by creating a Religion Corps (modeled after the Literacy Corps) and a group of Propagators of Religion. Despite their lack of vigor and their inefficiency, the religious leaders perceived these measures as a bid to liquidate the religious institution and annihilate Shi'ism altogether.

In the 1970s the predicament of the Shi'ite religious institution was aggravated by the pressure from the Iraqi Ba'thist regime on the resident Shi'ite *'ulama,* which did not relent after the Iraqi government's rapprochement with the Shah in March 1975. According to one informed estimate, the number of scholars and students at the Shi'ite centers of learning in Iraq had declined from 3000 to 600 in the years preceding the revolution in Iran.[46] All this convinced the *'ulama* that the secular state, in whatever form, was the chief enemy of Islam.

With their backs to the wall, the hierocracy within Iran increasingly heeded Khomeini's incessant appeals, and the latter's position among the grand ayatollahs was strengthened. Clerical reaction to the Shah's aggressive encroachments was to prove decisive. In an interview conducted in 1975, a prominent cleric spoke of "the awakening of Iran's religious community after the frontal attack of His Majesty." He went on to boast about the hierocracy's new political maturity: in the 1960s the eligibility of women for voting was a major preoccupation of the religious leaders; now with their increasing political savvy, they would not lose any popularity by incautiously opposing women's electoral rights.[47]

Having examined the *'ulama*'s relationship with the monarchy, let us now turn to their relations with the intellectuals. For the Shi'ite hierocracy, the ex-

perience of the period of the Constitutional Revolution—most notably the execution of Nuri and the assassination of Behbehani—left a bitter feeling of
distrust toward the Westernized intelligentsia generally and of alarm and resentment toward its radical faction, the Democrats.[48] It thus created a rift
between the religious and the secular intelligentsia, a rift reflected in the
acrimonious debates of the Fourth Majles which preceded the rise of the first
Pahlavi to power.[49] The mutual antipathy between the clerical and the lay
intelligentsia persisted after World War II, and reinforced by the hierocracy's fear of the growing influence of the leaders of the Tudeh Party, eventually wrecked Mosaddeq's nationalist government. Ayatollah Kashani, the
most prominent clerical politician of the period, fully supported Mosaddeq in the summer of 1952 but, not having received favors that he
had wanted in exchange for this support, orchestrated Mosaddeq's downfall
a year later.[50] Behind Kashani's personal dissatisfaction, there was a more
fundamental cause of tension in the relationship between Mosaddeq and the
hierocracy. The great popularity of Mosaddeq with the bazaar and his nationalist platform was making the bazaar-mosque political alliance obsolescent by
providing the former with viable alternative secular leadership. Years later, the
clerics resented being let down by the National Front (Mosaddeq's followers)
during their uprising in 1963.

The oppositional hierocracy came to regard the Westernized intelligentsia
as an integral part of a virtual xenocracy that had betrayed the Islamic tradition and double-crossed its custodians at critical junctures. When Khomeini
embarked on his bid for the overthrow of the Pahlavi regime around 1970,
he had in mind to settle not one but two old scores: to avenge himself and the
Shi'ite hierocracy against the two Pahlavis, and to turn the tables on the
Westernized intellectuals who, according to him, had cheated the hierocracy
in all the important nationwide movements of the preceding century. As we
now know, having ejected the Pahlavis, he wasted no time in initiating a
massive *Kulturkampf* against the Westernized intelligentsia.[51]

As a result of the dislocation caused by excessively rapid social change and
a mismanaged economic policy, popular discontent mounted while petrodollars sapped the vigor and commitment of the upholders of the regime.
When the first crack suddenly appeared and the imposing edifice of the
Pahlavi state began to crumble from within, the autonomous heirocracy could
rejoice at the prospect of defeating and subjugating the impiously arrogant
state.

II

FROM TEMPORAL TO
THEOCRATIC ABSOLUTISM

Thus was my turning from the King (*Shah*) to the Canonist (*faqih*):
Fleeing the ant, I entered the mouth of the dragon.

Naser-e Khosraw (d. 1072–77)

5

Khomeini and the Islamic
Revolutionary Movement

The Resurgence of Islam

The facile presumption of a worldwide "secularization of culture" is one of the more important commonly held misconceptions convincingly dispelled by the recent upheaval in Iran. In view of an uncontestable historical association between urban strata and congregational religiosity, especially of the ethical type, the contemporary Islamic revival in the wake of rapid urbanization in the Islamic world should not have generated the surprise it has. The connection between congregational religion and urban life is at least as firm, if not firmer, in Islam as in Christianity and other congregational religions. In its classic pattern, cities with their mosques and *madrasas* (seminaries) constitute the centers of Islamic orthodoxy and rural and tribal areas constitute the periphery, comprised of those who, according to the Koranic phrase, are "most stubborn in their unbelief and hypocrisy."[1] Movement from the rural periphery into urban centers has thus been historically associated with increasing religious orthodoxy and a more rigorous adherence to the legalistic and puritanical central tradition of Islam.[2] The contemporary revivalist movement in Shi'ite Islam follows this classic pattern.[3]

Between 1956 and 1976, the urban population of Iran nearly trebled while the rural population increased by just over one-third (see Table 11, Appendix). Internal migration accounts for a substantial proportion of this growth in urban population (see Table 12, Appendix; note that the offspring of the migrants born in the cities are *not* included in the figures). Furthermore, it is worth noting that the percentage of rural to urban internal migrants is considerably higher for the most developed and urbanized Central Province (capital, Tehran), 46.6 percent in 1972.[4]

In the late 1960s and 1970s, these expanding urban centers of Iran witnessed an increased vitality in a variety of religious activities.[5] With the spread of literacy and the creation of a public sphere in the period after World War II, Shi'ite traditionalism, advocated through the writings of a number of clerics and laymen, became a distinct trend. From the 1960s onward, traditionalism gathered impressive momentum.

Religious periodicals gained progressively wider circulation and religious books became more and more popular. A survey in 1976 found forty-eight publishers of religious books in Tehran alone, of whom twenty-six had begun their activities during the 1965 to 1975 decade. Other indicators, such as the number of pilgrims to Mecca or visits and donations to religious shrines, support the assumption of the increased vitality of traditional religious sentiments. During the summer of 1977, while conducting interviews with prominent clerics in Tehran and the provincial towns, I was struck by the number of times the interviews had to be interrupted because of requests for *estek-hareh* (Koranic bibliomancy), usually over the telephone. In 1973, *Mafatih al-Jenan* (Keys to the Garden [of Heaven])—a book singularly maligned by the modernist Shari'ati for representing the most other-worldly aspects of fossilized traditional Shi'ism—sold 490,000 copies and was second only to the perennial bestseller, the Koran (about 700,000 copies).[6]

Furthermore, a large number of religious associations mushroomed among laymen. From 1965 onward, there was an astonishing growth in their number. Often associated with the groupings of humbler occupations or of poorer city quarters, the associations met mostly during the religious months of Moharram and Ramadan but occasionally at other times. By 1974, there were 322 *Hosayniyyeh*-type centers for commemoration of the martyrdom of Imam Hosayn and other religious events in Tehran, 305 in Khuzestan, and 731 in Azerbaijan. In addition, there were over 12,300 religious associations in Tehran alone, most of which were formed after 1965. Of these associations 1821 designated themselves formally by a title. These titles typically refer either to the guild or profession of the members, to their geographical town or region of origin, or to their aspirations. As such, they are highly revealing of the social background of their members and of the type of religious sentiment motivating them to form these associations.

The unmistakable impression given by the titles is that their members fall into two quite possibly overlapping social groups: lower middle class guilds and professions associated with the bazaar economy, and recent migrants from the provinces. Here are some typical examples: religious associations of shoemakers, of workers at public baths, of the guild of fruit-juicers (on street corners), of tailors, of the natives of Natanz resident in Tehran, of the natives of Semnan, of the Desperates (*bicharehha*) of [Imam] Hosayn, of the Abjects (*zalilha*) of [Imam] Musa ibn Ja'far. Furthermore, there can be no mistake that their religiosity is solidly traditional. A survey of the immigrant poor in Tehran in the 1970s found these religious groups to be the *only* secondary voluntary associations of significance among them. As one young squatter told the researcher: "Nothing brings us together more than the love for Imam Husayn. My personal view is that these *hey'ats* (associations) have a positive aspect in uniting us and keeping us informed about each other's affairs."[7] The religious associations thus constituted a major agency for the social integration of the migrant population.

With the spread of the religious associations, demand for preachers and cantors outstripped the supply in the 1970s. The unsatisfied demand created a market for religious tapes and cassettes. By the mid-1970s, a survey reported some thirteen centers for recording and distribution of tapes.[8] The contributions of this organizational network to the success of the Islamic revolution was of crucial importance. Their organizers distributed Khomeini's taped messages and carried out the task of planning the massive demonstrations during the winter of 1978 and of enforcing order and discipline during those demonstrations.

A religiously inclined section of the rapidly expanding middle class, often from traditional bourgeois parents, also took part in the Islamic revival. From the early 1960s onward, an increased number of religious societies were formed in the universities, abroad, and by engineers and physicians, and their activities became interlinked with those of clerical publicists. Mahmud Taleqani was the most notable of this group of clerical publicists, which included such younger figures as Mortaza Motahhari and Mohammad Hosayn Beheshti, who combined a concern for Islamic reform with formulation and dissemination of an Islamic traditionalist ideology attractive to the intelligentsia.

The first Islamic society in the University of Tehran had been founded by Mehdi Bazargan in 1942, and he remained a major influence throughout the 1960s. But the 1960s witnessed the emergence of a more charismatic advocate of Islamic reform who had a much stronger bent for revolutionary ideology, 'Ali Shari'ati. At the same time, one Islamic society that had originally followed Bazargan was developing a revolutionary ideology of its own and constituted itself as a radical political group: the Mojahedin-e Khalq.

Above all else, the novel phenomenon of the late 1960s and early 1970s was 'Ali Shari'ati's advocacy of Islamic reform. Shari'ati found Islam the most perfect of the Abrahamic faiths, and at the present time, the most decayed or dilapidated. This decay was due to the fact that the Koran, containing the blueprint for the perfect social life, had been taken from the city, the center of life, to the cemetery, the abode of the dead, and the *Book of Prayers* (*ketab-e du'a*) (to secure supernatural succour and intercession) had been brought to the city from the cemetery. Shari'ati's attempt to recover Islam by going back to the book results in the creation of a radically populist theology of revolution. In a manner reminiscent of Shatov in Dostoevsky's *The Possessed,* Shari'ati equates, as regards matters social, God and the People. He insists that the central Islamic principle of *tawhid* (unity of God)—should correspond to a monistic or classless social order and fulminates against all social and economic stratification as the consequence of coercive (military), economic, and spiritual (clerical) domination. Throughout history this 'trinity' of the forces of domination is held responsible for corrupting the pristine Abrahamic monotheism and the corresponding classless monism of the social order. In this leveling, populist interpretation of Islam, Shari'ati naturally champions the cause of the people, doubly oppressed by the internal forces of domination and by the external force of imperialism. Shari'ati revives the graphic Koranic term *mostaz'afin* (the disinherited) to refer to the oppressed masses and renders Franz Fanon's *Les Damnés de la Terre* in Persian translations as the Disinherited of the Earth, a term that was to occupy a central position in the Islamic revolutionary rhetoric.

Shari'ati's ideas contributed directly to the revolutionary outbreak through his influence on Iranian students and young intellectuals, especially the highly organized and motivated Mojahedin-e Khalq, who did some of the decisive fighting in the fateful days of February 1979. His ideas also had an important influence on the writings of the clerical pamphleteers and preachers, who were quick to take up the rhetoric of social justice and the cause of the Disinherited. Furthermore, Shari'ati's writings won over a substantial part of the lay intelligentsia to Khomeini's side by leading them to believe the Islamic revolution under his leadership would be a "progressive" one. Presumably a model to be followed by himself as reformer, Shari'ati had written that Prophet Mohammad had preserved the *form* of traditional norms but had changed their *content* in a revolutionary manner. Had he lived to see the revolution, which he had projected as the this-worldly enactment of Shi'ite millenarianism, he might have appreciated the irony of the fact that it was the clerical party, one of his forces of domination, who succeeded in preserving the form of modern revolutionary rhetoric that he had introduced while changing its content in a strikingly traditionalist manner.[9]

Revolutionary Politicization of the Islamic Movement

During the Pahlavi period, individual Shi'ite clerics who did not subscribe to Nuri's rejection of the Majles, such as Sayyed Hasan Madaress and Ayatollah Kashani, gained prominence during periods of lively parliamentary politics, such as the 1920s, the 1940s, and the early 1950s. Outside the Majles, too, feeble beginnings of a politicized Islamic movement can be detected during Reza Shah's reign.[10] But religion first emerged as an important factor in Iranian politics in the late 1940s owing to the activities of a terroristic group of young clerics, the Feda'iyan-e Islam (Devotees of Islam) who engaged in the assassination of "corrupt" pro-Western Iranian statesmen, most notably Prime Minister Razmara. The Feda'iyan were protected by Ayatollah Kashani, who as the Speaker of the Majles secured the release of Razmara's assassin by what amounted to a bill of attainder in November 1952. They also enjoyed the support of Khomeini, Taleqani, and other radical members of the hierocracy and have the distinction of being the first group to publish a blueprint for Islamic government in 1950. It is interesting to note that Khomeini, who had gradually established himself as the political advisor of the foremost religious leader, Ayatollah Borujerdi, by the early 1950s, reportedly broke with the latter for Borujerdi's refusal to intercede with the Shah on behalf of the condemned Feda'iyan in 1954.

The period 1962 to 1963 witnessed the birth of a movement led by the militant clerics who remained faithful to Khomeini, continued to protest against his detention and exile, and subsequently kept in contact with him.

Many of the future rulers of the Islamic Republic had their first bitter political experience and embarked on clandestine political journalism and organization while using religious sermons as a political platform. They were, first and foremost, Khomeini's students, and secondly, the middle-aged clerics who collaborated with him in the agitations of the 1960s. Some of the *ad hoc* organizations and groupings, which had come into being to distribute Khomeini's proclamations in Tehran and other cities and to organize demonstrations, continued their existence underground after June 1963. In these clandestine associations elements from the bazaar, the religious youths, and the militant *'ulama* cooperated intimately. One such association, Hey'at-e Mo'talefeh-ye Eslami, had four "clerical advisors" (chosen by Khomeini) attached to it. It is claimed that on a single day in 1964, its 500 young men, simultaneously in Tehran and Isfahan in a matter of a few minutes, distributed 40,000 leaflets containing Khomeini's declaration on the granting of extraterritorial jurisdiction to the United States over its military personnel in Iran. In January 1965 one of its members succeeded in assassinating Prime Minister Hasan-'Ali Mansur, allegedly after obtaining an injunction (*fatva*) to do so from the group's clerical advisers, who included the late Ayatollah Beheshti. It was disclosed in the trial of its arrested members that the association had also planned to assassinate the Shah and set up a "unified Islamic government."[11]

Furthermore, the events of the mid-1960s marked the beginning of cooperation between the militant clerics and petty bourgeois intellectuals, which was to culminate in the Islamic revolution of 1979. After the destruction of the Hey'at-e Mo'talefeh in 1965, Mohammad-'Ali Raja'i—son of a haberdasher who had lost his father at the age of four, worked his way up from peddling to becoming a high school teacher, and was destined to become the prime minister and the president of the Islamic Republic of Iran—joined a young cleric, Mohamad-Javad Bahonar, in picking up the pieces. Bahonar was destined to replace Ayatollah Beheshti as the leader of the Islamic Republican Party and to serve briefly as prime minister during Raja'i's presidency before they were both blown to pieces by a bomb that exploded in his office on August 30, 1981. It is also worth noting that the third collaborator in their effort to reconstruct the demolished secret association was Jalal al-Din Farsi, a chief lay ideologue of the future Islamic revolution.[12]

The agitation of 1963, and its aftermath after Khomeini's exile at the end of 1964, also produced the opportunity for some instant journalism by the young clerics and seminarians and added this modern publicity tool to the traditional repertoire of preaching and inserting political commentaries into the elegiac sermons on the Imams. In his trial for his part in the events of June 1963, Taleqani reportedly boasted that they had been hoping to make the clerics of Qom newspaper readers in fifty years but Ayatollah Khomeini has made them journalists within a few months. Instant clerical journalism continued after Khomeini's exile. On December 20, 1964, the birthday of the Hidden Imam, a newspaper, *Revenge* (*Enteqam*), was issued by the clerical activists of Qom. Its aim was stated as "the enlightening of drugged minds, the awakening of dormant religious sentiments . . . and mobilization of all with the ideological [*sic*] weapon of faith."[13]

Meanwhile, as we have seen, certain social trends in Iran were creating a receptive audience for the militant Shi'ite *'ulama*. Rapid urbanization in the 1960s and 1970s was going hand in hand with increased vitality of religion. The vitality of traditional sentiment among the recent migrants into cities made them receptive to the propaganda of the traditionalist preachers and pamphleteers. These groups remained marginal and excluded from political processes until the early days of 1978, when with dramatic suddenness, they were massively mobilized against the Pahlavi regime by Khomeini's traditionalist party. It should be noted, however, that Khomeini's militant party did face immediate competition for mass audience from other religious leaders. After Khomeini's exile, the Grand Ayatollah Shari'at-madari set up a *Dar al-Tabliq* in Qom to pursue traditional apolitical missionary activities by using modern communications media. There were immediate clashes between the two groups, which produced a rift in the Shi'ite hierocracy between Khomeini and Shari'at-madari.[14]

Clerical publicists and orthodox reformists, such as Ayatollahs Taleqani, Motahhari, and Beheshti, who were bent on the purification of the Shi'ite tradition as much as its ideologization, became associated with the more serious religious associations of laymen and university students. In 1970, Ayatollah Motahhari, for instance, gave a series of talks to the Islamic Association (*Anjoman*) of Engineers, which the latter published under the title of *Khadamat Moteqabel-e Islam va Iran* (Mutual Services of Islam and Iran). In the preface to this volume, the Islamic engineers offer a fairly sophisticated analysis of increasing political awareness and national consciousness in the Third World, which is said to unfold in three stages: In the first stage, the intellectuals of underdeveloped countries engage in pure imitation of Western patterns of life, which are assumed to lead to prosperity. In the second stage, they become aware of the nation to which they belong, but turn to its past history, folklore, and myths; nationalism, imitative of Western nationalist ideology, that is, according to the classic Western definition, is typical of this stage of national awareness among the intellectuals representing the colonial masses. In the third stage this is to be replaced by Islamic consciousness appropriate to a unified Islamic community of believers. In this last stage, which is about to begin, the intellectuals will stop daydreaming about folklore and imitating their Western masters, and they will pay heed to the real factor making for unity in the sentiments of masses. A new and unified Islamic community, a "new nationality of *tawhid* (unity)" (*melliyat-e jadid-e tawhid*), as foretold by the Muslim thinkers of the late nineteenth and early twentieth centuries, will be born.[15]

The growth of Islamic associations among the intelligentsia had far-reaching consequences for the politicization of the Islamic revival, and their control offered the clerical activists a much greater challenge than did the political associations of the recent migrants. Unlike the apolitical migrants who were content with integration into a new religious societal community, the highly politicized intelligentsia in addition demanded political enfranchisement and inclusion in the political system. Any integrative social movement that hoped to gain their adherence would have to offer them the prospect of membership in a new political order in place of the Pahlavi regime from which they were excluded. Furthermore, because of their education, the intelligentsia were less prone to accept the intellectual authority of the 'ulama unquestioningly than were the recent migrants. The major achievement of the clerical activists was to meet both these challenges of offering an ideology attractive to the intelligentsia and of maintaining their intellectual authority and leadership over the latter.

The primacy of their political motive and their self-conception as modern intellectuals account for the obsession of the highly politicized segment of the new Iranian middle class that frequented the religious associations with the creation of an Islamic *ideology*. This meant the arrangement of readily accessible maxims constituting the sources of the Islamic tradition, the Koran and the sayings of the Prophet and the Imams, in accordance with a new pattern suggested by the Western total ideologies such as communism and fascism. A number of clerics took up the challenge of constructing the requisite Islamic total ideology. They already possessed thorough familiary with the source of the Islamic tradition and were quick to learn the art of constructing an ideology from the lay intellectuals. They learned this art both from their opponents—most notably the ideologues of the Tudeh party—and their allies—lay Islamic reformists such as Bazargan and Shari'ati. Here, the importance of intense ideological debates between the Tudeh ideologues and the militant clerics in the Shah's prisons in the 1960s should be noted.[16] It must also be noted that Shi'ite clerical publicists were greatly aided by their knowledge of Arabic, which gave them access to the internationally current Islamic fundamentalist ideology. Writings of Sayyid Qutb, the theoretician of the Muslim Brotherhood, and of Mawlana Abo'l-A'la' Mawdudi were translated into Persian and avidly read. With these writings came the invidious contrast between "Islamic government" as the rule of God on earth and the secular state as the rule of "idolatry" (*taghut*). In his preface to Sayyid Qutb's *Social Justice in Islam,* its clerical translator, Hadi Khosraw-shahi, reveals the basis of the book's appeal to the young intelligentsia:

> Sayyid Qutb is one of the few who . . . have been able to offer to the world Islamic issues *in the style of today,* and in the form of systematic books, *as a living and invaluable ideology* as against (other) social, political, economic schools of Communism, Imperialism, Socialism and Capitalism.[17]

The intelligentsia could have accepted the fundamentalist Islamic ideology transmitted to them by laymen as well as clerics without accepting clerical leadership. But in fact, they accepted clerical leadership too. Two factors seem to account for this. First, the Islamic intelligentsia were aware of the influence of the Shi'ite hierocracy on the masses and sought to use it against the Pahlavi regime. Second, they were impressed by the oppositional role of the *'ulama* to the state in recent Iranian history and full of unrequited admiration for them.[18]

Thus, Iranian Islamic intelligentsia, unlike their counterparts in Egypt and Malaysia, constituted the followers and not the leaders, the consumers and not the producers, of the traditionalist ideology because of the vigor shown by the publicists of the Shi'ite hierocracy. The clerical publicists who appeared to have lost the initiative to Shari'ati in the late 1960s, recovered it after the latter's death in 1977. They formulated the constitutive values and ideas of the movement in the light of their ideal and material interests as the custodians of the threatened religious tradition.

In any event, Khomeini did not restrict his attention to the intelligentsia by any means. He put considerable emphasis on preaching, an activity shunned by the *mojtaheds* and confined to the lower ranks of the hierocracy, and instructed a large number of preachers, who were to spread his message among the common people in Iran and among the Iranian pilgrims abroad.[19] In this, he was aided by the Shah's renewal of his father's policy of the internal banishment of dissident clerics:

> Ironically, exile to isolated provincial towns only allowed these clerics to spread their teaching further afield. The 1970s found Mahdavi-Kani teaching and preaching in Bukan; 'Ali Khamene'i in Iranshahr; Montazeri in Tabas, Khalkhal, and Saqqez; 'Ali Tehrani in Saqqez; and Mohammad-Mehdi Rabbani (a member of the Assembly of Experts that, after the revolution, wrote the new constitution) in Shahr-e Bakak.[20]

It was claimed in the mid-1970s that Khomeini had trained 500 *mojtaheds* throughout his long teaching career and, with less exaggeration, that 12,000 students attended his lectures in the years immediately preceding his exile in 1964.[21] The leading militant *'ulama* who have occupied the highest positions of power since the revolution have been, with rare exceptions, drawn from the large group of former Khomeini students. Socially, they are in all likelihood from the traditional urban background typical of the members of the Shi'ite hierocracy in the second half of the nineteenth and first half of the twentieth centuries. They were keenly aware of the dispossession of the Shi'ite hierocracy by the Pahlavi state and bent on the recovery of lost historical privileges. The younger militant clerics, on the other hand, were heavily drawn from humbler rural and small town backgrounds. For them, the Islamic revolution was to create avenues of rapid upward social mobility. However, the ideological weapon of the ranking militant ayatollahs for recovery of their lost privileges was the same as that of the younger clerics for safeguarding their rapid social ascent: Khomeini's theory of theocratic government or *velayat-e faqih*.

In the late 1960s, Khomeini began to think seriously about an Islamic government as an alternative to the Shah's. His theory of *velayat-e faqih* (Mandate of the Jurist), delivered as lectures in 1970 and published in the same year, is a bold innovation in the history of Shi'ism. He generalized the early *Osuli* arguments, which were designed to establish the legal and religious authority of the Shi'ite *mojtaheds*, to eliminate the duality of religious and temporal authority. Khomeini categorically stated that "the mandate of the jurist means governing and administering the country and implementing the provisions of the sacred law."[22] Having firmly rejected the separation of religion and politics, he argued that in the absence of the divinely inspired Imam, sovereignty devolves upon qualified jurists or the Shi'ite religious leaders. It is, therefore, the religious leaders, as the authoritative interpreters of the Sacred Law, who are entitled to rule. Furthermore, by encouraging his acclamation as Imam, Khomeini paved the way for the eventual restriction of *velayat-e faqih,* and thus the exercise of sovereignty, to Shi'ism's presumed supreme leader.

Throughout the 1970s, Khomeini's sense of alarm at the destruction of the Shi'ite culture and mores was increasing. In March 1975, he would refer to the Shah's White Revolution as "the revolution intended to spread the colonial culture to the remotest towns and villages and pollute the youth of the country."[23] Reacting to the public performance of explicit sexual acts by an avant-garde theater group in the Shiraz festival, he would be moved to say, in a speech made in September 1977 in Najaf, that the function of this government

> is tyranny and oppression and the spread of prostitution. You do not know what prostitution has begun in Iran. You are not informed: the prostitution which has begun in Iran, and was implemented in Shiraz— and they say it is to be implemented in Tehran, too—cannot be retold. Is this the ultimate—or can they go even further—to perform sexual acts among a crowd and under the eyes of the people?[24]

One final significant event, or rather its anticipation, should be mentioned as of utmost importance in turning politicized traditionalism into an intransigently revolutionary movement: the imminent beginning of the fifteenth century of Islam. The Shi'ite scholars' long-established tradition of the designation of a great *Mojadded* (Renovater) for each century, against a background of the sudden explosion of popular rage and the crumbling of the Pahlavi monarchical edifice, can safely be assumed to have changed the clerical estate's conception of time (and certainly Khomeini's) from a chronological to a kairotic one; the moment when time was to be pervaded by eternity seemed at hand, empirically and numerologically.

Certain specific features of Shi'ite Islam were highly suitable for the mobilization of the masses. The noncognitive character of social myth will be discussed in Chapter 6. In so far as the social myth of the Islamic revolution contained a concept embodying the desire to return to tradition and to preserve the threatened traditional norms of social relationships, this concept (i.e., "Islamic government") had to be linked in many ways with powerful images. Such images were drawn from the Shi'ite theodicy of suffering, which centered around the martyrdom of the Third Imam, Hosayn, and his family in Karbala in the year 680. This theodicy had constituted a repertoire of highly emotive imagery used for mobilization of the Shi'ite masses. For over a century, the struggle of the Imam against the Umayyad caliph Yazid had, from time to time, been transfigured into the archetype of the conflict between justice and tyranny. When Khomeini compared the Shah to Yazid in 1963, and much more effectively in 1978, he was stepping along a well-trodden path. Edwards noted that an enormous development of "oppression psychosis" precedes the major revolutions.[25] The Shi'ite theodicy of suffering provided the Islamic party with an armory of emotive images for expressing the oppression psychosis in terms of primeval tyranny (*zolm*) and for articulating the appropriate response in its glorification of martyrdom.

Charismatic Leadership and Revolutionary Traditionalism

Khomeini's charismatic leadership was undoubtedly a major factor both in the revolutionary politicization of Shi'ism and in the success of the Islamic revolution in Iran. Khomeini's courage and unswerving determination in challenging the Shah were indeed extraordinary personal qualities that could and did generate charisma. It would, however, be wrong to conceive charisma too restrictively as the extraordinary quality of the individual to whom it is attributed. Charisma is also much in the eye of the beholder and is determined by his or her cultural sensibilities. Khomeini's embodiment of Islam, which most of his followers considered endangered, had as much to do with his charismatic appeal as did his heroic stature and resolution.

As an individual, Khomeini had the making of a revolutionary transformer of tradition. He was an orphan and highly conscious of this fact.[26] His father had been killed by bandits before he was one year old, and he lost his mother and the aunt who had raised him when he was fifteen.[27] He had been named Ruhollah (the spirit of God), and can be presumed to have attached considerable significance to his name, given his intense interest in the mystical discussion of the Great Names of God as the constituents of reality.[28] He was deeply imbued with the culture of Qom, a holy city and center of Shi'ite learning, yet his intellectual passion, *'erfan* (mystical philosophy) had been truly marginal to Shi'ite learning and barely tolerated after the triumph of *Osuli* jurisprudence in the nineteenth century. On a number of important occasions Khomeini has recalled the pressure he felt under when going almost secretly to the house of his professor *'erfan*. To the marginality of *'erfan* as a subject of study should be added the possibility of millenarian interpretation of Shi'ism from its perspective, a possibility that had been realized by some of the claimants to Mahdihood who had launched "extremist" Shi'ite movements in the fifteenth century.[29] Both his early personal life and untypical intellectual formation thus poised Khomeini to be the author of a revolutionary transformation of Shi'ism, which he has in fact carried out since 1979. As we shall see in Chapter 9, this revolutionary transformation was to take the form of the extension of Shi'ite tradition to the sphere of political culture and political organization.

Khomeini was able to carry out this revolutionary transformation of the Shi'ite tradition in part because of his charismatic authority. Once again, it should be emphasized that this charismatic authority did not derive solely or even primarily from Khomeini's extraordinary qualities as an individual but had a strong normative foundation in the Shi'ite culture. Inspired generally by the Egyptian Muslim Brethren's notion of supreme leadership, by the activist followers of Ayatollah Khalesi (d. 1963), and more immediately by the example of their fellow countryman, Musa Sadr (whose Arabic-speaking Lebanese followers called the "Imam"), Khomeini's militant followers were calling him the Imam *in Persian* by 1970. The acclamation of Khomeini as Imam by his followers was a startling event in Shi'ite history in Iran.[30] Never since the majority of Iranians had become Shi'ite in the sixteenth century had they called any living person Imam. The term had hitherto only been used in reference to one of the twelve holy Imams and its connotations in the mind of the Shi'ite believers as divinely guided, infallible leaders undoubtedly worked to build up Khomeini's charisma. It did so especially by being suggestive of a link between Khomeini and the Hidden Imam of the Age, the Lord of Time.

An important feature of Shi'ite Islam is its millenarianism. We have noted that Shi'ism was established in Iran by the supreme leader of an aberrant millenarian warrior order. Though millenarianism was *contained* by the orthodox interpretation of the belief that the last Imam, the Mahdi, had gone into hiding, it could not be eradicated. The Mahdistic tenet remained inescapably chiliastic and would from time to time be activated (the most notable instance being the rise of the Bab in the mid-nineteenth century). As part of the general revival of religion in the late 1960s and 1970s, there was a marked increase in the popularity of *du'a-ye nodbeh,* the supplication for the return of the Hidden Imam as the Mahdi; and special sessions were being arranged for its recital.[31] Without claiming to be the returning Mahdi, Khomeini ingeniously exploited this messianic yearning by encouraging his acclamation as the Imam from about 1970 onward. An unmistakably apocalyptic mood was observable during the religious month of Moharram 1399 (December 1978) among the masses in Tehran. Intense discussions were raging as to whether or not Khomeini was the Imam of the Age and the Lord of Time. Those who answered in the affirmative were undoubtedly among the millions who massed in the streets of Tehran to welcome the returning leader in February 1979 and whose frenzy was to be televised across the globe. But many of those who answered in the negative were also ready to accept him as the Mahdi's precursor: Khomeini could not be the Lord of Time himself, since the Lord of Time would liberate the entire world, and Khomeini was going to liberate only Iran. Khomeini's face was allegedly seen on the moon in provincial cities, and those who had been vouchsafed that vision duly proceeded with the sacrifice of lambs on the ensuing days. As is usually the case, expectations of material gain were woven into the messianic yearning. When the Aqa (Master) came, the Pahlavis and the Rockefellers would be stopped from robbing the oppressed, and every family would have a Mercedes-Benz.

Khomeini used the charismatic title of Imam with its subtle millenarian connotations in his bid to oust the Shah, to destroy the new, Westernized middle class, and to set up a theocracy in Iran. On November 5, 1978, when the Shah announced the installation of a military government as a temporary measure while emphasizing that "your revolutionary message has been heard, Khomeini retorted, "Once more, the Shah has clung to the [usual] two means to have himself: trickery and the bayonet," ending his speech with the following words:

> If you give this fellow a breathing spell, tomorrow neither Islam nor your country nor your family will be left for you. Do not give him a chance; squeeze his neck until he is strangled.[32]

Khomeini destroyed his lesser opponents with the same single-minded determination and with much greater ease. On January 27, 1979, Mehdi Bazargan and other members of the Revolutionary Council who had been negotiating with Shahpur Bakhtiar secured Khomeini's acceptance of a plan according to which Bakhtiar would leave a letter of resignation with them before flying to see Khomeini in France. Bazargan was also to accompany him, and to offer him a position in the cabinet once Bazargan was appointed provisional prime minister by Khomeini.[33] Khomeini, however, changed his mind after midnight, insisting on Bakhtiar's open resignation before receiving him.[34] The next evening, Khomeini declared, with characteristic disdain, that he was using the term resignation for the sake of appearances: "But what does 'resignation' mean? You [*taw* (second person singular, contemptuous form of address)] are not at all Prime Minister! But let this [fellow (contemptuous reference)] say I am Prime Minister."[35] In the same way, during the ten or eleven days after his return to Iran, Khomeini rejected several compromise solutions for a transition from Bakhtiar to an Islamic Republic against the advice of virtually all his advisors, lay and clerical, because they might have implied that the new regime derived its legitimacy from the old.[36]

Khomeini's determined charismatic leadership was equally decisive in setting the direction of the Iranian revolution after the overthrow of the monarchy. In mid-April 1979, while harassed by separatist insurrections of the Kurds and the Turkmans and demoralized by the poor show of the army he called upon to demonstrate in his support, Khomeini confronted Ayatollah Taleqani, who favored an Islamic democratic republic as opposed to Khomeini's proposed theocracy, and was backed by the well-armed and disciplined Mojahedin-e Khalq, "with no strings attached." Khomeini used the weight of his charismatic authority as the leader of the Islamic Revolution and the sheer strength of his personality to force Taleqani to back off and suffer public humiliation.[37]

Last but not least, we have Khomeini's determined leadership in approving the taking of hostages at the American Embassy—in all likelihood at the planning stage—and in vetoing a speedy solution to the hostage crisis. He did not flinch at the serious risk of war with the United States to achieve the two remaining goals of his revolution: the political destruction of the Westernized intelligentsia and the establishment of Islamic theocracy.[38]

6

The Revolution of
February 1979

The Myth of the Islamic Revolution

The revolutionary mobilization of the masses, so Sorel tells us in his *Reflections on Violence*,

> could not be produced in any very certain manner by the use of ordinary language; use must be made of a body of images which, *by intuition alone*, and before any considered analyses are made, is capable of evoking as an undivided whole the mass sentiments which corresponds to the different manifestations of the war undertaken by Socialism against modern society. This problem [is solved] perfectly, by concentrating the whole of Socialism in the drama of the general strike; there is thus no longer any place for the reconciliation of contraries in the equivocations of the professors; everything is clearly mapped out, so that only one interpretation of Socialism is possible. This method has all the advantages which "integral" knowledge has over analysis.[1]

Substitute "Islam" for "socialism" and couple "general strike" with "revolt against tyranny" and you can understand the efficacy with which "Islamic government" acted as a social myth in the spiritual dynamic of the Iranian Revolution, producing a general strike of unprecedented tenacity that lasted some five months and put an end to twenty-five centuries of monarchical rule. The condensation of social and political reality was brought about by a social myth of Islamic government as the restoration of the Golden Age: the reign of Prophet Mohammad and of the first Shi'ite Imam, 'Ali. But it was not so much this utopia per se as its stark juxtaposition to the Shah's regime that primarily accounted for its effectiveness. The militant clerics contrasted the "monistic" (*tawhidi*) Islamic order to the Shah's tyrannical regime, described by the emotive Koranic term *taghut*, a term for idolatry that denotes tyrannical earthly power arrogating to itself that absolute authority over the lives of men that is God's alone. The term *taghut* capitalized on the conspicuous presence of the Shah as the embodiment of impious earthly power setting itself up against God's majesty. The vilification of the Shah and his regime were more important than the glorification of Khomeini. For every one slogan for Khomeini, there were probably more than two against the Shah.[2]

103

The famous headline *Shah raft, Imam amad* (The Shah has gone, the Imam has come) pithily captures the substance and the outcome of the Islamic revolutionary struggle in Iran. It contains the same rigidly binary juxtaposition between absolutist political power and divinely ordained religious authority depicted in so many other revolutionary shibboleths: the Pharaoh versus Moses, the Prophet of God; the corrupt and tyrannical Umayyad Caliph Yazid versus the martyred Imam Hosayn, the grandson of Mohammad and the third Shi'ite Imam. Let us cite just two examples of slogans from the Islamic Revolution:

> Oh brave soldier, do not kill Moses for the sake of the Pharaoh
> For Yazid's sake, do not kill the son of the Prophet; fear God

And

> Three were the idol-breakers
> Abraham, Mohammed, and Ruhollah [Khomeini].

Adoption of the modern political myth of revolution by the Islamic movement led by Khomeini was a crucial factor in the success of its mass mobilization against the Shah. This adoption cannot be accounted for by the plight of the embattled Shi'ite hierocracy in Iran but depended on the international context of the Islamic movement. Appropriation of the myth of revolution and development of an Islamic revolutionary ideology had already taken place in the Sunni world, notably in Indo-Pakistan and Egypt, and Khomeini's followers imported it and later modified it to suit Shi'ite clericalism.

Already in 1926, in a work that anticipates most of the ideological developments of the last two decades, the youth Indian Muslim journalist Abo'l-A'la' Mawdudi (d. 1979) had declared: "Islam is a revolutionary ideology and a revolutionary practice, which aims at destroying the social order of the world totally and rebuilding it from scratch . . . and *jihad* (holy war) denotes the revolutionary struggle." Mawdudi conceived the modern world as the arena of the "conflict between Islam and un-Islam," the latter term being equated with pre-Islamic Ignorance (*jahiliyya*) and polytheism. Modern creeds and political philosophies were assimilated to polytheism and Ignorance. Their predominance necessitated the revival of Islam.[3] Decades later, the Egyptian Sayyid Qutb (d. 1966) took the contrast between Islam and un-Islam from Mawdudi and made it the cornerstone of his revolutionary Islamic ideology, in which contemporary Muslim societies are branded as societies of Ignorance for accepting secular states. To extirpate Ignorance from these societies and make them truly Islamic, it is necessary to establish an Islamic government and to apply the Sacred Law. To establish an Islamic government, that is, to establish the rule of God, Islamic revolution is necessary.[4]

The distinctively clericalist Shi'ite idea of Islamic government, realized after the revolution of 1979, was not directly influenced by this trend in Sunni Islam. It draws on Shi'ite jurisprudence and is best understood in the context of the contest between the Shi'ite hierocracy and the centralizing monarchy discussed earlier. Khomeini's idea of Islamic government, though a radical innovation in Shi'ite history, is nevertheless stated within the traditional Shi'ite frame of reference and does not betray any influence of the ideological innovations of Mawdudi and Qutb. Nevertheless with the onset of revolutionary agitation, these innovations provided an indispensable ideological supplement to Khomeini's clericalism which was not emphasized at the beginning and did not even surface for a few months after the revolution. In fact, Mawdudi and Qutb were read avidly, in both Persian translation and Arabic, by Khomeini's militant followers; their influence is unmistakable in the revolutionary slogans and pamphleteering, most notably in the application of the term *taghut* to the Pahlavi political order.

In the latter part of the 1970s, with the massive influx of foreign civil and military technicians and the avalanche of foreign products, the Shah's propagation of the alien culture of the West could easily be contrasted to Khomeini's embodiment of Iran's traditional culture and authentic identity. The antithesis between legitimate religious authority and impious political power was thus neatly compounded and became doubly effective.

Not everyone who decisively contributed to the fall of the Pahlavi regime was moved by the myth of the clerical party or believed in or even knew about their political theory. "All revolutions," writes Dunn, "are supported by many who would not have supported them had they had a clear understanding of what the revolutions were in fact to bring about."[5] Such undoubtedly was the case with many in Iranian society who withheld their support from a compromise with the Shah and suicidally supported Khomeini in the autumn and spring of 1978.

For Khomeini and his supporters, Islamic government was to be established by means of an Islamic *revolution*. This meant that the proponents of Islamic traditionalism had appropriated the most potent myth of modern politics, the myth of the revolution. Symbolic reality usually comes in a multilayered cake. For those who did not care for the *Islamic* layer, the revolution was still a revolution of the Iranian *nation* against U.S. imperialism. Above all, it was *the revolution*. As a perceptive observer of the Russian revolution has remarked:

> The decisive part in the subjugation of the intelligentsia was played not by terror and bribery (though, God knows, there was enough of both), but by the word "Revolution," which none of them could bear to give up. It is a word to which whole nations have succumbed, and its force was such that one wonders why our leaders still needed prisons and capital punishment.[6]

The widespread belief in the modern myth of the revolution motivated a variety of social groups that did not share Khomeini's vision of an Islamic theocracy to gather nevertheless under his banner to overthrow the Shah.

The Revolutionary Alignment of Social Classes

The small groups of organized revolutionists—the original propagators of the myth of revolution—hardly needed the utopia of Islamic government to be mobilized for action, nor could they take this utopia seriously. For these groups, the Tudeh Party, the Marxist Feda'iyan-e Khalq (Devotees of the People), the Maoists, and perhaps even for some of the Islamic Mojahedin-e Khalq (Fighters for the People), the revolution itself was the supreme redemptive act and would automatically produce the ideal society through their agency in the vanguard.

The professional revolutionists came from the intelligentsia and were overwhelmingly the product of the universities and other institutions of higher learning. During the decade of the 1970s, the enrollment in these institutions had more than doubled to 175,000 (see Table 13, Appendix).

Furthermore, many high-school graduates who were unable to gain admission into universities must be considered members of the intelligentsia. Taking 1974 as a typical year, about 100,000 persons obtained high school diplomas, while 42,000 persons were accepted in the institutions of higher learning for the 1974 to 1975 academic year.[7] Some of the new graduates went directly into employment or compulsory military service, but many would join the growing pool of aspiring young intelligentsia who would spend a year or two or three just preparing for the competitive examinations to gain entry into universities. In 1977, 290,000 persons applied to universities and only 60,000 were accepted.[8] The young intelligentsia supplied not only regular demonstrators but also the urban guerrillas who acted effectively in early 1979.

However, one of the most striking features of the Iranian revolution is that the professional revolutionists were not very conspicuous when the revolutionary movement gathered momentum and did not gain control over the state apparatus. Instead, the revolution was marked by the astonishingly widespread participation of other social groups. To examine this wide participation, Table 14 presents a rough picture of Iran's class structure as inferred from the composition of the active labor force on the eve of the revolution (see Appendix).

In Chapter 5, we encountered the social groups who were genuinely moved by the myth of the Islamic government and Islamic revolution as proposed by Khomeini and the militant clerics. The most important of these were the traditional bourgeois sectors of the population, which as Table 14 suggests, expanded in absolute but *not* in relative terms during the two decades prior to the revolution. The bazaar consistently opposed the monarchy in the postwar period. Though it benefitted from the general prosperity created from the mid-1960s onward, there were no governmental policies designed to further its interests. On the contrary, developmental policies of the government were often detrimental to the interests of the bazaar and, especially, of the guilds. The encouragement of chain supermarket and department store development created direct competition between the modern and the traditional sectors of the urban economy, as did the introduction of machine-made bread and shoes, which adversely affected the guilds of bakers and shoemakers. The plight of the bakers supplies us with a good illustration of the consequences of the Pahlavi state's ill-conceived approach to industrialization. In 1972, the government high-handedly decided to replace the traditional fresh-baked bread with bland machine-made, Western-style bread. Despite the fierce resistance by the bakers and the sensible refusal of the Iranian consumers to switch to the tasteless new processed bread, some 6,000 bakery workers lost their jobs (the loss projected by the planners had been 12,000 jobs).[9] Furthermore, the bazaar became increasingly excluded from access to the huge amounts of credit for developmental projects that became available in the 1970s. Last, but by no means least, the Shah launched an anti-profiteering campagin in 1975. Some 10,000 students were hired to inspect the shops. Their work was continued by the guild courts set up by the government. Some 8,000 merchants and retailers were imprisoned and a very large number exiled and fined.[10] All this was certainly incentive enough for the bazaar to reaffirm its historical coalition with the hierocracy in opposition to the Shah. The bazaars were frequently closed and the bazaaris were most active in the anti-Shah demonstrations. Furthermore, they also gave financial support to those striking in the private sector, especially to the journalists' strike in November 1978.[11]

The second group, the recent migrants into towns—the "disinherited" of the Islamic revolutionary ideology—were to be found mainly in the construction and nonindustrial working-class categories of Table 14. The Shah did not integrate this group into his political system. By contrast, as we have seen, this group was successfully integrated into a societal community in their new surroundings by local or regional networks of self-help, in chained migration, and by religious associations.[12]

In curbing all independent group formation, and in following the principle of divide and rule within Iranian society, the Shah failed to build a constituency capable of concerted action on his behalf among any major social group. This failure was most glaring in the case of the middle peasantry who had acquired land through the Shah's agrarian reform and were now the dominant class in the countryside. He did not capitalize on this important source of potential support. In the arrogant, over-confident mood of the seventies, the Shah completely forgot about the peasantry, never doubting that he could dispense with their help indefinitely. By 1975, in pursuit of an abstract and disastrously misconceived strategy for economic growth, and in reckless disregard of political consequences, he would go so far as to state: "Iran's small and relatively unproductive farmers are an extravagance that the country can no longer afford."[13] It is not surprising that the land-owning peasants thus characterized by their ruler were conspicuous by their absence from the political scene during his crisis except in one staged pro-Shah demonstration on November 12, 1978.

Ironically enough, the only group whose support the Shah could successfully buy was the industrial working class. With the exception of the government-employed workers in the oil and tobacco industries, the industrial working class largely did not participate in the revolutionary movement.[14] This latter group was in any case not very important in Iran. As indicated in Table 14, Iran's industrial labor force amounted to only some 7 percent of the active population (see Appendix).

We finally come to the group that was by and large the product of the modern state created by the Pahlavis and that had expanded considerably both in absolute and in relative terms in the two decades preceding the revolution: the new middle class. Their decisive rejection of the Shah sealed the fate of the Pahlavi dynasty. With the religious party's agitation firmly under way, the new middle class acquiesced in the demise of the monarchy. However, what it wanted was not an Islamic government but "national sovereignty and popular democratic government."[15]

In short, the groups and social classes that were aligned against the Shah in 1978 had different motives for opposing him and proposed different alternatives to the Pahlavi regime. They did, however, share a basic political motive for their opposition: the desire for political enfranchisement and for inclusion in the political system. They all lacked political power and were denied any meaningful political participation. One of the most serious shortcomings of the Pahlavi regime was its failure to increase political participation in the face of rapid social change.[16] The two-party system the Shah had set up in 1957 was never taken seriously, but was at least beginning to serve as a mechanism for the recruitment of the political elite under Prime Minister Amir 'Abbas Hoveyda when the Shah abruptly decided to replace it by a one-party system in 1975. He set up the Rastakhiz (Resurrection) Party and offered all those who did not wish to be members of it passports to leave the country they were thus betraying! The attempt to integrate the socially mobile groups into the political system through the agitprops of the Rastakhiz Party was a fiasco. In fact, the headquarters of the Rastakhiz Party were among the main targets of the urban riots of 1978 to 1979—an eloquent comment on the Shah's failure to increase political participation.

The New Middle Class and the Revolution

It is true that the renewed political activities of the writers and intellectuals—chiefly in the form of open letters addressed to the Shah and, later, in politicized poetry readings—predated the crisis by at least a year. Furthermore, it is also true that Khomeini was encouraged in launching his own agitation by observing, in November-December 1977, the activities the secular opposition could engage in with impunity.[17] The relative vigor of these activities, and the relaxation of repression by the SAVAK were due to two factors: The Shah's project to make the political system more democratic as preparation for the succession of his son, and President Carter's Human Rights pronouncement on Iran in January 1977. The Shah's promises of liberalization, seen as his bowing to American pressure on the issue of human rights, did create a good deal of expectation among lawyers, university professors, intellectuals, and artists, though none envisaged an abrupt end to the monarchy. The Writers' Association was revived in June 1977 and a Group for Free Books and Free Thought formed in early July. The fall of 1977 witnessed the formation of the Iranian Committee for the Defense of Freedom and Human Rights, the Association of Iranian Lawyers, and the National Organization of University Teachers. Chronic disturbance in the universities began in October 1977 and continued thereafter. In the same period, Karim Sanjabi revived Mosaddeq's National Front, and Mehdi Bazargan, the Liberation Movement of Iran. Early in August 1978, the Shah invited the opposition to participate in the next elections; and throughout the summer and fall of 1978, these groups contemplated the prospect of making the constitutional monarchy a viable reality. In October and November 1978, the situation changed suddenly. A friend has told me how a group of professors at the University of Tehran decided to form a society for freedom of speech in November 1978. Within two weeks they realized they were so overtaken by events that they changed the formal purpose of their society to the abolition of the monarchy and establishment of a republic! But such an *ex post facto* adjustment did not enable Tehran's professors, any more than the liberal or socialist intellectuals in general, to establish a grip on the tumult that had gathered momentum under the direction of the clerical party.

Against this background, one is astounded by the behavior of Iran's sizable new middle class, the class that could be expected to assume the nation's political leadership by virtue of its educational and economic resources. How can one comprehend that this unprepared middle class could not think beyond the instant gratification of regicidal vengefulness and rise to the most elementary of the long-term political calculations? Why, instead of wringing concession after concession from a desperate Shah and a frightened military elite, did they choose to become subordinate allies of a man who treated them with haughty contempt and rejected their principles of national sovereignty and democracy? How can one account for the abject surrender to the clerical party of one after another of the feeble, middle-class based political factions: liberals, nationalists, and Stalinist communists alike?

Let us begin our analysis of the reaction of the middle class to sudden revolutionary explosion on the psychological level. The self-deception that was necessary on the part of the new middle class to make their surrender to the religious party psychologically acceptable was made easier by the politicization of their national cultural identity. Khomeini, who said little to disabuse them, was seen as the embodiment of Iranian tradition, totally uncontaminated by that strange condition of cultural alienation that unfortunately did afflict the Westernized bourgeoisie. During the first massive peaceful demonstration in September, Karim Sanjabi, the leader of the National Front who was soon to discover his "Islamic identity," was moved to remark that during the march "there was no longer an I but only a We." Better still, consider the middle-class feminist, in one of the massive December demonstrations, who was chanting, "Death to Shah, Long Live Khomeini!" Tears were streaming down her face because, as she explained to the French reporters covering the event, Khomeini was making her rediscover her "Iranianness."[18] Was it *Nostalgie de la boue?* Perhaps, but not entirely. Like the rest of Iran's new middle class, educated women were too inebriated by a newly discovered national communion and too seduced by the myth of the revolution to think clearly. As Monnerot aptly put it, in its nostalgia for social consensus and in its longing for unity, "the psyche will feed upon the grossest of counterfeits: the golden age, or the day of glory."[19]

Psychological insecurity usually underlies the politics of identity, and the case of Iran's parvenu middle class was no exception. As Durkheim pointed out in his classic study of suicide, "crises of prosperity" generate disorientation and anomie by disturbing the collective order. In the decade preceding the Organization of Petroleum Exporting Countries (OPEC) price rise of 1973, Iran's record of industrialization and economic growth was very respectable, with the gross national product (GNP) increasing 9 to 10 percent annually. The sudden increase in the price of oil generated a boom for the next three years while seriously distorting the path of economic development. The axis of the Iranian economy shifted from production to distribution (of the oil revenue), and economically unproductive activities mushroomed. The GNP went up by 30.3 percent in 1973–74 and 42 percent in 1974–75. Then came the economic debacle, despite the massive oil revenue: severe bottlenecks created by the shortage of skilled manpower and the inadequate infrastructure halted economic growth. In 1976, production remained stagnant or declined in all sectors of the economy except construction.[20] The curvilinearly minded sociologists of revolution might find in this a confirmation of the "J-curve" (continuously rising expectations followed by sudden frustration).[21] For my part, I think the source of malaise lay deeper and therefore prefer the old-fashioned Durkheim, who stressed normative disorientation. Persistent sense of malaise and anomie caused by rapid social change was a much more basic cause than the sudden downturn in economic growth that triggered it. It was not so much sudden frustration as constant disorientation resulting from a decade of rapid change that finally—after a period in which the *nouveaux riches* had sought to allay their insecurity by conspicuous consumption of the gaudiest kind—heightened the concern of the middle class for their fundamental national and cultural identity. For the salaried middle class, an important catalyst was also at work to heighten their nationalist sentiment: resident foreigners. The Shah's massive purchases of sophisticated weaponry brought a large number of highly paid American paramilitary technicians to the country, while his capital-intensive growth strategy pulled in a large number of foreign skilled technicians, chiefly Europeans, some of whom received higher salaries than their Iranian counterparts.

However, foremost among forces that could crystallize this sense of disorientation into a definite political stand was the presence of the Shah. Palpable even in the remotest corners of his realm, the imperial personage was the only entity that could be blamed for everything and thus give focus to the anomic malaise and frustration of the middle class, integrating its members into what would have been an inchoate political movement. Their discontent soon found a second point of reference with the meteoric emergence of Khomeini. Thus they readily accepted the religious party's portrayal of the Shah as the Anti-Christ and concomitantly took refuge in the comforting discovery of Khomeini as his messianic counterimage (and the only available one).

Let us look more closely at the upper layer of the new middle class. The Shah had alienated the industrialists and businessmen by his anti-profiteering campaign of 1975 and by his various profit-sharing schemes on behalf of the workers, and had also alienated the bureaucratic elite by the arrests in November 1978 of a considerable number of former ministers and high officials who had served him. But the political importance of this point should not be exaggerated. Far more important was the fact that the oil bonanza had played havoc with the mentality of the moneyed middle class. Corruption among the high civilian officials became phenomenal and spread to the generals as billions of dollars were being siphoned off through government and army contracts. When I was in Iran in July and August of 1978, I was astonished at the utter lack of any moral commitment to the Shah's regime among those who had a stake in it, the top civil servants and well-to-do entrepreneurs. (The two groups were linked by the distribution of the oil revenue, which was partly funneled out of the state treasury at the discretion of the former into the pockets of the latter.)

Petrodollars had sapped the last drop of public spirit, which could occasionally be detected during the early years of the decade. Both groups were filling their pockets as fast as they could, expecting the whole thing to collapse—a self-fulfilling prophecy, as it turned out. It was obvious they would not lift a finger to save the regime when the crunch came, and they did not. Many of them transferred their bank balances. In mid-1976, the British Embassy estimated that $1 billion a month of private capital was leaving Iran;[22] by the end of 1976, about 20,000 Iranians had bought houses in or near London, not to mention those who had acquired property in California. A much more massive flight of capital preceded the Shah's departure in the last months of 1978. The inglorious *haute bourgeoisie* followed its money.

More critical for the fate of the monarchy was the withdrawal of support by the middle and lower layers of the new middle class and of the government-employed workers in the oil industry. For these groups, living mostly on fixed incomes, the massive inflow of oil revenue from 1975 onward meant more inflation rather than additional income.[23] The rents in the major cities increased by over 100 percent annually. Real estate prices and construction costs soared similarly. Thus, the purely economic grounds for dissatisfaction were by no means lacking. This group *was* affected by the turn of the J-curve.

The Shah did not help matters by his prolonged indecision and by his unrealistic hope to drive a hard bargain with the opposition, which could leave him fully in control of the army.[24] Nevertheless, being totally unprepared, representatives of the new middle class within the opposition were for their part not forthcoming either.

Eventually, in December 1978, the Shah begged the representatives of the new middle class to inherit the state and the country. Meanwhile, Sanjabi, the leader of the National Front, had capitulated to Khomeini. Once again, fearing the wrath of the populace, he and other middle-class political figures had self-deceivingly rationalized their lack of preparation and courage as a heroic refusal to compromise, and sought additional solace in utopian revolutionary faith and in the wishful thought that they would get the better of the mollahs after the revolution. When another member of the National Front, Shahpur Bakhtiar, did courageously step forward, they disowned him, disguising their pusillanimity as a refusal to collaborate with the tyrannical dictator. The person they were talking about bore no resemblance to this description. The Shah was not the megalomanic of 1975, but a shaken, dejected prisoner of Niyavaran Palace who was now only insisting on a respectable farewell. Be that as it may, the middle classes took leave of their political senses, wished away their differences with the clerical party, and abdicated their historic political responsibility. They prostrated themselves before the contemptuous Khomeini and passed on the gift the Shah was begging them to inherit to the Grand Ayatollah who had already presumed the lofty title of Imam.

Though distressing, it is hardly surprising that Iran's new middle class proved incapable of concerted political action and was in no way ready to assume political leadership of the nation. In consistently following his policy of divide and rule, the Shah had succeeded in atomizing the new middle class to a far greater extent than was the case with the bazaar and the related lower middle class. The intelligentsia had been kept under close surveillance by the SAVAK. With the exception of the Bar Association, no significant professional associations and, needless to say, no real political parties were tolerated. Even after he had embarked on his liberalization policy in 1977, the Shah could not overcome his distrust and apprehension of all the middle class opposition figures. He disregarded the suggestions put to him for a serious compromise to bring the moderate middle class opposition into the political process. In November 1977, his press began vilifying the former reformist premier, Dr. 'Ali Amini as "the CIA boy" to discredit him as a potential rival and SAVAK forces attacked a gathering of the National Front. These moves did keep the middle class opponents of the Shah from becoming united but they also destroyed the chance of their subsequent co-optation.[25]

The Shah thus kept the new middle class divided and emasculated. Under these circumstances its political immaturity and lack of responsibility were hardly surprising. Nor were its negative politics: the refusal to go beyond denunciation of an oppressive regime, which had come to an end anyway, and the irresponsible, intransigent posturing of the middle-class politicians whose immediate interest appeared to be served not by solving any of the concrete problems but by inflammatory rhetoric. Furthermore, the ability of the new middle class to act was seriously impaired by the isolation of army officers from the rest of their class. Political representatives of the new middle class were seriously hampered by the impossibility of a coalition with the army, which had become too closely identified with the Shah and his regime after the coup of August 1953 and the ensuing purges.[26]

As we shall see in the following sections, the moderates and the militants within the Shi'ite hierocracy, headed respectively by the Grand Ayatollahs Shari'at-madari and Khomeini, were united in their opposition to the Shah by the bungling and ineptitude of the government. It was easy and rational for the political representatives of the new middle class to enter a coalition with Grand Ayatollah Shari'at-madari, who adopted their political demand for the full observance of the Constitution.[27] Such a coalition was forged by political figures from Azerbaijan such as Moqaddam-Maragheh'i and Hasan Nazih, the head of the Bar Association whose visit to Shari'at-madari in Qom on September 20, 1978, attracted considerable attention.

What was less easy, and less rational, was forming a coalition with the militant wing of the Shi'ite hierocracy led by Ayatollah Khomeini. What made formation of such a coalition difficult and unwise was the profound distrust of the secular national democrats by these clerical leaders.[28] The first move toward this coalition was made with difficulty[29] in November 1977 during the memorial services held for Khomeini's son who had died unexpectedly, purportedly at the hands of the SAVAK. The totally unexpected weakness of the Shah in the fall of 1978 made the unification of the opposition forces all the more attractive, and a coalition was formed between Bazargan's Liberation Movement of Iran and the leading pro-Khomeini *'ulama* such as Ayatollahs Motahhari, Beheshti, Mofattah, and Mahdavi-Kani. The National Front and Sanjabi toed the line behind Bazargan, with only a few figures such as Sadiqi and Bakhtiar refusing to do so. In December, Khomeini set up a secret Revolutionary Council that consisted of the fifteen representative members of the coalition, taking good care to appoint a majority of clerics.

The sad truth of the matter was that because of twenty-five years of systematic political sterilization, the new middle class had produced no notable figures with a sense of political vocation and the requisite political experience. The seedy old politicians of the Mosaddeq era had to be pulled out of the closet. We have already met Dr. Karim Sanjabi who was the first to distinguish himself by his act of prostration at the threshold of the Grand Ayatollah near Versailles. Along the same lines came, on January 21, the ignominious resignation of Sayyed Jalal al-Din Tehrani, Chairman of the Regency Council sent by Bakhtiar to Paris to negotiate with Khomeini. Khomeini demanded that he resign as a condition for receiving him. Tehrani abjectly complied; all he got in return was a ten-minute dismissive audience with the Imam.[30] Let us also consider the inglorious appearance of some other new middle class figures on the political scene after the revolution of February 1979. Enter, shortly after the revolution, Dr. 'Ali Shayegan, another senile politician of the Mosaddeq period, immediately acclaimed as the first prospective president of the Iranian republic. Exit, before the month was out, Shayegan, into immediate oblivion, with a frustrated remark that the reactionary mollahs should stick to their prayer rugs. Exit, in April, Sanjabi from his post as Foreign Minister after being slighted and humiliated by Khomeini's protege Ebrahim Yazdi (who then succeeded him in that post, only to be dumped in turn and equally unceremoniously before long). Daryush Foruhar, the other National Front man who had turned up in France to pay his homage to Khomeini and had been rewarded with the portfolio of the Ministry of Labor and Social Affairs, was soon deprived of his portfolio despite his agreeably resigning from the National Front. Nazih, the only member of this group to act with determination and integrity, was contemptuously dismissed from his post as the director of the National Iranian Oil Company and summoned to trial for having the courage to criticize, in respectful terms, Khomeini for wanting to turn modern Iran into a theocracy, and for not displaying the requisite obsequiousness toward the mollahs. Bazargan, the ablest septuagenarian of the Mosaddeq era, held out the longest, but at the cost of capitulating to the mollahs on every major point and in exchange for giving the mollahs the crucially necessary time to organize for the final takeover of the state. He, too, cut a pathetic figure beside the Ayatollah with his oft-repeated threats of resignation, the first barely three weeks after taking office, and long before Khomeini made him the archetypical example of "Bazarganization."

Paralysis and Collapse of the State

The mammoth edifice of the state became hollow as a result of the complete withdrawal of moral commitment to its preservation. The Shah stated in an interview from exile that he watched the top officials of the state run away like rats. "Rat" seems to be the telling word. During his trial, General Rabi'i, the last commander of the Imperial Air Force, related his disillusionment when (the American) General Huyser came to Iran, took the Shah by the tail, and threw him into exile like a dead rat.[31] One may or may not take exception to the language, but the truth is that the decisive person to withdraw his commitment to the preservation of the state was no other than the Shah himself. Once he gave up the will to fight for survival, under the impact of the initial shock in September, the state crumbled from within and out of its own momentum.

The Shah's contradictory impulses and policies did much to wreck the state apparatus and sever its bureaucratic and military wings. On September 8, 1978, martial law was imposed in Tehran; early in the evening the troops fired on a group of demonstrators in Jaleh Square. The casualties were heavy and the event became known as the massacre of the Black Friday.[32] The opposition was horrified and scared; the Shah wept before the cabinet and the generals whom he sharply rebuked. From the Black Friday onward, a three-way struggle among the Shah, the prime minister, and the army became constant. Sharif-Emami had formed his "government of national reconciliation" in order to implement the Shah's liberalization policy,[33] under which thirty-three senior officials of the SAVAK were dismissed in September and some 1000 political prisoners freed in October. On the other hand, martial law was declared in the major cities, and the leading hard-liner, General Gholam-'Ali Oveisi, was put in charge of the administration of martial law in the capital. As he retained command of the Ground Force (the Army), and as most of the administrators of martial law in other cities served in the Ground Force and were thus his subordinates, Oveisi in command of the administration of martial law was in command of a state within the state and at odds with government and the prime minister.[34] However, the Shah did not give Oveisi a free hand: Troops were only allowed to fire into the air, the Shah frequently scolded Oveisi for fatal shootings, and commanders in Qazvin, Mashhad, Tabriz, and elsewhere who disobeyed Tehran and ordered their men to shoot to kill in self-defense in October were reprimanded.[35] Meanwhile, friction between the prime minister and the general became especially intense after October 17, when the prime minister won the battle over the removal of censorship.[36] During the last days of October, General Moqaddam, Director of the SAVAK, complained to a CIA representative about the gradual weakening of the armed forces, reined in by a Shah who "fully believes in an open political atmosphere, democracy and the implementation of the Constitution." Sharif-Emami's government, he continued, kept blaming the SAVAK for everything—not without cause[37]—and had forced him to remove two of his officials who occupied sensitive posts. General Moqaddam ended his plaint as follows:

> I declare that the Shahanshah has tied our arms and the hands of the armed forces and has left the affairs of the country to the Prime Minister. We are of course astonished as to why the Shahanshah follows these policies.[38]

In the fall of 1978, the release of political prisoners and a new freedom of the press were among the Shah's first liberalizing measures. His absurd insistence at continuing these liberalizing measures in the hope of placating the opposition alongside martial law, even after installing a military government, and his refusal to authorize massive arrests introduced a major, indeed fateful, inconsistency in his handling of the revolutionary crisis. Day after day following the end of censorship, the newspapers came out with harrowing reports of the SAVAK's torture chambers as related by released political prisoners.

These tales of horror intensified the moral indignation of the new middle class groups over the seemingly avoidable Black Friday massacre, which had occurred only a few hours after the imposition of martial law and before its announcement had sunk in, while the weakness shown by the Shah in handling the crisis emboldened them to join his foes. Group after group agitated in step with the rhythm set by Khomeini and excitedly went on strike: bank clerks and journalists in the private sector; power plant engineers and technicians; Customs officials; and Iran Air, National Iranian Oil Company, and Central Bank employees in the public sector. Late in November, the oil workers joined the movement, resuming their strike and causing shortages in the towns. Strikes and slowdowns also paralyzed many of the government agencies and ministries. Iran plunged into anarchy in the closing months of 1978.

The recurrent clashes between Prime Minister Sharif-Emami and General Oveisi came to a head on November 4, when ten or twelve students were killed in demonstrations at Tehran University and Sharif-Emami wanted the responsible officers disciplined. On November 5, the day of Tehran's worst riot, the army was present in the street but stood by and did not intervene. Tehran burned. Banks, hotels, cinemas, showrooms, the Ministry of Information and other buildings were destroyed, and the British Embassy was occupied. While seventeen police stations were attacked by guerrillas, the SAVAK detachments settled old scores.[39]

General Oveisi had ordered the army not to intervene,[40] hoping thus to force the Shah to bring in a military government and appoint him prime minister. The Shah, as usual, could not make up his mind. Two days earlier, on November 3, National Security Advisor Zbigniew Brzezinski had called him to urge him to take a hard line; the Shah sought but did not receive confirmation of the message from Ambassador William Sullivan on the following day. He then did the worst thing possible. He brought in a military government, but one that was military in name only. The Shah bypassed the hard-liner Oveisi and instead appointed an ineffectual parlor general, Azhari, who soon earned himself the rhyming sobriquet *"eshali"* (diarrhea-ridden) and physically broke down in his conciliatory back bendings, suffering a mild heart attack in December. On November 6, in the very speech announcing the installation of military government, the Shah told the Iranian nation, "I have heard the message of your revolution," and vowed to continue his policy of liberalization. The military government was not given the freedom to apply martial law, especially a provision allowing the massive arrests that the SAVAK wanted to make.[41] On the contrary, scores more of political prisoners were freed. Nor was the military government given immunity, as they usually are, to questioning and ridicule by the parliament. (The Shah had not dissolved his handpicked parliament, whose members were now trying to catch up with the popular mood.)

It is revealing that, according to one report, when the Shah finally made up his mind to have a military government, he telephoned General Oveisi and told him: "Begin applying the martial law tomorrow, but I do not want anybody's nose to bleed!" Putting the receiver down, he turned to General Tufanian who was present and said: "Now what more will the Americans say?"[42]

The installation of military government did produce a lull of two weeks in the strikes and demonstrations, until Khomeini decided to start the next wave of agitation in anticipation of the holy month of Moharram. Meanwhile, the paralysis of the state was becoming definitive. Resignations of high officials became quite common, as did their flight abroad.[43] Furthermore, a number of very prominent members of the political elite, mostly Baha'is, were forced to resign, fourteen former ministers and high officials were arrested, and many more arrest warrants were issued under the martial law code in order to appease the opposition.[44] The employees of the Central Bank published a list of 183 prominent persons and high government officials who had allegedly transferred large sums of money abroad. It outraged public opinion and devastated the morale of the political elite. It is highly revealing of the disunity among the elite of the Pahlavi regime that its members, mindless of the danger threatening the survival of all of them, still sought to destroy their personal rivals with any weapon supplied by the opposition. The owners of the two major newspapers, *Ettela'at* and *Kayhan*, for instance, each circulated partially forged lists of money transmitters containing the name of his competitor.[45]

Much more important than the defections and demoralization of the high officials was the development of contacts and coalitions between the government employees and the oppositional organizations and personalities. The employees of such important organizations as the state radio and television network were won over by the opposition. Furthermore, these contacts produced a massive wave of strikes in government offices, notably the Customs, and among the workers in the state-owned oil industry, railways, and tobacco factories in late November and early December 1978, which continued until the fall of the monarch.[46] Bakhtiar's ministers were to find the doors to their offices locked and could not get in until January 21, 1979![47]

The Shah had painstakingly expanded the state machine and restructured it into a neo-patrimonial dictatorship around his own person. There can now be no doubt that the collapse of the man preceded the collapse of his neo-patrimonial machine. Many a visitor who met him in his palace in the fall of 1978 described him as a shattered man on the verge of a nervous breakdown.[48] Had it not been for the Shah's mistakes, his debilitating dejection, bungling, and indecisiveness, the collapse of the monarchical regime would not have happened so soon. The government had made unnecessary, inadvisable moves ever since publishing in the daily *Ettela'at* a slanderous letter calling Khomeini, among other things, a British agent of Indian descent. It had triggered open agitation of the clerical party in January 1978 and provoked the outrage of the students at the religious colleges in Qom, the unrest that produced the first group of martyrs and thus set in motion the 40-day cycles of mourning and violence that led to the Islamic revolution. Forty days later, on February 18, there was the first major riot in Tabriz in which, according to the government figures, 27 were killed and 262 wounded. The Tabriz riot, in which symbols of Westernized Iran such as cinemas, liquor stores, restaurants, banks, hairdressing salons, and the headquarters of the hated Rastakhiz Party were the targets of violence and destruction, set the model for the subsequent periodic unrest that culminated in the burning of the Shams brewery and the entire red light district of Tehran on January 29, 1979.

Even more serious than the bungling over the letter was the Shah's failure to come to terms with the moderates within the Shi'ite hierocracy. He was totally insensitive not only to the radicalizing pressure of the seminarians on the Grand Ayatollah Shari'at-madari, which forced the latter to declare days of mourning for demonstrators killed by the troops,[49] but also unresponsive to Shari'at-madari's desperate approach in the fall of 1978. In fact, the troops had attacked Shari'at-madari's home in May 1978 and killed two of his aides. The Shah did not respond to Shari'at-madari's search for an understanding in the autumn,[50] nor to his direct calls in the last days.[51] Last but not least, Khomeini was relatively inconspicuous in his residence in Iraq in September 1978 and was inadverantly pushed into the limelight of Versailles by the Iraqi government at the Shah's ill-advised instigation.

Already in July the Shah had disappeared from the public eye. A variant of the King Faisal story, to the effect that he had been shot by a nephew, widely circulated. In response to this rumor, the national radio and television network came up with a brief strip showing a Shah, stiffened by gout, walking beside the Empress on the Caspian coast. I heard the film being referred to as the "We-have-not-been-shot-and-are-alive movie." In August came the serious riots of Isfahan, Shiraz, Tabriz, Ahwaz, and Tehran, with banks, restaurants, and cinemas being burned. On August 27, Prime Minister Amuzegar resigned after complaining about the existence of "a government within the government," a remark taken to refer to the activities of the SAVAK's *agents provocateurs,* which were designed to persuade the Shah to reinstate repression by fanning unrest. But the SAVAK had miscalculated. The Shah did not order it to clamp down on the opposition, even though he was falsely but effectively blamed for the worst of the incidents: Cinema Rex in Abadan, in which 480 persons died as a result of arson on August 19 (the anniversary of the royalist coup d'etat of 1953), was renamed the Kebab House of the Sun of the Aryans (*Aryamehr;* the title the Shah had pompously given himself in imitation of Louis XIV). The Shah then appointed Sharif-Emami, whom he apparently did not trust and would allow very little latitude in dealing with the crisis. As for his response to the strikes: all carrot and no stick. The strikers in the public sector secured large pay increases and continued to be paid in full while on strike or slowdown. There were thus substantial rewards and no penalty for going on strike and joining the anti-Shah political festival. It was not the Shah but Bakhtiar who eventually announced the rule of no work–no pay after the monarch's departure.[52]

Insofar as it is possible to reconstruct the psychodynamics of the Shah's collapse, we may say that in the summer of 1978 his well-known disposition to *folies de grandeur* understandably turned into paranoia. Contrary to the image he so successfully projected in the 1970s, the Shah, unlike his father, was a weak person, and had failed to act decisively in all the major crises of his reign. In his last book, the Shah admitted that in his confrontation with Mosaddeq in July 1952, when faced with large crowds of demonstrators, "I refused to order my troops to fire and was forced to recall Mosaddeq."[53] In August 1953, while the coup d'etat was in preparation, he lost his nerve and fled the country. The next crisis came in June 1963. In this instance, the troops did fire on crowds and bloodily suppressed the protest movement led by Khomeini. However, it is now clear that it was Prime Minister Asadollah 'Alam who acted decisively *despite* the Shah's hesitations.[54] (Incidentally,

'Alam was the ablest of the Shah's trusted advisers and his death, together with that of Manuchehr Eqbal, another trusted aide, in the fall of 1977 increased the Shah's helplessness.) The Shah's conspiratorial view of the world is amply documented in his last book.[55] Especially after a visit by Chairman Hua Kuo-fang at the end of August, he became convinced that the United States, Russia, and China had reached an agreement to force him to preside over a blood bath and then get rid of him. He halfheartedly decided to go but to avoid the blood bath. He ordered the army to behave with restraint. It is said that he would at times have no real ammunitions issued to the soldiers who were parading the streets to enforce martial law with plastic bullets.[56] In the end, the Shah decided simply to pack his bags and go. His fear of being taken prisoner by his generals did not materialize, and he did obtain the respectable exit he was seeking.

Psychological considerations apart, two features of the revolutionary movement of 1978 to 1979 made it very difficult for the Shah to guide the unwieldy apparatus of the state in a clear direction during the crisis. First, it should be remembered that national integration and improved nationwide communications were the essential preconditions of the revolution of 1979. Unrest was extraordinarily widespread. The Tabriz riot occurred on February 18, 1978. There were disturbances in Reza'iyeh, Varamin, Isfahan, Yazd, and Arak in April and in Qom in May. The first major clash in the renewed wave of violence occurred in Mashhad on July 22, 1978. From August onward, unrest rocked Isfahan, Shiraz, and Tabriz as well as many of the hitherto stagnant peripheral towns. On October 29, a Commune came into being in the town of Amol and lasted for forty-eight hours, with sixteen-year-olds forming their own police force. Forty persons died in the unrest in Kerman. Some fifty provincial towns, notably Kermanshah, Zanjan, Gorgan, and Sanandaj, witnessed similar insurrections. All this was brought to the immediate attention of the entire nation by the mass media, raising the nation's political fever to the level of delirium.

Second, though Islamic activists and professional revolutionists were intensely committed to their cause and willing to risk their lives, it was not their activism per se but the widespread, low-level participation of virtually all urban social groups precipitated by their activism that brought about paralysis of the state. This widespread, persistent participation in a loosely structured and amorphous movement without a single center made it extremely difficult for the state to apply its coercive means.[57] In other words, it was a massive campaign of civil disobedience, both within the state bureaucracy and state economic enterprises and outside of the state among other urban strata, against which the army was of little avail.

The Armed Forces and the Revolution

It is now time to turn to the most important pillar of the Pahlavi state, the one on which its survival critically depended: the armed forces. The modernized standing army, Reza Shah's proud achievement, had disintegrated in a few days when faced with the Allied invasion in 1941. The young Mohammad Reza Shah, himself a product of his father's military academy, had taken good care to reorganize it and update its armament. In the 1970s, he made it into one of the dozen best equipped armies in the world. In the mid-1970s, the flowers of Reza Shah's military academy—those who had been selected as classmates for the crown prince in 1936—had attained the highest rank of four-star general, and together with the rest of the elite of their generation, dominated the army that was to face the revolution.

"How could he then miscarry, having . . . so many trained soldiers . . . and diverse magazines of ammunitions in places fortified?"[58] The question Hobbes asked himself about Charles I of England is all the more puzzling as regards the Shah, who had not only a much greater quantity of ammunitions and many more soldiers than Charles, but also all the money in the world with which to pay them. The answer is as follows: the Shah miscarried because he refused to use the army effectively to repress the revolutionary movement, hoping, in vain, to quell the mounting popular rage by the threat or semblance of its use. Unlike the czar's army in 1917, the Shah's remained intact and loyal until he departed on January 16, 1979. It was the Shah himself who muffled the army to the outrage of the hard-liners among his generals. He simply did not dare take the ruthless measures required by the logic of his neo-patrimonial dictatorship to assure its survival. This was reflected in the relatively small number of casualties for an upheaval of major proportions. Notwithstanding wild rumors on the number of martyrs, being echoed by the media, the number of persons killed between October 1, 1978, and January 15, 1979, was between two and three thousand.[59]

The first act of sabotage in the army was the explosion of a helicopter in Isfahan on October 10, 1978,[60] but the incident was insignificant, and the army was intact and remained so until the end of the year. The hard-liners among the military wanted the Shah to allow them to clamp down on all opposition activities, and General Oveisi began to feel out the foreign powers concerning their reaction to a military takeover. In mid-October, the American and British ambassadors called on him to let them know they favored progressive democratization as a solution to the current crisis and were not well-disposed toward his project.[61]

During the last ten days of November, with the commemoration of the martyrdom of Imam Hosayn during the month of Moharram approaching, Khomeini began to put the army to its severest test in its confrontation with the civilian population. This was intensified at the beginning of the month of Moharram, on December 1, when the curfew was defied with cries of *Allah-o akbar* (God is great) from the rooftops. The army, which the Shah, unlike his father, had never thought of using against the population, completely lacked training in crowd control, and the casualties were relatively heavy (probably about 1500 between November 20 and December 10.)[62] The strains of confrontation were aggravated by pro-Khomeini propaganda and distribution of leaflets among the soldiers. Nevertheless, the military machine remained largely intact. This made the opposition leaders in Iran, both lay and clerical, extremely apprehensive that the demonstrations planned for December 10 and December 11 in defiance of government orders might turn into a bloodbath.[63] However, the Shah did not have the nerve to use the army and backed down. Instead, he allowed the demonstrations and instructed Generals Moqaddam and Qarabaghi—the director of the SAVAK and the minister of the interior, respectively—to enter into negotiations with the organizers of the demonstrations in order to avoid incidents and bloody encounters.

The truly massive demonstrations of December 10 and December 11 (*Tasu'a* and *'Ashura;* the eve and the day of martyrdom of Imam Hosayn) marked the end of the Shah, who had not used his last and most formidable weapon. Signs of breakdown of discipline in the army itself appeared in the wake of the first massive demonstration. On December 11, the day of 'Ashura, a small band of enlisted men and noncommissioned officers opened fire on the officers' mess of the elite Imperial Guard, killing tens of officers.[64] On December 13, martial law administrators organized "spontaneous" pro-Shah demonstrations, which resulted in some bloodshed. The local commanders responsible were dismissed and threatened with court martial. On December 18, a number of soldiers in Tabriz deserted their posts and joined the demonstrators, and others refused to obey orders to fire. In the second half of December, there were further signs of breakdown of discipline, including sabotage at a number of bases and increased desertions. An army division in the shrine city of Mashhad broke down, and its commander had to be replaced. Desertions among the conscripts continued to increase until the end of December. However, discipline in the ranks was maintained and the army was still largely intact.[65]

The news of the imminent departure of the Shah on the last days of 1978 had the predictable effect on the morale of the army. The towns of Dezful and Andimeshk exploded and the army was forced to retreat.[66] Desertions increased in early January 1979. According to General Huyser there were 100 desertions daily before the Shah's departure and 100 to 200 by January 20.[67] Furthermore, the generals, including Prime Minister Azhari, began to criticize the Shah openly in the second half of December. A senior officer even went as far as telling his American counterpart that the Shah must go.[68] General Oveisi fled the country in the early days of January 1979; some others started packing their bags more discreetly. Nevertheless, the incidents of breakdown of discipline were isolated and the desertions did not amount to more than a fraction of the army of 400,000 men by the day the Shah left Iran (January 16, 1979). The army had held together very well, perhaps even better than could have been expected. And the opposition in Iran knew it, as did Khomeini in Paris.

Fascinating new evidence shows that Bakhtiar offered to resign on the same day as the Shah left. The collapse of the Pahlavi regime might have happened even faster than it did had it not been for Khomeini's apprehension of the army and his conviction that elements from the army had to be won over before the revolution could succeed. When Bakhtiar's immediate offer of resignation on January 16 was conveyed by Bazargan to Khomeini, he did not respond, telling his entourage that it was too soon because no serious contacts with the military were yet established—it would be inconvenient for Bakhtiar to resign now; let that be done later when the revolutionaries reach an understanding with at least a few of the key generals. Indeed, it was too soon; the new approach to the military—flowers in the gun muzzles and garlands over the tank guns—had only been launched during the demonstrations of January 15.[69] From January 17 onward, contacts with the generals by Bazargan and Ayatollah Beheshti in Tehran were intensified with Khomeini's full endorsement.[70] Khomeini himself began making conciliatory appeals to the army while the Revolutionary Council in Tehran regularly remained in touch with the generals, especially the director of the SAVAK, General Moqaddam, and General Qarabaghi, now the chief of staff in the Shah's absence.[71]

Until now it has been assumed that it was Bakhtiar who needed time and sought to postpone Khomeini's return. This is true, but he was not the only one who needed time. Khomeini and the Revolutionary Council also needed to slow down the astonishingly rapid crumbling of the headless and hollow monarchical state.[72] Despite this mutual interest, while both sides were courting the generals, negotiations between Bakhtiar and the Revolutionary Council continued, and by January 25, 1979, Bakhtiar had written a letter of resignation and made several alterations in his own handwriting, to be left behind as he flew to Paris for a meeting with Khomeini.[73] As we have seen in Chapter 5, Khomeini wrecked the arrangement by insisting on Bakhtiar's public resignation. Bakhtiar reacted by planning to intercept Khomeini's airplane, which in the end he dared not do.

Meanwhile, the opposition was putting all its efforts into working on the army. This time, the efforts were not confined to penetrating the lower ranks but also included negotiating with the senior officers and the commanders. On the day of the Shah's departure, pictures of the civilian demonstrators and the clerics embracing the soldiers appeared in the newspapers. In the subsequent days, centers were set up to give army deserters civilian clothes and the equivalent of $30 in bus fare to their town or village of origin.[74] General Huyser reports that desertions increased after January 26,[75] General Qarabaghi mentions the figure of 1000, adding that it gradually increased to 1200 desertions a day in the last days of the monarchy (February 9–11).[76] Hand in hand with the effort to persuade the military to join the revolution went harassment and intimidation of the loyalist officers. The military were boycotted by physicians and pharmacists, and by oil company distributors and railroad employees; notices and signs were attached to their doors.[77] Some army personnel began to take part in the demonstrations in civilian clothes, and a number of police personnel wrote to their commanders that as followers (*moqalleds*) of Khomeini in the Sacred Law, they could not disobey him.[78] The number of generals requesting retirement or offering their resignations increased.[79] Particularly demoralizing for the high-ranking officers was the fact that the Shah had not wanted the commanders of the armed forces to see him off at the airport and that, despite the insistence of the alarmed generals, he had refused to leave any instructions for being contacted by the chief of staff.[80] The situation deteriorated further at the end of January with instances of violence, at times fatal, against army officers and their families.[81]

Most serious of all, there was the *en masse* defection of some 800 non-commissioned officers of the Air Force, known as Homafaran, in Dezful, Hamadan, Mashhad, and Isfahan a week after the Shah's departure. Orders were issued for their detention and 450 of them were transferred from their original bases.[82] The Homafaran was a 12,000 man corps of technicians and maintenance workers who held a sense of deprivation relative to regular officers, who could rise much higher in rank, with less education. They were singly the most important unit in the armed forces won over by the revolutionaries and were decisive in precipitating the final split in the army that sealed its fate on February 9 to 11.

After Khomeini's return on February 1, the presence of the army in Tehran and elsewhere progressively decreased, and at least half a dozen key generals in the Air Force, the Army, the Navy, and the police defected.[83] During the military parade of February 1, following the Imperial Guards chanting "Long Live the Shah!" were twenty-eight truckloads of riot police with carnations in their rifle muzzles and several truckloads of Air Force personnel carrying pictures of Khomeini and shouting "Death to the Shah!" The dialogue between Bazargan, Beheshti, and the military (including the Shah's sinister director of the Imperial Bureau of Intelligence, General Fardust,[84] General Moqaddam, and General Qarabaghi) continued while scores of retired or purged officers were declaring their willingness to serve under Khomeini.[85] The most important of the above-mentioned figures was General Fardust, who had played a major role in the Shah's neo-patrimonial control of the army. A former schoolmate and trusted personal friend of the Shah who had consistently held the highest positions in the intelligence system created after the 1953 coup, Fardust had inordinate influence over the generals. It would be no exaggeration to say that his defection was a major factor in the eventual paralysis of the army. On February 6, the pledge of loyalty to the Shah was omitted from the oath administered to the cadets of Tehran's military academy at their graduation ceremony, which was presided over by General Qarabaghi. The ceremony was rebroadcast by the national radio and television network the next day. The loyalist commander of the Ground Force, General Badre'i, to be assassinated six days later, was reportedly furious over the incident; he put the Imperial Guards at the Lavizan base on alert, but was kept at bay by Qarabaghi.[86]

We have already pointed out that the Shah restrained the army as long as he remained in Iran. It is worth emphasizing that he could not overcome his fear of a strong military leader even after he had decided to pack his bags and go. In the early days of January 1979, he could not be pressured into appointing the hard-liner, General Ja'farian, as the commander of the Ground Force. On the contrary, he appointed Qarabaghi, a general not of the army but of the gendarmerie, a soft-liner who had not been present at a gathering of the top brass at the Shah's palace to obtain his authorization for a military takeover a few days earlier.[87] Furthermore, the Shah refused to give control of the armed forces to General Jam, a man known for his independence whom Bakhtiar had proposed as the minister of defense in order to boost his chance of survival. Hoping against hope for a return to Iran when things calmed down, the Shah still feared a *coup de grace* from the army and persisted in perpetuating a divided military command: instead of Jam, he appointed General Shafaqat who, as he knew full well, was resentful and hostile toward Chief of Staff Qarabaghi.[88]

All this, as one observer has put it, "was wholly consistent with the Shah's tenacious refusal throughout the crisis to transfer any real power from himself to the government."[89] It was also fully consistent with the Shah's method of controlling the army and with the neo-patrimonial command structure the Shah had set up and maintained in his otherwise modernized army.

The appointment of Bakhtiar and the announcement of the Shah's departure caused a predictable stir among the military hard-liners. They made a feeble attempt to pressure him into authorizing a military takeover in the early days of January and, with encouragement from American General Huyser, began working on a contingency plan for complete military takeover if the Bakhtiar government collapsed. According to one general involved, the Shah knew about this but would not acknowledge it.[90] General Ja'farian, the commander of the divisions in Khuzestan, favored a military takeover and allowed his men to have their way on the days following the Shah's departure. On January 17, tanks with troops shouting "The Shah must return" marched through the city of Ahvaz and began shooting. On January 20, two armed units in Ahvaz and Dezful broke ranks and savaged the people.[91] After Khomeini's return, Admiral Habibollahi, who had been working on the plan for a coup, sided with Chief of Staff General Qarabaghi in opposing a military takeover while Generals Rabi'i, Badre'i, and Tufanian remained "ready at the drop of a hat for military action," or so General Huyser thought.[92] Be that as it may, little headway had been made with the contingency plan by the time decisive defections in the Air Force in February eliminated the feasibility of military action.[93]

What remains for us to emphasize is that the Shah's army would not act without him. Its eventual split and paralysis should therefore not have come as a surprise. The neo-patrimonial command structure was the result of the application of the principle of divide and rule to the highest echelons of the army. To assure the loyalty of the army to his person, the Shah had sought to maximize rivalry and mutual resentment among the generals, and placed personal enemies alternately in the chain of command whenever possible. The same went for the three commanders of the Armed Forces and the commanders of the police and of the gendarmerie, who each reported directly to the Shah, circumventing the chief of staff. He did not change this policy during the crisis, and there was a great deal of mutual antipathy among the generals in key positions: between Oveisi and Qarabaghi; between Qarabaghi and Shafaqat; between Shafaqat, the minister of war, and Tufanian (his senior in age if no longer in rank, who resented having to serve under him as deputy minister of war); between Fardust and Tufanian; and between Commander of the Air Force Rabi'i and Azarbarzin (second in command in the Air Force, to become its commander under Bazargan); etc.[94] Such a fragmented command structure, perpetuated by the Shah to prevent a coup against him, made it very difficult for the senior army officers to carry out the complicated coordination necessary for any form of concerted action in a crisis.[95] It is not surprising that General Huyser found that generals who were talking about a military coup in early January 1979 had no plans at all and did not even know where the key installations were located.[96] Nor did Huyser's arrival do anything for the generals other than keep them trapped. He already found the generals deeply suspicious of the United States' motives for both real and imaginary reasons.[97] They had a real basis for their suspicion in Ambassador Sullivan's contacts with the opposition and they imagined the extensive coverage of the activities of the opposition by the BBC to reflect the attitude of Western governments. General Huyser's refusal to discourage the Shah's departure at their persistent request could only have confirmed this suspicion.[98] So did General Huyser's contradictory instructions which, in his own words, amounted to asking for a "tightrope act":[99] They were to support the Bakhtiar government fully but also be prepared to carry out a coup d'etat against it at short notice if the situation looked desperate. To make matters even more complicated, they were to keep in contact with the opposition leaders.[100] General Huyser could do little to allay their suspicions;[101] General Gast, who remained in Iran after his departure, could do even less.[102]

During the two weeks beginning on January 10, General Huyser and the American military advisors helped the generals draw up a contingency plan for the military takeover of the vital services: the Army was to take over communications and the distribution of food and water, and the Navy the oil fields. One of the simplest steps in the plan was the takeover of Customs, whose employees remained on strike. There were twenty-six customs posts operated by 6000 employees. Alarmed by the unrestricted entry of Khomeini's tapes and guns and ammunitions that went to the opposition forces while all else was blocked, Huyser insisted on the implementation of the plan for the takeover of Customs two days after the Shah's departure. Qarabaghi agreed, with Prime Minister Bakhtiar's approval. On the appointed day, January 20, the plan was called off. On the same day, Qarabaghi offered his resignation, while Grand Ayatollah Shari'at-madari announced that customs officials would release perishable and essential goods![103] General Huyser made no headway in having the plan implemented subsequently. Nor could he get the generals to move on the plan for the takeover of the oil fields.[104] In fact, despite working since January 18 with the full might of the Iranian army on the infinitely more modest task of unloading the fuel from a U.S. tanker, Huyser did not have the satisfaction of seeing it unloaded when he left Iran on February 3.[105]

Nevertheless, when reporting to President Carter on February 5, 1979, General Huyser still considered the Iranian army capable of a military take-over if given a signal from Washington,[106] presumably on the basis of the maintenance of discipline in the bases he had frequently inspected and the manageable number of desertions. Within a week, the events in Iran were to prove him wrong. I submit that he was wrong because there is more to the capability of an army than the maintenance of discipline among its sol-diers. The generals, too, need to be united and capable of concerted action in crisis. This was not the case with the generals of the Shah's headless army.

The split within the army became evident immediately before the "three glorious days" of the revolution, February 9 to 11. Significantly, the trouble began with the Air Force cadets and the Homafaran, many of whom were sympathetic to Islamic reformism as advocated by Bazargan and Shari'ati. When state television rebroadcast film of Khomeini's return on Friday eve-ning, February 9, their mutinous demonstrations of support for him pro-voked a punitive attack by 50 to 200 Imperial Guards stationed at the Dawshan Tappeh Base in eastern Tehran. Fighting continued around the air force barracks until early Saturday afternoon when the Homafaran seized 2000 rifles and distributed them to the people. By then arms were being distributed in mosques of Tehran, and special phone numbers for calling to receive arms were posted on placards. Meanwhile, Isfahan had fallen into the hands of Khomeini supporters. A meeting was arranged between Bazar-gan, Qarabaghi, and Bakhtiar for the following day in which the latter was to hand in his letter of resignation.

At this stage, the forces that were to seal the fate of the revolution triumphantly enter the arena: the populace and the guerrilla groups formed among university students. The two guerrilla groups ideologically committed to revolutionary armed struggle and aided by the PLO, the Marxist Feda-'iyan-e Khalq and the radical Islamic Mojahedin-e Khalq, had been re-ceiving considerable quantities of arms and ammunitions since the Shah's departure and had been responsible for the outbreaks of violence at the end of January.[107] They now joined the defenders of the mutinous air force barracks. A dusk-to-dawn curfew was decreed on February 11, the anni-versary of Siahkal, in which a small band of Feda'iyan had been overcome by the Shah's forces after heroic resistance. But Tehran did not sleep; the curfew was ignored.

The incident of the night of February 9 was all that was needed to break up the army. The day before a considerable number of military men, many of them in uniform, had gone to Khomeini's headquarters and openly pledged their solidarity.[108] With no forwarding address or phone number from the Shah and no unity and no experience of teamwork, some of the key generals secretly declared their solidarity with the revolutionaries in desperation. The most important defections of the last days were those of the commander of the Air Force, General Rabi'i, on February 8 and of the commander of the Imperial Guard, General Neshat, on February 10.[109] On February 10, Rabi'i allowed the distribution of arms among the mutinous cadets and Homafaran while Neshat refused to send in the reinforcements requested by the Imperial Guard unit fighting at Dawshan Tappeh, pretending he had special orders from the Shah. In all probability, however, Neshat was acting according to instructions from General Fardust, who controlled most of the high-ranking officers of the Imperial Guard. With some difficulty and after hours of delay in putting together the equipment and ammunition, General Badre'i dispatched a column with thirty tanks to Dawshan Tappeh late in the evening. It was blocked by the sheer mass of people in eastern Tehran and its commander was killed around midnight. Only a handful of the tanks returned to their base the following morning; the rest joined the revolutionaries.[110] Another army unit sent from Qazvin was similarly blocked in the western outskirts of the city. Meanwhile, some of the martial law commanders on their own initiative ordered their units to return to the bases. The revolutionaries began attacking police stations. One police station after another fell after saturation of the surrounding area by the populace and well-executed attacks by the guerrilla units. Many of the stations had been left empty as a result of collusion with the revolutionaries of the senior officers.[111] The turmoil continued unabated into the following morning and afternoon. The fallen targets included an arsenal captured in the early morning hours of February 11.

The uprising was decidedly spontaneous. The sense of impunity resulting from the disunity in the armed forces emboldened the populace, releasing its political will and energy, but without a specific direction. That was supplied by the Mojahedin and the Feda'iyan. The clerical party did not actively lead the uprising. On February 4, a spokesman for Khomeini had said that the transfer of power could be carried out within the framework of the 1906 Constitution. On Saturday, February 10, Khomeini spoke to the nation, condemning the attack on the air force barracks as unbearable savagery, but affirming that he would still like to finish the matter in a "peaceful and legal (a word rarely encountered in the Grand Ayatollah's vocabulary) way." Early in the afternoon of February 11, Ayatollah Mofattah, speaking for Khomeini, reiterated that statement, adding that no order for Holy War but only for preparation had been issued by the Imam.[112]

Demoralized by the miserable performance of the armed forces, their lack of preparation, logistic failure, and fragmentation, Qarabaghi gave up hope of any bargains, however small, that he had planned to drive during his afternoon meeting with Bazargan and convened a meeting of all the senior officers of the armed forces. The meeting began at 10:30 A.M. and the twenty-seven generals present drafted a declaration of neutrality which was signed by all of them around 1 P.M. It was conveyed to the revolutionary leaders through a liaison and was read out over the radio at 2 P.M., shortly before the radio station was captured by the revolutionaries.[113]

The Bahktiar government had fallen and a slow-motion collapse of the monarchical edifice was completed twenty-six days after the departure of the monarch. The revolution in the name of God had triumphed.

The United States and the Iranian Revolution

The United States was the major power behind the Shah. He bore this in mind in all his decisions and indecisions. After suspecting the CIA of instigating the popular unrest, during the last two months of his reign the Shah pathetically turned to the United States, begging to be told what to do. The opposition, for its part, considered the Shah a lackey of the United States and was constantly mindful of the American position in its tactical calculations. It is therefore worthwhile to examine the role of the United States in the unfolding drama of the fall of the Shah.

In the United States, the unexpected suddenness of the popular explosion in Iran immediately resulted in soul searching over the American "intelligence failure" and the major policy disagreement within the government at the highest level of decision making, and set in motion a search for the villain "who lost Iran." Such quests by the American public, however understandable, overlook a fundamental question: Did it all matter? Would the outcome have been significantly different if there had been no intelligence failure and no policy disagreements within the United States government? To offer a reasonable answer to these questions, a subsidiary question also needs to be addressed: Did it seem *to the Iranians,* the Shah included, that there had been an intelligence failure and that the United States had no policy?

Intelligence failure was a minor contributory factor to the success of the revolution but not in the above-mentioned sense. The Iranian intelligence services were organized with the assistance of the CIA after the 1953 coup, and took on the CIA's ingrained bias for seeing red. One is startled by the obsession of the SAVAK and military intelligence with communists even as late as January 1979. The Shah and the generals fully shared this obsession.[114] This inability to conceive of any kind of agitators other than the communists, or the communists' unsuspecting tools, at times made the Shah and his intelligence officers deny the evidence of their senses, and could be presumed to have distorted their diagnosis of the revolutionary situation. Beyond this, however, it would be simpleminded to assume that the revolution could have been prevented had it not been for any intelligence failure on the part of the United States, deplorable though this failure may have been for other reasons.

The fact that the first high-level meeting in Washington on Iran took place as late as November 2 is usually taken as a sign of disarray in American policy making. In retrospect, the United States was doing *better* when there was no high-level discussion of Iran. Until the end of October 1978, no contradictory message reached the Iranians. The United States did not like the military option and the American ambassador had dissuaded Oveisi and, on October 24, the Shah from considering that option. In the last days of October, Ardashir Zahedi, Iran's flamboyant ambassador to Washington, won the National Security Advisor Brzezinski's ear, who became set on pushing the military option.[115] On November 3, the day after the first high-level meeting on Iran, Brzezinski called the Shah and assured him that the United States would back him if he decided to bring in a military government. As we have seen, two days later, General Oveisi ordered the army to stand by as Tehran's worst riot broke out, and the Shah felt compelled to appoint General Azhari prime minister. On November 9, Ambassador Sullivan urged Washington to "think the unthinkable," that is, an Iran without the Shah. By mid-November, most officials in the State Department considered the Shah finished, though Secretary of State Cyrus Vance did not share this view. While Vance was preoccupied with other matters during the last days of November, George Ball was brought to Washington to prepare a report on Iran for the president. Ball completed his report in two weeks with heavy input from the State Department.[116] Brzezinski later regretted commissioning the Ball report and contended that it delayed the basic choices.[117] The two weeks Ball was at work covered the crucial period in which the Shah decided not to use the army and to allow the massive marches of Tasu'a and 'Ashura. Meanwhile, on the basis of his and the State Department's assessments of the situation, which were being incorporated into the Ball report, Ambassador Sullivan actively negotiated with the organizers of the marchers and relayed their more general proposal for a Regency Council to Washington. The proposal was adopted by George Ball, who recommended setting up a Council of Notables to take over from the Shah.[118] In the two weeks after the submission of the Ball report, there was a stalemate between the National Security Council and the State Department. President Carter rejected the Ball report. On the other hand, at the urging of his staff, Vance had a showdown with Brzezinski who was told by the president not to communicate with Tehran directly.[119] It is worth stressing that Sullivan was not removed and Brzezinski could not interfere with his conduct of affairs in Tehran in this critical period. Sullivan was thus able to foil Zahedi's attempts to urge the Shah to clamp down by withholding confirmation of Brzezinski's messages of support carried by Zahedi upon his return to Tehran. Furthermore, he was able to maintain his dialogue with the opposition. When Secretary of State Vance told the ambassador to get in touch with the opposition on December 29, the latter replied that he already was. In fact, by the early days of January 1979, Sullivan had a list of some eighty to a hundred generals to leave the country with the Shah and their replacements.[120]

How did the people in Iran perceive U.S. policy in the fateful year of 1978? Until the end of October 1978, the Shah knew the United States favored liberalization and was opposed to a military government. In September and October, the Shah kept seeking reassurance that the United States was not plotting against him.[121] In November and December, he received contradictory messages from Sullivan and Brzezinski, and was confused, suspecting a well-planned American policy as a part of a global grand design. On December 26, he gave up trying to guess what the grand design might be and declared his readiness to execute blindly his part in it, only to be told by the American ambassador that he was the king and the decisions were his.[122]

The opposition, on the other hand, did not feel particularly confused about the U.S. policy. It attributed the Shah's liberalization measures to President Carter's Human Rights policy and was becoming increasingly convinced that the Americans did not mind seeing the Shah replaced. In September, Bazargan's Liberation Movement submitted transition plans for a change of regime that included the setting up of a Regency Council. It is true that Washington did not react to this plan despite Sullivan's prodding.[123] Nevertheless, the opposition must have known the plan was transmitted to Washington. When Bazargan visited Khomeini on October 20, he emphasized the importance of an agreement with the Americans and warned Khomeini not to underestimate the United States.[124] Dr. Naser Minachi, the spokesman of the Liberation Movement, continued to push for the Regency Council plan throughout November. The plan was conveyed to the Shah by Dr. Amini on the night of December 2 to 3. The Shah accepted it at first but was prevailed upon by Zahedi to change his mind twenty-four hours later.[125] Against this background, it is not surprising that President Carter's statement on December 7 that he liked the Shah personally, but that it was up to the Iranian people to make up their minds about him, was widely interpreted to mean that the Shah was abandoned. Everyone who mattered in Iran thought the Americans were dumping the Shah, and his decision to back down and allow the Moharram demonstrations was attributed to the U.S. policy. They also knew—as the officials in Washington evidently did not—that the constant statements of support for the Shah by the White House spokesmen were doing him in by discrediting him even more in the eyes of his outraged people. The corrspondents of the French newspaper, *Liberation,* perceptively reported on the festive air of the December 11 demonstration: The cry of "Death to the Shah!" no longer had the venom of the preceding months and was uttered lightly, as if it meant "He must go!"[126]

It was only after Brzezinski read about Sullivan's refusal to tell the Shah what to do on December 26 that the disagreement over policy in Washington actually resulted in doing contradictory things and multiplying the channels of communication with Tehran. On December 28, after an acrimonious debate between Vance and Brzezinski, a telegram was sent to the Shah purporting to state the preferences of the United States for his guidance. The fact that both Brzezinski and Vance claim victory in their memoirs for the compromise formulation of the message is enough to suggest how it must have compounded the Shah's confusion. The Shah was told to set up a coalition government but also to think of resorting to military action if this were not possible.[127]

Disagreements in the U.S. government sharpened and its policy became a muddle in the new year of 1979. Rumors were reaching Washington that the generals were about to jump ship as the Shah had appointed Bakhtiar and was leaving Iran. On January 3, General Robert Huyser was sent to Iran to persuade the generals to stay on and prevent the disintegration of the army after the Shah's departure.[128] He found that the rumors were well founded and the top generals were ready to go.[129] As has been pointed out, Sullivan had been negotiating with the opposition precisely on the departure of the generals along with the Shah and on their replacement by other officers acceptable to the opposition and was predictably furious to learn that Huyser's mission was to tell the generals to *stay*. Sullivan had also suggested that a ranking U.S. diplomat visit Khomeini, and arrangements for such a visit on January 8 proceeded. The visit, however, was cancelled at the last minute by President Carter. Sullivan characterized the president's decision as insane. He was not removed but was no longer trusted by the president, and competing channels of communication with Tehran increased.[130] However, the idea of establishing contact with Khomeini was pursued, and an American diplomat of lesser rank met Khomeini's aide, Dr. Yazdi, on January 14.[131] Meanwhile, by the end of his first week in Tehran, General Huyser had started a group of Iranian generals on a contingency plan for the complete military takeover of the country.

As his stay in Iran lengthened, General Huyser gradually began to see Ambassador Sullivan's point of view and instructed the new chief of staff, Qarabaghi, to contact the opposition. The latter proceeded to do so with the Shah's permission.[132] Meanwhile, a transition plan for the setting up of a Regency Council (which would change its name to the National Council, dissolve the Majles, dismiss Bakhtiar, and appoint a man of Khomeini's choice as prime minister) was being shaped in the opposition circle, apparently with the approval of the American ambassador.[133] On January 23, Bazargan put the head of the SAVAK directly in touch with Ayatollah Beheshti.[134] On the same day, Huyser and Sullivan submitted to Washington a request for change of instructions to make possible a military-Khomeini coalition.[135] The request, however, was not granted, and Huyser was instructed to secure the general's support for the Bakhtiar government and to keep them prepared for a military takeover in case it failed. Early in February, the urgent request that General Huyser be relieved of his mission was granted and he left Tehran two days after Khomeini's arrival.

How did all this look to the Iranians? The Shah knew that Sullivan was negotiating with the opposition and was told about the meeting with Khomeini scheduled for January 8. It is only when the proposed meeting was abruptly cancelled that he claims to have realized the United States had no policy at all. But by then his bags were being packed. The generals who, despite Brzezinski's best efforts, had never been told to take over, took Huyser's contradictory instructions and pointless intervention to indicate that they were being left to their own devices after the Shah. As for the opposition, nothing indicated any need to revise the assessment that the Americans were dumping the Shah. The empty talk of a coup was discounted in view of their developing contacts with the generals and the chief of staff's repeated assurances, at Bakhtiar's request, that none was in the making. Furthermore, "the Imam has come, the American general has departed" seemed a natural conclusion to "The Shah has gone, the Imam has come." As Ayatollah Taleqani typically commented, the Americans had dumped the Shah, now they would do the same with Bakhtiar.[136] And they did.

The Iranian perception that the United States had abandoned the Shah was a significant factor contributing to the outbreak of revolution. As Hobbes emphasized three centuries ago, the inability of the government to instill awe is decisive for the occurrence of rebellions.[137] With the awe-inspiring SAVAK disabled by its master, the erosion of the fear of U.S. intervention removed the last noteworthy inhibition to revolutionary action against the Pahlavi regime.[138]

The account given above suggests that the United States *did* follow a consistent line down to the last days of 1978 despite the disagreement of the National Security Advisor, and it was seen that way in Iran. From December 28, 1978, the U.S. policy became a complete muddle and contradictory measures were taken by the U.S. government. But by this time, it was too late to do anything effective, and therefore the disagreements did not matter anyway. Let us review the Shah's last four months to see if alternative American decisions could have made a difference.

What if the Carter administration had abruptly forgotten its Human Rights Policy and given General Oveisi the green light for a coup in October 1978? Nothing we know contradicts Sullivan's judgment that it would have been a "delayed disaster." Given its fragmented command structure and the disunity of its generals, the Iranian army without the Shah could not have prevailed in a revolutionary crisis. What if the United States had urged the Shah to use the iron fist? In his memoirs, Dr. Brzezinski stated that the Shah's reluctance to clamp down was due to the fact that he wanted the United States to take full responsibility for the use of the army against the civilian population.[139] Although this statement is true with regard to the hopeless last days of December, it is doubtful that he would have done any more than he did in November and early December even if Ambassador Sullivan had been fully in line with Brzezinski. It should be recalled that the Shah was deeply suspicious of the United States and constantly sought reassurance that it was not plotting against him. Nor was he particularly comfortable with Zahedi—a man who bragged about his influence in Washington and was the son of the general who had restored him to the throne in 1953—as a liaison with the American National Security Advisor. His above-cited remark to Generals Oveisi and Tufanian[140]—"Now what more will the Americans say," meaning, what more do they want—suggests that he considered the military government he brought in on November 6 to be a concession to the United States—probably the most he was prepared to do in response to the U.S. pressure so long as he perceived any margin of hope. As for a military coup after December 28 for which the United States would take full responsibility, given the lack of high-level preparation attested to by General Huyser's account, it would in all probability have been a disaster without delay.

What if the United States had pushed for a "centrist coalition" *before* the massive demonstrations of December 10 and 11? What if the Ball report had been written and accepted two or three weeks earlier and the United States had forced the Shah to give up his tenacious insistence on the control of the army and to set up a Regency Council while the army was still intact and the opposition in awe of it? In retrospect, this seems to have been the best option for the United States and was feasible. Nevertheless, its chances of success were small owing to the following factors: the lack of support for the Shah in the middle class itself, the lack of ability on the part of middle class politicians, the small likelihood of concerted political action by the military and civilian politicians, and last but not least, the opposition of Khomeini. It should also be noted that such a plan could probably not have been implemented fast enough. In fact, while he was negotiating with Sadiqi and Bakhtiar in the second half of December, the Shah still considered Amini's proposal for a Regency Council to be on the table as a third option.[141] This option would not have been all that different from what the Shah eventually did when he appointed Bakhtiar. All one can say is that either Bakhtiar or a Regency Council would have had a better chance a month earlier.

Finally, what if the United States had pursued consistently and with greater vigor Sullivan's plan for a Bazargan-army coalition? What if General Huyser had not been sent to Iran, and the American diplomat who visited Khomeini's headquarters in France had had the ambassadorial rank and done so six days earlier as originally planned? Would the outcome for the United States-Iranian relations have been different? I do not think so. Sullivan's dialogues with Bazargan and Beheshti was enough to secure the initial good will of the new government toward the United States, and any closer ties would only have gotten the Bazargan government into trouble with the leftist revolutionaries in the early days.

Indeed, the United States did not do badly. The proof: nine months later the radicals had to resort to the taking of hostages to stop the return of U.S. influence, and thus ousted Bazargan for his friendly handshake with Brzezinski in Algiers.

To conclude, an intelligence failure and disagreements among the U.S. policy makers certainly existed, but they did not matter. Only two things could have made a real difference: (1) the thorough revamping of His shattered Imperial Majesty's personality to enable him to use the army and the secret police with decisiveness and brutal severity; (2) the endowment, overnight, of the atomized Iranian new middle class with a sense of unity and the political maturity and responsibility that would have enabled it to take over the modernized bureaucratic state and its army from the Shah. I am inclined to think that the United States could not have performed either miracle even if it had had perfect intelligence and complete consensus among its high-level policy makers.

7

Revolutionary Iran:
February 1979 – December 1982

Disintegration of the Pahlavi
Regime and Establishment of Dual Power

On September 16, 1978, there was a massive earthquake in the remote town of Tabas in which some 15,000 people lost their lives. The occasion gave the hierocracy the opportunity to make a bid for the allegiance of the people by organizing relief work in direct and pointed competition with the state.[1] The event foreshadowed the emergence of dual power a few months later. With formation of the coalition between Bazargan and his associates and Khomeini and his militant clerics, dual power became a reality. By the end of November 1978, the shrine cities of Qom and Mashdad were in the hands of the Islamic militants, with an Islamic republic declared in the latter.[2] Dual sovereignty can be said to have spread to the rest of the country on December 11 when millions of demonstrators approved a 17-point revolutionary program that included recognition of Khomeini as Imam, abolition of the monarchy, and establishment of an Islamic government.[3] This was taken by Khomeini and the opposition to constitute "the referendum against the Shah." At the end of December, Qom and Mashhad were still in the hands of the Islamic militants, and Isfahan, Khorramshahr, and Tabriz were in part controlled by them.[4] To these cities were added Dezful and Andimeshk on January 1, 1979, and Shiraz fell into the hands of the revolutionaries on January 11, two days after the lifting of martial law. Even in Tehran, the police disappeared and food and fuel were distributed from the mosques.[5]

On January 12, 1979, Khomeini set up a Council of Islamic Revolution with the task of establishing a transitional government. The names of its members were not disclosed, but it was promised that a provisional government would be presented to the nation as soon as possible.[6] According to Yazdi, Bazargan was sent to the striking oil workers to see if a provisional government could take control of the oil industry and start selling oil as a first step in replacing the monarchy. He did not succeed. However, there was a report of the oil workers seeking representation on the Revolutionary Council. On January 24, the council succeeded in persuading the postal employees to end their strike.[7] Khomeini's return to Iran was planned for after the appointment of a provisional government, when dual sovereignty would be fully in place. But fear of a military takeover prompted Khomeini to return to Iran before he had had the chance to appoint the provisional government.[8]

In any event, with Khomeini's return to Iran on February 1, 1979, the balance of dual power overwhelmingly shifted in favor of the revolutionary side. When he arrived in Tehran and millions gathered from all over the country to welcome him, the army stood back and security was handled by the revolutionary forces with the agreement of the chief of staff. This was the first clear recognition of the existence of dual sovereignty by the state and indicated the decisive change in favor of unofficial sovereignty. On February 4, he appointed Bazargan provisional prime minister while Bakhtiar's government was still constitutionally in power. The ceremony of appointment was broadcast by the state radio and television network, as was the subsequent speech of Bazargan delineating the program of his government.[9] Meanwhile, with the progressive disintegration of the monarchical state, *ad hoc* committees (*komitehs*) were being set up in some of the important cities. These committees controlled the distribution of gas and kerosene. By January 25, individuals designated as "police of the Islamic government" were performing a number of functions.[10] By the first week of February, much of the city of Isfahan had come under the control of the *komiteh* set up by Ayatollah Khademi in the last week of January.[11]

Like other revolutions before it, the Iranian revolution of February 1979 immediately produced a period of dispersion of power or "multiple sovereignty." By the decree of February 4, on whose wording Bazargan had insisted, Khomeini appointed him provisional prime minister "upon the recommendation of the Revolutionary Council and according to the lawful (*shar'i*) right deriving from the opinion of the decisive and nearly unanimous majority of the Iranian nation."[12] The cabinet he subsequently formed consisted of men from the new middle class and did not include any clerics. His government took over the administration, the police and the remnant of the army. However, it remained subordinate to the clerically dominated Revolutionary Council, the identity of whose members was in fact not disclosed. Furthermore, a "parallel government" of revolutionary committees, courts, and guards also developed spontaneously. This parallel government came increasingly under clerical control, thus gradually transforming multiple sovereignty into a new system of dual power. Meanwhile, the period of multiple sovereignty witnessed increased violence and leftist-inspired autonomist agitation in peripheral ethnic regions, notably among the Turkmans, the Kurds, and the Arabs of Khuzestan during the spring of 1979.[13]

In Tehran alone, there were 1500 revolutionary committees. As Bazargan put it, "the committees are everywhere, and no one knows how many exist, not even the Imam himself."[14] Late in February, Khomeini put Ayatollah Mahdavi-Kani in charge of the committees. Many were abolished, those remaining thoroughly purged, and major district committees headed by clerics set up.[15] Revolutionary courts had also come into being from the first day of the revolution. Their creation was the first step by the militant clerics to revive their expropriated judicial authority while at the same time avenging themselves against the representatives of the expropriating Pahlavi state. The clerics soon put an end to the executions ordered independently by the committees and thereafter did not share, nor did they ever relinquish, their full control over the formidable revolutionary courts. Furthermore, Khomeini almost at once began to appoint *imam jom'eh*s in every town, and emphatically reinstituted the Friday congregational prayers. He also treated the property confiscated from the Pahlavis and their associates as booty according to the Sacred Law and gave it to the Foundation for the Disinherited (*bonyad-e mostaz'afin*) to be set up under clerical control in place of the Pahlavi Foundation.

In March and April 1979, Bazargan missed a real chance to capitalize on the considerable legitimacy of the post-revolutionary state and put an end to dual power, or at least alter its balance decisively in favor of central government.[16] Alarmed by the separatist insurrections of the Kurds and the Turkmans in March and apprehensive of the armed strength of the Marxist groups, Khomeini was well disposed toward strengthening central government and, indeed, criticized Bazargan for being weak. Furthermore, Bazargan enjoyed the unqualified support of Ayatollah Taleqani, the second most popular figure in the Islamic revolution. This chance existed even after the clash between Khomeini and Taleqani in mid-April in which the latter was humbled. In fact, on April 25, Ayatollah Mahdani-Kani and the chief of police, Colonel Mojallali, announced a plan for the incorporation of 4000 guards and militiamen into the regular police force.[17] The assassination on May 1, 1979, of Ayatollah Motahhari, the chairman of the Revolutionary Council and Khomeini's close aide, alarmed Khomeini and made him opt for an alternative proposed by the activists of the clerical party. In the following week, the armed revolutionary bands under very loose control of the Texan Islamic modernist Dr. Ebrahim Yazdi, were purged and reorganized into the carefully recruited, 6000-man Corps of the Guardians of the Islamic Revolution, set up to serve as the armed wing of the Islamic Republican Party.[18] The corps was placed under the clerical supervision first of Ayatollah Lahuti and subsequently of two rising clerical stars: Hashemi-Rafsanjani and Khamene'i.

On July 19, 1979, having taken stock of the realities of post-revolutionary dual power, Bazargan invited four militant clerics to join the cabinet: Mahdavi-Kani as minister of the interior and Hashemi-Rafsanjani, Khamane'i, and Bahonar as deputy ministers of interior, defense, and education. The direct clerical invasion of the debilitated state had begun at Bazargan's insistence. Meanwhile, Ayatollah Beheshti was organizing the Islamic Republican Party (IRP) on the basis of the new fundamentalist Islamic ideology and Khomeini's principle of *velayat-e faqih* to replace all others as the sole party of the Islamic ideological state of the near future.

The Clerical Coup d'Etat of November 1979

"Not the 'republic of Iran,' nor the 'democratic republic of Iran,' nor the 'democratic Islamic republic of Iran,' just the 'Islamic Republic of Iran,'" the Ayatollah Khomeini told the nation imperiously before the referendum of March 1979. On behalf of the nation, his prime minister, Bazargan, had submitted to the will of Khomeini, the Imam, the Supreme Leader of the Revolution: Islam was not to be denigrated by the adjective "democratic." Late in May 1979, prefacing what he considered his most important speech since the February revolution with "O God, witness that I gave your message," Khomeini reaffirmed that Iran was to remain an *Islamic* republic; anyone wishing Iran to be just a republic, or a democratic republic, or a democratic Islamic republic was the enemy of Islam and of God. In a speech on June 5, marking the anniversary of the 1963 uprising as the beginning of the Islamic movement, Khomeini sternly warned the intellectuals not to oppose the *'ulama* who endowed the nation with a God-given power:

> Those who did not participate in this movement have no right to advance any claims. . . . Who are they that wish to divert our Islamic movement from Islam? . . . It was the mosques that created this Revolution, the mosques that brought this movement into being. . . . So preserve your mosques, O people. Intellectuals, do not be Western-style intellectuals, imported intellectuals; do your share to preserve the mosques.[19]

Like so much else, the significance of Khomeini's insistence on the categorical exclusion of any reference to democracy was still lost on Iran's timorous middle classes, especially on its illusion-prone intelligentsia: the students, bank clerks, teachers, government functionaries, and junior army or air force officers who were still heady with the unprecedented freedom of expression and engrossed in a prodigious torrent of discussions. In his first press conference, Dr. Bakhtiar, whose outlook if not his courage was shared by a large section of Iran's educated middle class, would still consider it appropriate to remark that, unlike Lenin, Hitler, Nasser, and Castro, Khomeini did not know where he was leading the nation [*sic*]. This, alas, was but a wishful thought on the part of this representative of the new middle classes, which by the end of the summer could unmistakably be identified as the losers.

Shortly before his fall, in a remarkable interview with Oriana Fallaci, Bazargan stated:

> Something unforeseen and unforeseeable happened after the revolution. What happened was that the clergy supplanted us and succeeded in taking over the country. . . . If, instead of being distracted, we had behaved like a party then this mess wouldn't have occurred. . . . In that respect, all the political parties . . . went to sleep after the revolution. And that included the parties of the left, which have never been able to attract the masses in Iran and have always remained on the fringes of reality. Yes, it was the lack of initiative by the laity that permitted the takeover by the clergy.

Nevertheless, like Bakhtiar, Bazargan subscribed to the erroneous view:

> In fact, it cannot even be said that they had it in mind to monopolize the country. They simply seized the opportunity offered by history to fill the vacuum left by us.[20]

However fantastic his program must have sounded to Westernized ears within or outside Iran, Khomeini knew full well, and had repeatedly said, where he was leading the nation. It was his enemies and idolizers who chose not to notice. Both before and after the revolution, he had, with utmost clarity, stated his twin aims: the establishment of an Islamic theocracy and the *complete eradication* of Occidentalism, or Western cultural influence, that, according to him, had ravaged Iran for nearly a century. His audience pretended—or needed to pretend—that he was not serious. While the intelligentsia and middle-class liberals wishfully continued to support him, Khomeini, in the consistent pursuit of his twin aims, dealt with them pitilessly and with utmost disdain. The more sycophantically the middle-class political factions idolized him, the more scornfully he pushed them aside. Until August, the National Front, the Marxist Feda'iyan, the radical Islamic Mojahedin, and, finally, the Tudeh (Communist) Party, which outdid all others in servility, patiently continued to receive insults according to the degree of their opportunistic miscalculations, their thirst for humiliation, or the magnitude of their dashed hopes. By the end of August, however, the losing parties' illusions about a unity of purpose with the true victors of the February revolution were finally dispelled. Unnoticed by other groups enraptured by daydreams, teach-ins, and sit-ins, Khomeini's traditionalist clerical party launched its bid for political domination.

Immediately after the effective clampdown on the press and the National Democratic Party, the Feda'iyan, the Mojahedin, and the Tudeh Party in August 1979, Khomeini ominously announced that, having allowed the opposition six months to show their true faces, the clerical party would now break their poisonous pens and crush their conspiracies, together with those of the Kurds who were restive. A day later, on August 25, he resumed his fulminations not only against the Kurdish Democratic Party, but also against the National Front, secularist liberals and intellectuals, "lackeys of the West," and journalists, whom he would not allow "to drown the people in corruption and prostitution in the name of freedom."

Following Khomeini's speeches, Public Prosecutor of the Islamic Revolutionary Tribunals Ayatollah Azari Qomi announced the expansion of the tribunal's jurisdiction to "counterrevolutionary activities," including those in commerce and industry. August thus came to a close with the consolidation of clerical hegemony, setting the stage for the attainment in September of the destiny Khomeini had chosen for Iran as the first Twelver Shi'ite theocracy in history. In his inaugural message to the Assembly of Constitutional Experts, Khomeini told his hand-picked mollahs that he expected them to create "a 100 percent Islamic constitution."

Meanwhile, Khomeini himself pointedly seized the opportunity offered by the death of the democratic Ayatollah Taleqani in September 1979 to propagate the good tidings of the instauration of the rule of the *'ulama* on behalf of God: did the people kiss Taleqani's hand because he was a democrat, or because he was a liberal? No! They kissed his hand because he was the deputy of the Holy Imams, because he was the deputy of the Prophet.

Following the taking of hostages by an Islamic student group at the American Embassy on November 4, 1979, the Bazargan government fell, and the clerical coup d'etat took place, in the form of a direct takeover of the state by the clerically dominated Revolutionary Council. The coup d'etat unfolded amidst mobilization of mass opinion against the United States by the Tudeh Party, who were bent on the elimination of the liberals as a political force, and by the Mojahedin, who put the struggle against U.S. imperialism before all else. This was masterfully exploited by Khomeini for his own purposes: anti-American resentment, coupled with a cry for vengeance against the dying tyrant, lent itself far better to Khomeini's demagogic manipulation than had the anti-Kurdish and anti-intellectual sentiment in August. It assured, perfectly in accord with Khomeini's design, the smooth passage of Iran's theocratic constitution in the referendum of December 2 to 3, which made him, *de jure,* the ruler of Iran on behalf of the Hidden Imam. The coup d'etat of November 1979 also knocked down one of the two poles of the system of dual power, leaving the state under the supervision of the clerically dominated Revolutionary Council, and paved the way for the further clerical subjugation of the state. It is no wonder that the episode has been subsequently referred to as the Second Islamic Revolution.

The taking of the American hostages marked the onset of Ayatollah Khomeini's phantasmagorical struggle with the imperialist Satan. The first advantage Khomeini drew from this struggle was to assure the ratification of the clericalist constitution. While Bazargan was in office, only one general clause on the mandate of the jurist (Article 5) had been passed by the Assembly of Experts. After Bazargan's dismissal, the Assembly passed the substantive Articles 107 to 110, which granted the jurist extensive governmental power. Bazargan, Bani-Sadr, and others had been grumbling about the "improved" draft of the proposed constitution, which gave the Imam absolute power without the slightest responsibility. But now, who could oppose the constitution when doing so could only mean siding with Satan against God, with the imperialist plotters against the long-suffering nation and its revolutionary leader? Bazargan and most political groups, notably the Tudeh Party, declared themselves constrained to support the new constitution in order not to jeopardize "the ongoing anti-imperialist struggle and the political line of the Imam"!

The disinherited would undoubtedly have voted for the new constitution upon Khomeini's order, and their vote would have been enough to secure its ratification with a comfortable margin. What showed Khomeini's political acumen much more clearly was the sequence of events from the first week of December 1979 until January 11, 1980, beginning with an insurrection in Tabriz.

The insurrection represented the reaction of the coalition of moderate Shi'ite *'ulama* and national democratic politicians to the clericalist constitution making and coup d'etat of November 1979. Everyone expected trouble in Tabriz during and after the constitutional referendum. The Grand Ayatollah Shari'at-madari had spoken out on the eve of the referendum against Khomeini's constitution, calling it an incoherent and self-contradictory piece of legal work and indicating that he would therefore abstain from voting. Precipitated by an attack on Shari'at-madari's house in Qom and an abortive attempt to assassinate him on December 5, a general strike and demonstration in Tabriz were planned for Thursday, December 6. Tabriz rose in insurrection; the radio station and government buildings were taken over by Shari'at-madari's partisans. The following days witnessed demonstrations, counter-demonstrations, and numerous clashes between supporters of the two ayatollahs, all pervaded by what Hobsbawn has called "populist legitimism"[21] on the part of the protesting party: the insurrectionists wanted the director of the National Radio and Television Network executed because of the vicious censorship he had imposed; they wanted the dictatorial constitution repealed, but Khomeini, they said, was their Imam!

Of the troubled December days in Tabriz, I should like to choose the bleakest for close scrutiny: Friday, December 7, the day that proved the faintness of the voice of democracy in Iranian revolutionary politics and the deafness of the ears upon which it fell. This one day vindicated the political astuteness of Khomeini's decision to prolong his anti-imperialist struggle by refusing to release the American hostages.

Shari'at-madari's followers had risen because of loyalty to the person of the Grand Ayatollah. Shari'at-madari's attempt to broaden the support for his protest movement by creating a democratic platform failed. On Friday, December 7, an appeal from Shari'at-madari's political ally, Hasan Nazih[22]— to those who stood for democracy and freedom "to continue the struggle until the elimination of personal power, and of the despotism institutionalized by the new Constitution"—was broadcast three times from the captured radio station but fell on deaf ears.

By contrast, the same day witnessed the triumph of political unreason at the service of Khomeini. Taher Ahmad-zadeh,[23] a supporter of Masaddeq and a revolutionary with impeccable credentials, addressed a large pro-Shari'at-madari gathering. He reminded the crowd that Khomeini was the champion of the disinherited against the imperialist oppressors. For this reason, just like Mosaddeq before him, the Imam was being unconsciously vilified by the imperialist press and by the agents of American capitalism in Iran. Mosaddeq, too, had been compared to Hitler and accused of being a dictator when he nationalized the oil industry, thereby expropriating the British. The crowd was manifestly moved. (Not one among them raised a voice to remind the speaker that Mosaddeq had no comparable constitution introducing institutionalized despotism; nor was the speaker asked in what way the incarceration of unarmed hostages was comparable to the nationalization of the British-owned oil company in 1951.) With whatever rational power they may have possessed under calmer circumstances totally eclipsed and with their emotions reigning supreme, the crowd took to the streets, repenting the betrayal of the Imam and accusing themselves of abandoning David in his struggle against the imperialist Goliath, as the people of Kufa had abandoned the martyred Imam Hosayn in his struggle against the tyrannical Yazid.

The ruthless counteroffensive by the pro-Khomeini forces began immediately after the repentance of the neo-Kufan betrayers of the just Imam. On Monday, December 10, the pro-Khomeini militants, aided by the leftist guerillas, attacked and recaptured the radio station. What was left of the insurgence was bloodily suppressed in the ensuing weeks.[24]

In the early months of 1980, the clerical party began its systematic use of the documents captured at the U.S. Embassy to discredit Bazargan and his associates in the Liberation Movement with devastating effect. Securing the passage of the theocratic constitution may have been his primary aim in exploiting the hostage crisis but he also had unsettled business with the liberals who now represented the new middle class.[25]

I have dwelled on the insurrection in Tabriz at some length because it was decisive for excluding from the teleology of the Islamic revolution what I consider most valuable in the Shi'ite tradition and in modern political thought respectively: the separation of religious and political authority, and liberal democracy. The tradition of separation of religious and political authority was routed in the person of Ayatollah Shari'at-madari, the idea of liberal democracy in the person of Hasan Nazih. This left the field to the worst elements in the Shi'ite tradition and in modern political thought: clericalism and the idea of a totalitarian ideological state. These were espoused and amalgamated by Khomeini and his followers with passionate intensity. Their Islamic ideology carried the day and the Iranian revolution became the third major revolutionary force of the century in succession to communism and fascism.

Termination of Dual Power and Direct Clerical Takeover of the State

In January 1980, thanks largely to the inexperience of the Islamic Republican Party and confusion over its candidate,[26] Abo'l-Hasan Bani-Sadr, running as an independent, won the presidential elections by a landslide. Reacting to the fierce exclusivism—or "monopolism," to use Bazargan's term—of the militant clerics and the Islamic Republican Party, both the Islamic reformist and the nationalist and liberal, democratic elements began to rally behind Bani-Sadr. The stage was set for an intense power struggle between the presidency, with the broad support of the new middle class, and the clerical party. The first Majles, which was elected in March and held its first session in July 1980, was overwhelmingly dominated by the clerics and the Islamic Republican Party (IRP). The political struggle between the president and the Majles, whose first manifestation was the prolonged disagreement over the choice of a prime minister and then of the cabinet, was greatly complicated by the Iraqi invasion of Iran in September 1980. At this stage, neither side was well organized and the power struggle resulted in a stalemate that lasted for a year. A small number of important *'ulama* sided with Bani-Sadr in this period and were forced to recant or retire from the political arena after Bani-Sadr's defeat. It is important to note that Khomeini himself did not, as a rule, directly intervene in the struggle between Bani-Sadr and the IRP. He preferred to remain "above politics," as he apparently had for a few months after the revolution in 1979. Khomeini was at this stage opposed to the occupation of the highest offices of the state by the ayatollahs (the matter was being vigorously debated among the militant *'ulama,* and such influential figures as Ayatollah Mahdavi-Kani, the president of the important Society of Militant Clergy [*Jame'eh-ye Ruhaniyyat-e Mobarez*], were opposed to the idea, as was Khomeini, who had vetoed Ayatollah Beheshti's idea of running for president).

After the presidential elections, Khomeini appeared to have decided in favor of consolidating the enormous gains of the Shi'ite hierocracy as embodied in the Constitution. He relented on his Islamicization campaign in order to give the newly elected President Bani-Sadr and his team a chance to run the state and revive the economy for him. Khomeini's position did not change in February and much of March. The pressure from below might also have acted as a factor in inducing the aged patriarch to strengthen the hand of the state. The temporary inability of the clerical structures of power to absorb the radicalizing pressure from leftists at the grass roots was clearly demonstrated in mid-February by strikes of the employees at the Revolutionary Public Prosecutor's office and the Foundation for the Disinherited, which forced Khomeini to react by publicly stressing the need for strengthening the central government.[27] However, he was not rash enough to allow Bani-Sadr to form a cabinet and dissolve the Revolutionary Council.

The situation changed rapidly in the new Persian year that began on March 21, 1980. In his new year's message, Khomeini lashed out against the syncretic mixers of Marxism and Islam (the Mojahedin) and warned the intellectuals "with links to the East or the West" that they would be excluded from the new Islamic order.[28] Emboldened by its massive electoral victory, which secured it some two-thirds of the seats in the Majles, the Islamic Republican Party impatiently resumed the program of "Islamicization" to reshape the Iranian state and society. In addition to their doctrinally defensible legal authority, the hierocracy also had had *de facto* control of the educational system prior to the twentieth century. To restore the traditional pattern, therefore, the clerical party set out to desecularize the national educational system, and to bring it under clerical control. Furthermore, the desecularization of the educational system was essential for the creation of an ideological state advocated by the new Islamic fundamentalist ideology, and took the form of a "cultural revolution" designed to eradicate all traces of Western cultural influence from the universities and high schools. It was launched by the

Islamic Republican Party in April and was intensified after the abortive U.S. rescue attempt. The universities were closed indefinitely and only the faculties of medicine were to reopen in October.

The Islamic cultural revolution, instituted by Khomeini himself[29] in April 1980, represents an interesting extension of the modern myth of revolution due to Mao Tse-Tung and the repercussions of the Chinese revolution. As the modern myth of revolution was increasingly appropriated by Khomeini and his clerical followers in 1979 and 1980, it was natural for them to look at the latest model of revolution with added features. Khomeini therefore ordered the creation of the Commission (*setad*) for Cultural Revolution to take charge of the Islamicization of universities. It is also interesting to note that a revolutionary ideologue such as Bani-Sadr found it impossible to resist the temptation to capitalize on the appeal of cultural revolution. On April 22, three days after the Party of God had ransacked offices of leftist students in the University of Shiraz and other provincial universities, and a day after bloody clashes which claimed twenty lives, Bani-Sadr led a large crowd into the University of Tehran and proclaimed the beginning of the cultural revolution.[30] However, in this move, unlike Khomeini, Bani-Sadr was as usual mistaken. As Bakhash puts it, "by identifying himself with the attack on the universities, Bani-Sadr embittered the left and secured no credit with the clerical party."[31]

The dominant hierocracy turned its attention to the systematic desecularization of the judiciary system in a seminar convened in April. During the last days of May, Khomeini exhorted the deputies of the Majles to implement Islamic justice "of which our nation has been deprived under the regime of the oppressors and the usurpers of the Pahlavi dynasty." The vigorous response of the IRP deputies was expressed in early July 1980 by the revolutionary public prosecutor, Ayatollah Musavi Ardabili. The first result of those endeavors, a penal code reviving atavistic practices that had been in abeyance for centuries, was presented to the Majles in early 1981.[32]

With the army in an advanced state of disintegration and the president deprived of a cabinet, the religiously significant month of Ramadan (July 1980) was chosen for the official inauguration of the new Islamic order, appropriately marked by the dismissal of a large number of female state employees for not wearing "modest Islamic attire," that is, a scarf—the modernized version of the veil and thick stockings. (The dress code was to be made more restrictive as time went on.) The extropunitive moralism of the militant clerics reached a new height in this period. Ayotallah Khalkhali, the Judge Blood of Iran's Islamic revolution, extended the bloodletting from the political remnants of the old regime to narcotics dealers. On July 5, two women and two men were stoned to death for illicit sexual intercourse, with the clerical judge who had passed the sentence throwing the first stones.[33] Intensive Islamicization could continue unabated throughout the month of August. On August 17, Khomeini felt confident enough to congratulate his clerical followers on their revolution's unprecedented success in installing all the basic institutions of the new Islamic order. On the same occasion, he referred to the hierocracy as the judiciary branch of the regime, completely ignoring the not-yet-dismantled secular judiciary apparatus.[34]

In Khomeini's Islamic order, the state was not only constitutionally weakened but also made Islamic ideologically, that is, manned, insofar as possible, by reliable Islamic personnel and brought under direct or indirect clerical control. The process had been formalized when Bazargan officially admitted clerical supervisors to a number of his cabinet ministries in 1979. With the success of the Islamic Republican Party, the process culminated, in late August 1980, in the installation of a group of servile, religiously minded laymen into the prime ministry and other high offices of state. Meanwhile, the IRP's infiltration of lower ranks of the state had been assured by the repeated purges of the Westernized elements during the summer of 1980. By the beginning of July, after less than ten days of purges, over 1000 functionaries had been discharged. There were some 150 purge committees operating. The purges continued throughout the month of July all around the country. During the last two days alone of major purges at the end of July, some 450 functionaries were discharged.[35] The purges were particularly extensive in the Ministry of Education and the army, which were considered strategically crucial from the ideological and military viewpoints. Some 20,000 teachers and nearly 8,000 officers were discharged.[36] The posts thus freed were distributed as spoils to Islamic activists, or to use the more recent shibboleth, *maktabi* (doctrinaire; belonging to the school [of ideological Islam]) activists. The highest cadres of state functionaries were purged even more completely. The cabinet of Prime Minister Raja'i consisted entirely of *maktabi* loyalists of traditional bourgeois origins and of clerics.

The Iraqi invasion temporarily halted the twin processes of weakening and appropriation of the state by the dominant *'ulama* and their lay agents. Nevertheless, it should be remembered that, as a result of the purges, by the fall of 1980, real power within the state apparatus was captured by the Islamic loyalists. Dual power came to pervade the state itself as a prelude to its clerical subjugation. In October, Prime Minister Raja'i instructed all government departments to deal with the president through his office only, and sought to block Bani-Sadr at every turn.

The war gave President Bani-Sadr and the beleaguered moderates clustered under his umbrella a serious chance to reverse the purges. The moderates' prospects for curbing the monopolism of the IRP and for regaining control of the state seemed good in November and December 1980. Public support for Bani-Sadr and his associates, Qotbzadeh and Salamatian, was organized in the bazaar. But in the long run, Bani-Sadr and the moderates around him failed to act decisively. Clerical hegemony was reestablished after what in retrospect appears to be only a three or four month breathing spell for the moderates.

As a consequence of the war, the army immediately became highly popular, and Bani-Sadr, as its commander in chief, became a national hero. The ruling clerics were blamed for the disintegration of a once powerful army, and found themselves on the defensive. Their rhetorical clamor subsided, and the voice of the moderates—mainly Bazargan and his associates, who had opposed the dismantling of the army during their tenure of office—was widely and appreciatively heard. One group in particular enthusiastically welcomed the reappearance of the moderates on the political stage: the disaffected bazaar merchants. The bazaar had constituted one of the main pillars of the Islamic Republic, and when the merchants sided with the president, the position of the IRP leaders was seriously weakened. In a defensive speech on November 16, 1980, Khomeini castigated the bazaar merchants for their defection, placing them in the same category as the maligned liberals and the minority faction of the Marxist Feda'iyan.

But Bani-Sadr was unable to capitalize on the popularity of the army and the discontent of the bazaar merchants. He failed to organize any concerted action by the sundry groups who supported him while the political climate was highly favorable, and contented himself with long and pathetic harangues, which even the state radio stopped broadcasting. The bazaar merchants continued to demonstrate their support well into March of 1981, but then demoralization apparently set in as they tired of waiting for a clear call from the president. And so the merchants resigned themselves and made a tacit accommodation with the ruling clerics and their lay servants.

The alert reaction of Khomeini and the IRP sharply contrasted with the ineffectuality of the president and the moderates. From the very beginning, Khomeini was careful to keep Bani-Sadr's exposure in the mass media to the minimum. He also set up a clerically dominated Supreme Defense Council to limit the president's power. By December, the ruling clerics had fully regained their composure and set out to secure their share of glory from the war effort. Clerical control of the Supreme Defense Council began to make itself felt through the appointment of permanent representatives, usually clerics, to supervise various fronts. More important, the role of the Guardians of the Islamic Revolution in the fighting was progressively enhanced, with its commander assuming the position of a full member of the Supreme Defense Council. On the last day of December 1980, the IRP pushed through a law envisioning a vast expansion of the Guardians' numbers and enabling its leadership to choose additional recruits from army conscripts. In January 1981, the ruling clerics stepped up their denunciations of Bani-Sadr's conduct of the war, and pushed him into undertaking a "major offensive" prematurely. The offensive, predictably, failed.

Meanwhile, Khomeini was wooing the bazaar merchants, inducing them to give up their support of Bani-Sadr and Bazargan and to come to an arrangement with Prime Minister Mohammad-'Ali Raja'i and his cabinet, who served the clerics. The Imam's Committee for Guild Affairs tried to mobilize the bazaar against Bani-Sadr but without much success. Early in March, clashes between the supporters of President Bani-Sadr and the IRP's club-wielding strike force known as the Party of God (Hezb Allah) broke out during the commemoration of the death of Mosaddeq—the symbol of Iranian nationalism. While initiating criminal investigations into his conduct, the clerical leaders demanded that Bani-Sadr be dismissed and put on trial for treason, thus presaging the scenario for what might be called the "Bazarganization" of Bani-Sadr.

The immediate events leading up to the fall of Bani-Sadr began with the broadcast on May 27 of a speech by Ayatollah Khomeini. The Imam made ominous pronouncements on the future of the first president of the Islamic Republic of Iran, suggesting that he would be deposed as had been the republic's first prime minister, Mehdi Bazargan. The Supreme Leader of the Islamic Revolution lashed out against the beleaguered president, telling him and his supporters to "go back to Europe, to the United States, or wherever else you want." On June 8, following the banning of Bani-Sadr's *Islamic Revolution,* Bazargan's *Mizan,* and other newspapers, Khomeini bluntly threatened Bani-Sadr with dismissal. Two days later, while touring the battlefront, the president was informed that he had been relieved of his duties as commander in chief of the armed forces.

Bani-Sadr was deposed by the Majles and went into hiding. A major demonstration in his favor on June 20, 1981, was bloodily suppressed, with at least 100 persons shot and another 150 arrested and summarily executed for "spreading corruption on earth." The brunt of the casualties was borne by the Mojahedin, but a few supporters of Bani-Sadr from the bazaar were also executed. Furthermore, the dismissal of the president was accompanied by strong threats against the corrupt profiteers of the bazaar. In the following month, Bani-Sadr fled Iran. The period of multiple sovereignty thus came to an end with the clerical takeover of the highest office of state. The subjugation of the modern bureaucratic state to clerical sovereignty became complete,[37] as was the defeat of the new middle class produced by that state in over five decades of modernization.

8

Consolidation
of Islamic Theocracy

Principles of Legitimacy of Theocratic Government
and the Constitution of the Islamic Republic of Iran

It is a truism, nonetheless a significant one, that the delegitimization of monarchy occurred prior to the legitimization of "Islamic government" (not to mention theocratic government) and was a far more important factor in contributing to the demise of the Pahlavi regime. The shaking of the foundations of the legitimacy of monarchy begins with the erosion of the traditional conceptions of kingship resulting from the introduction of democratic theories in Iran. The idea of republicanism gained some currency after World War I and gained even more after the abdication of Reza Shah in 1941. It remained in the background and, as one would expect, began to be espoused openly during the crisis of 1978. More significantly, the crisis of 1978 stimulated the rapid spread of the delegitimizing myth of revolution among the new middle class. In the eyes of the traditional bourgeoisie and the Shi'ite hierocracy, the Shah's failure to uphold Islamic morals,[1] the aggressive invasion of the religious sphere in the form of the Religion Corps and the destruction of seminaries, and finally specifically delegitimizing acts such as the time-saving alteration of the clock (seen as interference with the time of performance of the daily midday prayer), and above all, the foolish replacement of the Islamic calendar by an Imperial one, all combined to render the Pahlavi regime illegitimate.

In Khomeini's own writings, too, the delegitimization of the Pahlavi regime precedes the legitimization of theocracy by quite a few years. In his manual on practical jurisprudence, composed during his exile in Turkey from 1964 to 1965, in the section on "forbidden occupations" he forbade his followers from entering the service of "the tyrant" in positions in "the army, security forces, government of towns, etc." By clear implication, the Shah's tyrannical government was illegitimate. Furthermore, in another section he castigated the pseudo-clerics who, as devils in the clerical garb, had ties with the tyrannical rulers.[2] Like others, Khomeini had denied the legitimacy of the Shah's dictatorship, declaring it in violation of the Constitution in 1963, and continued to do so until at least as late as September 6, 1978. Furthermore, on

October 31, 1971, he countered the Shah's attempt to boost his image as the King of Kings with an extravagant celebration of twenty-five centuries of monarchy in Iran by stating that "Islam is fundamentally opposed to the whole notion of monarchy," citing the prophetic tradition that "King of Kings" is the most hated of all titles in the sight of God.[3] Unlike others, however, Khomeini did not stop at the denial of the legitimacy of Pahlavi rule and, as we have seen, proceeded to put forward a theory of theocratic government in 1970: the mandate of the Shi'ite jurist on behalf of the Hidden Imam (*velayat-e faqih*) included the right to rule.

In an interview with Ayatollah Khomeini in France on January 2, 1979, shortly before his return to Tehran, I asked him what room was left for parliamentary legislation in a government based on the mandate of the jurist as the authoritative interpreters of the Sacred Law. He did concede a role for legislation by a popularly elected parliament, but to the dismay of his aide Dr. Ebrahim Yazdi (subsequently to become Bazargan's foreign minister), who was present during the interview, he restricted its scope to matters which are "beneath the dignity of Islam to concern itself with." The relevant section of the interview follows:

> Q: How should one relate what you have written in *The Mandate of the Jurist* to the actual political organization of Iran and the management of the daily political affairs? What, briefly, are the principles of government in Shi'ism? To what extent and under what conditions can the constitutional regime and its Fundamental Law (of 1906 and 1907) be made to conform to them?
>
> A: *The Mandate of the Jurist* is about the principles of government. The organizational structure of government, the criteria for the appointment of political authorities are not treated in *The Mandate of the Jurist*. I concentrated on the fundamental principle [of the right to rule] because that was what I had been asked to do. There, I wanted to make it clear that government is the right of the religious jurists. The details of the matter [actual organization of government] will have to be dealt with by appropriate laws which will be enacted later.
>
> The Fundamental Law will have to be studied. Whenever it is in the interests of the nation, it will be accepted by us. Incidentally, the Supplementary Fundamental Law does state that [constitutional arrangements] should not be contrary to the principles of Islam. It is possible that some of the items of the Fundamental Law will be found contrary to [national] interests and will be abrogated. Some have already been abrogated by public referendum [in the course of the recent demonstrations] such as the monarchy.
>
> Q: You refer to the Supplementary Fundamental Law. The author of the relevant clause, Shaykh Fazlollah Nuri, did [subsequently] issue a ruling (*fatva*) declaring that legislation is contrary to Islam, because in Islam, law is the divine Sacred Law, which is only in need of interpretation, and whose interpretation is the function of the qualified doctors of Shi'ite jurisprudence. What is your view on the above ruling?
>
> A: It is correct. There are certain matters which are executive affairs such as urban planning and traffic regulations. These are not related to

[Sacred] Law, and it is beneath the dignity of Islam to concern itself with them; they are not related to basic laws. In Islam there is no room for the institution of basic laws and if an assembly is installed it will not be a legislative assembly in that sense, but an assembly to supervise government. It will deliberate [and determine] the executive matters of the kind I mentioned and not basic laws [which are already laid down by Islam].

It is highly significant that in Khomeini's book, *Islamic Government,* there is no mention of an Islamic *republic.* There is reason to believe that Khomeini considered the Islamic republic to be the appropriate form of government only for the period of transition to the truly Islamic government. In this final stage, sovereignty would belong to the hierocracy on behalf of God. There would be no room for sovereignty of the people nor for the supremacy of the state as the presumed embodiment of the national will. Khomeini's project required a drastic withering of the state to an appropriate size. The judiciary system was to be desecularized and brought under the control of the hierocracy. Beyond this, however, Khomeini had given little thought to the exact nature of a modern theocratic state.

The political and publicistic activism of the militant Shi'ite clerics in the 1960s and 1970s impressed upon its leaders, such as the late Ayatollahs Motahhari and Beheshti, the need for a distinct Islamic ideology. In this enterprise, while drawing their inspiration from the writings of the Arab and Pakistani fundamentalist ideologues, they were decisively aided by the Iranian Islamic reformists such as Bazargan, Shari'ati, and Bani-Sadr. These modernist laymen were their immediate masters in the art of formulating and elaborating a coherent ideology. Nevertheless, deep down the Shi'ite hierocracy was suspicious of the modernist lay ideologues and considered them somewhat contaminated by secular ideologies of liberalism, nationalism, and socialism. This is especially true of Khomeini himself who wanted his movement to remain purely Islamic in ordientation and membership. In 1972, in a typical statement that demonstrates his resolve on the creation of a theocracy, Khomeini warned that the problems of Iran would not be solved so long as "the nation of Islam" remained attached to "these colonial schools of thought [i.e., political philosophies] and compared them to Divine Laws [of Islam]." The differences between the militant Shi'ite *'ulama* and the Islamic "modernists" who variously accepted elements of nationalism, liberalism, and socialism did not take long to surface during the revolution, as the militant clerics attacked first the liberal nationalists and then the Islamic modernists.

Already in 1978 Ayatollah Motahhari had stressed the need for vigilance lest the nationalist and liberal intellectuals take the better of the clerical elite as they had done during the Constitutional Revolution. In May 1979, less than five months after the revolution, Ayatollah Beheshti considered the time ripe for openly fighting nationalism and liberal democracy in the person of Hasan Nazih, the president of the Bar Association and the chairman of the National Iranian Oil Company. In a speech demanding the trial of Nazih for treason, Beheshti referred to the years 1962 to 1963, and especially the clerically led uprising of June 1963, as the turning point in Iranian history at which the direction of "the pure Islamic revolution" was determined in clear contradistinction to nationalism and liberal democracy. A few months later, Beheshti incorporated this view of the militant *'ulama* into the preamble to the Constitution of the Islamic Republic.

> Although the Islamic way of thinking and militant clerical leadership played a major and fundamental role in [the constitutional and the nationalist/anti-imperialist] movements, these movements rapidly disintegrated because they became increasingly distant from the true Islamic position.
>
> At this point, the alert conscience of the nation, led by . . . the Grand Ayatollah Imam Khomeini, realized the necessity of pursuing the authentic ideological (*maktabi*) and Islamic path of struggle. . . .
>
> The plan for an Islamic Government based upon the concept of the Mandate of the Jurist (*velayat-e faqih*), which was introduced by Imam Khomeini . . . gave a fresh, concrete incentive to the Muslim people and opened the way for an authentic ideological Islamic struggle. This plan consolidated the efforts of those dedicated Muslims who were fighting both at home and abroad.

As one of the most articulate militant representatives of the Shi'ite hierocracy, Beheshti attacked reformist attempts to reconcile nationalism, liberal democracy, and socialism with Islam as "syncretic thought" (*elteqati*) and presented the theory of the Mandate of the Jurist, which was said to be the result of the researches of the militant *'ulama* on the issue of Islamic government since the 1960s, as the purely Islamic alternative. There can be no doubt that Khomeini and his followers had no clear plans for Islamic government in the 1960s. The only concrete proposal put forward by Khomeini in 1963 had been that the government hand over the responsibility for national education and the pious endowments to the hierocracy and allow them a few hours on the national radio.[4] Although Khomeini did put forward the idea of *velayat-e faqih* around 1970, as indicated in his interview with the author, he had not worked out the institutional and constitutional implications of the idea by January 1979. It is amply clear that he wanted the state to be subordinate to the hierocracy and was firm and careful in this regard. However, he attached little significance to constitution making and was prepared to accept in draft a constitution approved by the cabinet and the Revolutionary Council in June 1979 with only minor change. In fact, he proposed to bypass the promised constituent assembly and submit the draft directly to a referendum. It is highly significant that Bazargan and Bani-Sadr insisted on the election of a constituent assembly while Hojjatol-Islam Hashemi-Rafsanjani asked the latter: "Who do you think will be elected to a constituent assembly? A fistful of ignorant and fanatic fundamentalists who will do such damage that you will regret ever having convened them."[5]

It was decided to hold elections of an Assembly of Experts on August 3, and the draft constitution instantly became the subject of debate by various secular parties and organizations. These debates alarmed Khomeini. At the end of June, he told the Shi'ite clerics that the revision of the draft had to be undertaken from an Islamic perspective and that it was their exclusive prerogative.

> This right belongs to you. It is those knowledgeable in Islam who may express an opinion on the law of Islam. The constitution of the Islamic Republic means the constitution of Islam. Don't sit back while foreignized intellectuals, who have no faith is Islam, give their views and write the things they write. Pick up your pens and in the mosques, from the altars, in the streets and bazaars, speak of the things that in your view should be included in the constitution.[6]

And they did. At this point, a process, largely independent of the personal inclination of the participating ayatollahs, was set in motion: that of working out the full logical and institutional implications of Khomeini's theocratic idea in the farmework of the modern nation-state. This impersonal process, a novel rationalization of the political order,[7] unfolded in the form of the constitution making of the clerically dominated Assembly of Experts, which concluded its deliberations in mid-October 1979. Their proposed draft was ratified by the referendum of December 2–3, 1979.

The constitution of the Islamic Republic of Iran is an astounding document, perhaps without parallel since the writings of thirteenth century canonistic advocates of papal monarchy and Pope Boniface VIII's bull of November 1302, *Unam Sanctam*. It places the judiciary system under the exclusive control of the hierocracy, with a provision for extensive revision of the legal codes to render them Islamic. The constitution is also remarkable in its relating to Koranic verses and to Traditions as sources of the Shi'ite Sacred Law in an appendix. Furthermore, by putting a doctrinally novel emphasis on the continuous quality of Imamate (*imamat-e mostamarr*), it endows the jurist, as the representative of the Hidden Imam, with supreme power over men and responsibility only to God. Finally, it sets up a clerically controlled Council of the Guardians of the Constitution with inordinately extensive powers to represent the Shi'ite religious institution and to insure that the legislative and executive branches of the state remain within the straitjackets tailored for them.

Three Shi'ite norms of authority have been identified in Chapter 1: the Mahdistic (millenarian) norm, the juristic norm based on *ejtehad,* and the Akhbari (traditionalist) norm. The last norm was discarded in the nineteenth century and has not been revived since the revolution. In fact, the term *Akhbari* is used only as a pejorative label to designate the apolitical, "stagnant" and "superstitious" orientation of those *'ulama* who do not subscribe to the politicized and ideological Islam of the militant clerics and who reject the *velayat-e-faqih.* As has been pointed out, the *velayat-e faqih* consists in the extension of the Shi'ite juristic norm of authority from the religious to the political sphere, and thus the addition of the claim to political rule, to the prerogatives of the Shi'ite *'ulama.*

The Islamic revolution has also revived the Mahdistic (millenarian)

Shi'ite norm of authority, albeit in an implicit and modified form. The acclamation of Khomeini as "Imam" by his followers in the 1970s was an unprecedented event in Shi'ite history. The connotations of Imam as a divinely-inspired leader in sacred history in the mind of the believers has undoubtedly worked to enhance Khomeini's charisma. Already in 1978, with Khomeini's meteoric advent on the Iranian political arena, there were debates, especially among the uneducated, as to whether he was in fact the Mahdi or merely his forerunner. Millenarian yearnings and expectations were strengthened by the coincidence of the Islamic revolution in Iran with the turn of the fifteenth Islamic century. On at least one televised occasion, Khomeini was asked by a Majles deputy from Tehran, with a confirmed habit of comparing Khomeini with Abraham and other prophets, whether or not he was in fact the Mahdi. Khomeini conveniently observed noble silence. In 1982, side by side with the advocacy of *velayat-e faqih* and the campaign against dissident clerics, the tendency to attribute supernatural qualities to Khomeini was intensified. The influential late Ayatollah Saduqi of Yazd, for instance, reported a miracle performed by the Imam many years earlier (in the form of creating a spring in the middle of the desert under scorching sun).

Khomeini opted for the milder claim and let currency be given to the idea that he was the forerunner of the Mahdi. There was political wisdom in this decision—and a historical precedent. The founder of the Safavid empire in the sixteenth century had claimed Mahdihood and thus used the Mahdistic tenet for the purpose of millenarian mobilization of the tribes he led in the battlefield. Having completed the conquest of Iran, it seems that he recognized the inconvenience of political volatility stemming from millenarianism and modified his claim to that of being the forerunner of the Mahdi. Safavid scribes and historians subsequently attenuated the claim still further to the formula that the rule of the Safavid dynasty would continue until the Advent of the Mahdi.[8] A strikingly similar development has occurred in Iran since the revolution, this time with a modern revolutionary twist. The most frequently chanted slogan in demonstrations organized by the followers of the Line of the Imam has for some time been the following: "O God, O God, keep Khomeini until the Revolution of the Mahdi." In November 1982, *Sorush*, the intellectual journal of the Islamic militants, published an astonishing—though not untypical—article on "the Connectedness of the Two Movements" (that of Khomeini and that of the Mahdi) in which the above slogan was recommended to the reader as a constant prayer. The article referred to an interview published in the journal in June of the same year in which a man wounded on the front reported seeing the Mahdi and talking to him. The Mahdi reportedly told him, "Your prayer, 'O God, O God, keep Khomeini until the Revolution of the Mahdi,' has expedited my Advent by a few hundred years." In September 1982, a clerical deputy of the Majles predicted the imminent Advent of the Mahdi in Jerusalem and cited Traditions, especially those of the sixth Imam, Ja'far al-Sadeq, in support of this prediction.[9]

Khomeini knew that although millenarian expectations could motivate young believers to fight courageously and court martyrdom on the front, they were not a stable basis for the consolidation of his regime in Iran. Consolidation requires institutionalization of clerical rule, and could only be achieved on the basis of extension of the Shi'ite juristic norm of authority. Khomeini's charisma could not be transferred to his successor without routinization and institutionalization of the office of the supreme jurist, or to that of a council of supreme jurists. Therefore the legitimacy of theocratic government had to be established as a new article of Shi'ite faith, and procedural mechanisms brought into existence for the selection of a Jurist or a Council of Jurists to succeed Khomeini. As both these novel objectives lacked precedents in Shi'ite history, their attainment required all the weight of Khomeini's authority and the efficacy of his unique charisma. At any rate, the realization that resolution of the problems of legitimization and succession would be much more difficult after the death of Khomeini than in his lifetime prompted the Imam and the militant ayatollahs to take a series of concerted measures with uncompromising determination.

In 1982, propaganda on the question of *velayat-e faqih* was stepped up in the government-controlled media. One interesting means of promoting the theory was the publication of the wills of the "martyrs" of the war. Throughout May and June 1982 (and subsequently), the newspapers would regularly publish the martyrs' profession of faith in the *velayat-e faqih* and their praise for the Imam and the militant clergies. Statements to the effect that obedience to the clergy as "those in authority" (Koran IV. 59; a term hitherto invariably taken to refer to the twelve Infallible Imams in the Shi'ite tradition)[10] is incumbent upon the believer as a religious duty were often excerpted from the will and made into headlines in bold letters.

The anniversary of the explosion at the IRP headquarters (June 28, 1983) was used to tie the fate of the Islamic Republic to the institution of *velayat-e faqih*. The explosion was said to have inaugurated the Third Revolution, devoted to the complete installation of theocracy. (The Second Revolution was the occupation of the American Embassy, which resulted in the liquidation of the pro-American liberals.) Since then the *imam jom'ehs* (the Friday prayer leaders) have preached the doctrine of *velayat-e faqih* and have enjoined their congregations to obey the *'ulama* as a matter of religious obligation. A headline on the front page of the daily *Ettela'at* in the early days of December 1983, can be taken to represent the culmination of this trend. It was a statement by the prosecutor general and referred to Khomeini as *vali-ye faqih* (the ruling jurist) as synonymous with *vali-ye amr*—an astonishing phrase in view of the fact that the term *vali* has never been used in the Shi'ite tradition in this general sense except to refer to the twelve Imams.[11] From August 1983 onward, numerous conventions organized by revolutionary foundations and Islamic associations would pass resolutions endorsing and pledging full support to the *velayat-e faqih* and declaring obedience to the *faqih* to be a religious obligation. The belief in *velayat-e faqih* was often made a condition of eligibility for applying for vacancies in ministries and government-controlled organizations. But perhaps the most important measure taken to enshrine the novel doctrine of theocratic government has been to teach it in schools. The *velayat-e faqih* is now taught at schools throughout the country, most fully as a part of the compulsory course on Islamic ideology and world view in the third level of high schools.

Elimination of Opposition and Normalization

In less than two years after their direct takeover of the state, from November 1979 to mid-1981, the militant ayatollahs succeeded in removing secular nationalist and liberal elements from the political scene. The elimination of the rival *Islamic* forces took an additional year-and-a-half. By the end of 1982, they had destroyed the well-organized and highly dedicated Islamic radicals, the Mojahedin. They had also succeeded in putting an end to Ayatollah Shari'at-madari and the dissident ayatollahs as a political force, capable of throwing its support behind any oppositional political grouping or organization. Finally, in April 1983, the clerical regime destroyed the last remaining political organization of any consequence: the Tudeh Party. In sum, since their direct seizure of power in November 1979, the militant ayatollahs have ruthlessly dealt with all its organized political opponents, and have by and large succeeded in destroying them.

The period June 1981 to December 1982 was marked by the bloodiest power struggle of the Iranian revolution and ended with the elimination of all organized opposition to theocracy except in Kurdestan. A massive explosion at the headquarters of the Islamic Republican Party on June 28, 1981, killed nearly one hundred important members of the party, including its founder, Ayatollah Beheshti. It was followed by a similar explosion at the prime minister's office on August 30, which claimed the lives of the new president, Raja'i, and the prime minister, Hojjatol-Islam Bahonar. These events seem to have put an end to all doubts in Khomeini's mind. He withdrew his objection to the occupation of the highest offices of the state by clerics and resolved on the direct and full takeover of the state. He appointed a Presidential Council consisting of the Ayatollahs Musavi-Ardabili, Hashemi-Rafsanjani, and Mahdavi-Kani, with the latter as Acting Prime Minister. Ayatollah Khamene'i, having survived an attempt on his life, was elected President on October 2, 1981.

In the period June 1981 to December 1982, the Islamic theocracy passed decisive tests of survival: It survived the decimation of hundreds of its leaders as a result of the explosions of June and August 1981, and of numerous individual assassinations carried out over a period of fifteen months by the Mojahedin. It survived a serious setback in the war with Iraq, which involved heavy casualties in the summer of 1982. In this period, the bid for the full clerical takeover was accompanied by considerable radicalization of the regime. After the election of Khamene'i to the presidency and the dismissal of Mahdavi-Kani as the acting prime minister, every important state functionary was replaced by an IRP loyalist to assure the complete control of the state apparatus by the clerical hardliners. The move entailed a setback for "the Household of the Imam," his son-in-law, Ayatollah Shehab al-Din Eshraqi, who had been favorably disposed toward Bani-Sadr and who died in September 1981, and for his son Ahmad, who was forced to give a few re-cantatory interviews regretting his past support for the "Hypocrites" (*mona-feqin*)—that is, the Mojahedin—and join the chorus denouncing "the accursed Bani-Sadr" and "the accursed Qotbzadeh." The purges of the nonloyalist civil servants were carried out with particular thoroughness in the Ministry of the Interior where all mayors and provincial governors (both categories are ap-pointees of the central government) were replaced. The teachers and other em-ployees of the Ministry of Education also suffered particularly vicious and wide-spread purges. Meanwhile, the Revolutionary Guard was becoming much more homogeneous. Many of the "less Islamic" Guards did not report to work or were purged after the explosion at the IRP headquarters and in the subsequent months.

By January 1982, the clerics were feeling considerably safer and moving around with fewer or no bodyguards. Their perception that the Hypocrites were finished was perhaps somewhat premature, and violence erupted again in the summer of 1982. By the autumn of 1982, however, this assessment had come true. Statements by the Prosecutor General of the Revolutionary Courts that 90 percent of the organized networks of the Mojahedin were destroyed seem to have been fairly realistic. The definitive consolidation of the theocratic regime in Iran may therefore be dated from December 15, 1982, when Khomeini issued a decree promising the people of Iran a post-revolutionary era of security and stabilization. By this date over 10,000 Mo-jahedin had been killed or were awaiting execution and other organized armed opposition groups had been largely destroyed.

From April 1982, Khomeini and his followers began to devote them-selves fully to the resolution of the twin problems of succession and of the legitimacy of Islamic theocracy, both of which were crucial to the long-term survival of the regime. This resolution removed the most insidious obstacle to the survival of theocracy: clerical opposition to theocratic government and certain aspects of the legacy of the Shi'ite tradition itself. The theory of theocratic government, as propounded by Khomeini and incorporated into the Constitution of the Islamic Republic of Iran, extends the Shi'ite norm of juristic authority as elaborated in the nineteenth century into a new sphere. Leaving aside rival theories of government such as democracy and national sovereignty, Khomeini's theory of the Mandate of the Jurist is open to two forceful objections in terms of the *Shi'ite* tradition. The first, fundamental objection is that the mandate or authority of the Shi'ite jurists during the Occultation of the Twelfth Imam cannot be extended beyond the religiolegal sphere to include government. The second objection is that this mandate refers to the *collective* religiojuristic authority of *all* Shi'ite jurists and can-not be restricted to that of a single supreme jurist or, by extension, to a supreme council of three or five jurists (as envisioned in the Constitution of the Islamic Republic).

The above doctrinal objections to the *velayat-e faqih* have been voiced by the Grand Ayatollahs Kho'i, Qomi, and Shari'at-madri who, furthermore, pointed to its inconsistence with the principle of the sovereignty of the people, to which the Constitution of the Islamic Republic also pays lip service, and by Ayatollahs Baha' al-Din Mahallati, Sadeq Ruhani, Ahmad Zanjani, 'Ali Tehrani, and Mortaza Ha'eri Yazdi. The opposition of the last two ayatollahs, who were among Khomeini's favored students, and of Ayatollahs Mahallati and Qomi, who were his close associates in 1963 and were imprisoned with him, must have been particularly disappointing to Khomeini but did not deter him. Of the above-named, Mahallati died in August 1981, Zanjani in January 1984, and Ha'eri Yazdi in March 1986. Qomi has been under surveillance in Mashhad, Tehrani has fled to Iraq, and Kho'i resides in Iraq. Shari'at-madri was repeatedly humbled and, despite his subsequent acknowledgment under intense pressure of the legitimacy of theocratic government, subjected to a campaign of merciless vilification and character assassination until his death in Qom in April 1986. Other clerics who share the views of these Shi'ite dignitaries have been intimidated into silence or, whenever possible, obliged to declare their support for the *velayat-e faqih*.

The beginning of the rift between the militant followers of Khomeini and the *'ulama* who considered them overly politicized predates the revolution. Although active in protesting against the arrest of Khomeini and the other religious leaders and in securing their release from prison in 1963, Grand Ayatollah Shari'at-madri was uneasy about the primacy of political concerns in Qom and founded the Dar al-Tabliq for traditional apolitical missionary activity and learning. This was resented by Khomeini and his militant followers because it deflected clerical energies from political activity. There were even clashes between the two groups in December 1964. Although Khomeini and Shari'at-madari, who was by then the most influential of the Grand Ayatollahs residing in Iran, presented a united front against the Shah during the last months of his reign, differences between them surfaced soon after the revolution, and, as we have seen, resulted in serious violent clashes between the supporters of the two ayatollahs in Tabriz before the end of 1979. Against this background, one can see that the first obstacle to be removed was Shari'at-madari. In April 1982, in a move unprecedented in Shi'ite history, seventeen out of the forty-five professors of the Qom theological seminaries were prevailed upon to issue a declaration "demoting" Shari'at-madari from the rank of Grand Ayatollah. In May to June 1982, the leading pro-Khomeini clerics further decided on a purge of the pro-Shari'at-madari *'ulama* and of other "pseudo-clerics" reluctant to accept the *velayat-e faqih*. The Society of Militant Clergy was put in charge of confirming the true clerics.

Hand in hand with the demotion of Shari'at-madari and the silencing of clerical opposition came a sustained effort to promote the theory of *velayat-e faqih*. Ayatollah Khaz'ali, who presided over a series of seminars convened for the discussion of *velayat,* confirmed the principle that "the Jurist (*faqih*) is the lieutenant of the lieutenant of God, and his command is God's command" (March 1982). However, at this stage the ayatollahs were just beginning to address the issue of succession to Khomeini; opinions predictably varied as to the precise institutionalization of the *velayat-e faqih* and different positions were publicly aired by the militant ayatollahs themselves. There was a concerted attempt to address Ayatollah Montazeri as the "Esteemed Jurist" (*faqih*), Grand Ayatollah, and so forth. There were, however, dissenting voices in this regard, and the campaign of designation slackened in April and May.

Having pushed aside Shari'at-madari and the dissident pseudo-clerics in the spring and summer of 1982, the clerical rulers of the Islamic republic still had to reckon with another organized group of importance that was opposed to the doctrine of *velayat-e faqih,* the Hojjatiyyeh. Masters of identifying and isolating political problems and dealing with them one by one, the clerical ruling elite postponed the settling of the affair of the Hojjatiyyeh until the summer of 1983. The Hojjatiyyeh, or the Charitable Society of Mahdi, the Proof of God (*anjoman-e khayriyyeh-ye hojjatiyyeh-ye mahdaviyyeh*), was founded after the coup d'etat of August 1953 by Shaykh Mahmud Halabi, who has remained a close friend of Khomeini. Its aim was the "propagation of the religion of Islam and its Ja'farite [i.e., Shi'ite] branch, and the scientific defense of it." It was one of the relatively few centers of religious activity other than the *madrasas* that was allowed to function after 1963, and many clerics and lay Islamic activists took part in its readings and discussions. Khorasan was its strongest regional base. The society's efforts prior to the revolution were directed against Baha'ism as the chief enemy of Islam to be refuted and combatted. After the revolution, as the suppression of Baha'ism became the general clerical policy, the society turned to Marxism as the archenemy of Islam to be eradicated. As a society devoted to the Mahdi, the Hojjatiyyeh could not accept Khomeini's extension of the religiolegal authority to political rule, which it considered the nontransferable prerogative of the Mahdi. The founder and directors of the society insisted on this position and resisted the pressure from the younger, more politicized members to revise its charter, with the result that many of the members who were or hoped to become prominent in the theocratic regime left it to join the ranks of the IRP. The society supported the Islamic Republic, without considering it sanctioned by the Sacred Law, and accepted Khomeini's political leadership but refrained from designating him as Imam.

The Hojjatiyyeh first impressed the ruling powers in Iran with their organizational strength and disciplined control over the members in 1981, during the second presidential elections, in which Raja'i was elected. Four hundred thousand votes (about 2.5 percent of the total) were reportedly cast for the Twelfth Imam, the Mahdi.

From late 1981/early 1982 onward, the Hojjatiyyeh had been under intermittent fire from the IRP militants who were prodded into doing so by the Tudeh Party. The Tudeh ideologues, from whom the IRP cadre took many of their cues at the time, were anxious to isolate the economically conservative activists of the Islamic movement, a few of whom were affiliated with the Hojjatiyyeh. However, the hour of reckoning did not come until July 1983. The IRP followers of the Line of Imam mounted their full-scale attack and succeeded in persuading Khomeini to refer obliquely to the position of the Hojjatiyyeh as crooked and deviationist. The society wisely avoided a showdown and suspended all its activities indefinitely in deference to the opinion of "the esteemed leader of the Islamic revolution." The Hojjati Ministers of Commerce and Labor submitted their resignations.

The chief accusation leveled against the Hojjatiyyeh by the IRP was that they confined their missionary activities to the cultural level, the level of ideas; they were therefore stationary as opposed to dynamic and had a dry and empty view of Islam. This critique implied that they were not ideological and did not subscribe to the politicized ideological Islam of Khomeini and his followers. A second charge was that they did not accept the *velayat-e faqih* as legitimizing government by the hierocracy during the Occultation of the twelfth Imam, the Mahdi.

In a long series of polemical articles against the Hojjatiyyeh in the *Ettela'at* during September and October 1983, the society was vehemently attacked for being opposed to intervention in politics in the name of religion and for advocating—like the Baha'is whom they attacked in their apologetics for Islam—the separation of religion and politics as concocted by the imperialist propaganda machine. They were further attacked for their separation of religious authority (*marja'iyyat*) from political leadership, which enabled them to endorse Khomeini merely as a political leader and not "as a leader to whom obedience is obligatory [as a religious duty]." In October 1983, the author of the articles reacted sharply to the surreptitious use of the issue of the Hojjatiyyeh by the Tudeh Party to create division within the Islamic movement by attaching the label of "Hojjati" to prominent clerics and high government office holders. The Tudeh's attacks on the Hojjatiyyeh were said to have been hypocritical, stemming from the ulterior motive of creating divisions within the ruling clerical elite and the Islamic nation. Nevertheless, in the concluding article in the series, the author (inadvertently) repeated the Tudeh's chief argument against the Hojjatiyyeh: opposition to the Line of Imam in matters of economic policy.

Thus, while the above view of the Hojjatiyyeh primarily emphasized their rejection of the theory of *velayat-e faqih* and accused them of having swallowed the imperialist-inspired belief in the separation of religion and politics, the Tudeh view underscored the social and especially economic conservatism of the Hojjatis. The Tudeh Party had succeeded in giving wide currency to a scheme for dividing the ruling elite of the Islamic Republic into radicals and conservatives in terms of respective positions on socioeconomic policies and had astutely labeled the latter group as the Hojjatis. The schema contains elements of truth and was plausible enough to be seriously misleading: while the Hojjatis were socioeconomically conservative, not every socioeconomically conservative cleric was a Hojjati or necessarily sympathetic to the Hojjati's doctrinal position.

Western analysts, who are almost constitutionally indisposed to attaching any importance to doctrinal and cultural issues, have widely accepted the Tudeh interpretation of the clash between the Hojjatiyyeh and the followers of the Imam because it is stated in terms more familiar to them and more in line with their preconceptions on the presumed primacy of economic interests over religious and cultural factors. Nonetheless, the Tudeh interpretation was erroneous in discounting the doctrinal objection to *velayat-e faqih,* and it greatly exaggerated the importance of the Hojjatiyyeh as a political group.

The militant clerics of the IRP quickly mastered the art of ideological politics and succeeded in conducting politics in their own terms, shifting its foci to unifying rather than divisive issues. In fact, before long, they freed themselves from the tutelage of the Marxists in political analysis, as they had earlier freed themselves from the tutelage of the liberals and nationalists in the elaboration of a coherent world view and ideology. This liberation was solemnized in the early months of 1983. In a brusque and unexpected move in April 1983, the ruling ayatollahs dissolved the Tudeh Party, which had sycophantically applauded their revolution, and arrested its members en masse.

Though the Tudeh had acquired some influence among the intelligentsia immediately after the revolution, the party lacked popularity in revolutionary Iran. Nevertheless, it was systematically penetrating the army and the government bureaucracy, and the few thousand Tudeh members arrested in April included some 200 army officers. Furthermore, the party's ideological impact on the clerics while they were novices in Iranian politics was important. The militant clerics learned many of their political and journalistic tricks and tactics—first used during the antiliberal, antinationalist smear campaigns following the occupation of the American Embassy, their coining of political slogans, and their models for political analysis from the Tudeh ideologues.

The Hojjatiyyeh Society apart, the potential for a rift within the ruling clerical elite on socioeconomic policy has continued to exist and has indeed been highlighted by the acrimonious debates over land reform, nationalization of foreign trade, and most recently, over return of confiscated property to individuals associated with the old regime. The conservative ayatollahs wasted no time countering what they considered the excessive influence of socialism and since have determined an official position on some basic socioeconomic principles. In April 1982, the influential *imam jom'eh* of Qom, Ayatollah Meshkini, stated that any attempt to establish a classless society was a move against "the natural order of things," which was "itself a form of tyranny and oppression." In a typical statement, in September 1983, the minister of commerce (who had replaced the Hojjati former minister, 'Asgarawladi) stressed the regime's respect for the public, the private, and the cooperative sectors of the economy. More importantly, the official position has been propagated through government-printed textbooks. A textbook on the *Islamic World View (Binesh-e Eslami)* for the first level of high school, for instance, significantly uses the term "unitarian society" (*jame'eh-ye tawhidi*), popularized by Islamic modernists such as Shari'ati and Bani-Sadr, but empties it of all possible socialistic content. A unitarian society is a society based on negation of all worldly idols (*taghut*) other than the one God, and on "equal distribution of opportunities." "A society with no economic differences among its members is one of evident oppression and injustice [because rewards are not proportionate to effort]." The prime minister and the radical clerics, on the other hand, have continued to advocate redistribution and strengthening of the public sector of the economy at the expense of private enterprise.

The debate over economic policy did in fact intensify in the fall of 1985. The economically conservative ayatollahs and the allies of the bazaar in the Majles clearly had the upper hand and it took Khomeini's intervention to keep the radical prime minister Musavi in power when a vote of no-confidence seemed certain in November 1985. This, however, was partly due to the split between the prime minister and the president and their respective radical supporters. The debate has continued unabated and, in the words of the Majles Speaker, Hashemi-Rafsanjani, has divided the elite into two wings (*janah*).[12] Its latest manifestation has been the attack on the economic radicals by the newly founded conservative newspaper, *Resalat*, under the direction of Ayatollah Azari-Qomi, in the summer of 1986, which elicited a furious response from the radicals and the threat of repression from Khomeini himself.[13] It is interesting to note, however, that in contrast to the clash with the Hojjatiyyeh three years earlier, the clash with *Resalat* no longer brings into question the principles of legitimacy of the Islamic regime. It is a strong indication of the consolidation of the Islamic rule in Iran that questions of a constitutional nature are no longer raised in political debate. The disagreements over socioeconomic policy, however profound, do not challenge the constitutional framework of the regime but are rather expressed within it.

Khomeini has always been on guard to avoid divisions among his supporters. Even his attempts to force the dissident clerics into submission were accompanied by an emphasis on the imperative need for unity among the Shi'ite 'ulama. Already in October 1983, Khomeini was warning his supporters against the disunity whose seeds were being sown by hidden hands in the service of the superpowers. He vehemently attacked the constant atmosphere making (*javv-sazi*), labeling, poisoning the climate of opinion and the seditious division of the ayatollahs and their supporters into lines, such as the Line of the Imam. All these presentations were deceptions to create disunity. "There is no line in Iran except the Line of Islam."[14] The massively orchestrated drive for unity, however, did not come until December 1983, and it was timed to coincide with the nationwide convention of the *imam jom'ehs,* the pillars of the Islamic theocracy then under construction. Speeches on the need for unity and avoidance of factionalization by Khomeini, echoed by the Friday prayer leaders in the remotest towns of Iran, made it clear that the specter that haunted him was the division of the Shi'ite hierocracy during the Constitutional Revolution, which had enabled the secular intelligentsia to defeat and execute Shaykh Fazlollah Nuri and to oust the Shi'ite 'ulama from the political arena.

Khomeini realized that, in the long run, the possibility of debilitating rifts within the clerical regime could only be eliminated by a successful re-channeling of political energies from parliamentary debates and ideological discussions into institution building and construction of a theocratic state. Theocratic consolidation and institution building, by impressing upon the ayatollahs their common interest as the ruling stratum of the Islamic Republic, would encourage their unity.

Institution building cannot proceed as thoroughly in a situation of endemic political violence and insecurity as it can under conditions of peace and economic prosperity. In any event, revolutionary turmoil cannot continue indefinitely, and the return to normalcy and stability will sooner or later be attempted by any political regime that aspires to viability. Stability became the official policy of the Islamic regime in Iran on December 15, 1982, when Khomeini issued a decree guaranteeing all Iranian citizens security from arbitrary arrest and confiscation of property, and promising the restoration of law and order and vindication of wrongs.

The "prophet-like" decree of the Imam was immediately and widely hailed as a "rainfall of mercy," "the continuance of the guidance of mankind by the Prophets," and was taken to signal the inauguration of a new era of security, stability, and economic prosperity. The Commission (*setad*) for the Pursuance of the Decree of the Imam was soon set up. Its members, after being instructed by the Imam, made statements to the press reaffirming the sanctity of private property, the determination of judiciary authorities to combat arbitrary confiscations, and the country's need for internal peace and security. In the following two weeks, twenty-four special teams were sent to the provinces to investigate grievances, six revolutionary prosecutors, one commander of the Revolutionary Guards, and some other functionaries were dismissed (further dismissals and the reinstatement of a number of former employees of the Ministry of Education were reported in November 1983). In March 1983, as an act of mercy, a group of ordinary prisoners were released or had their sentences commuted following an earlier amnesty by Khomeini to mark the anniversary of the revolution. The Mojahedin and the leftist groups were explicitly excluded from the Imam's mercy in the decree. To assure this exclusion, Lajevardi, notorious Revolutionary Prosecutor and warden of Evin prison, immediately reacted to the news of its issue with a wave of summary executions.

On May 19, in response to "the people's concern and anxiety that the Decree of Imam is forgotten," the head of the commission affirmed the regime's continued commitment to implement the decree. It is a striking indication of the widespread discontent caused by revolutionary excesses, especially the purges, that over 160,000 complaints had been filed in the five months since the decree. On the same day it was announced that the hated Selection Committees in charge of the purges of governmental agencies had been dissolved by Khomeini's order. A few days later, May 31, 1983, the dissolution of the Commission for the Pursuance of the Decree of Imam was announced, and Khomeini agreed that the numerous cases of complaint submitted to it and pending investigations be referred to the Judiciary. Normalization had begun.

In 1984, there was talk of attracting the technical cadres and skilled manpower that had left Iran, and by the end of the summer, political prisoners were being amnestied in groups or released after repenting. Even the executions of drug-traffickers were reduced—hitherto well over 100 had been reported in the press each month. An important indication of normalization was the onset of pragmatism in Iran's foreign policy, marked by the arrangement of a state visit to Iran by the foreign minister of the Federal Republic of Germany in July. The new pragmatic policy was endorsed by Khomeini in a speech to the Iranian diplomats in October 1984.

Khomeini and the ayatollahs knew that normalization would rest on shaky foundations unless the question of succession could be successfully resolved. In August 1982, Khomeini asked the theology professors of Qom to put forward suitable candidates for election to the Assembly of Experts to determine the issue of succession. The Assembly of Experts was elected in December 1982, but did not meet until July 14, 1983. It was announced that it would ordinarily convene for one session each year and concluded its session for the first year on August 15. Its deliberations were not public. It received and sealed the political will and testament of the Imam without disclosing its contents. As a result, it was not clear whether there already was a clear majority for Ayatollah Montazeri as Khomeini's successor. Meanwhile, the public designation of Montazeri as Grand Ayatollah and the Esteemed Jurist has become routine. His portrait is often displayed with Khomeini's. Like Khomeini, he regularly grants audiences to various groups and has representatives in various organizations, agencies, and even foreign countries. Finally, like Khomeini, though less frequently, he issues decrees and makes various appointments.

The Assembly of Experts met annually but no significant decision was announced until a few months after its 1985 session. It is evident that the two-thirds majority necessary for the election of a successor to Khomeini could not be reached when the 1985 session adjourned. Hashemi-Rafsanjani and the other supporters had enough of a sense of urgency then to circumvent due process of law announcing the "election" of Montazeri in the interim late in November 1985. It was alleged that the list nominating Grand Ayatollah Montazeri already had fifty signatures. The remaining thirty-three honorable members of the Assembly who had not had the time or were otherwise unavailable for signing the list would be doing so at their convenience! Early in December 1985, Hashemi-Rafsanjani reported that they had all signed the list—presumably with arms other than the ones being twisted.[15] And so the road was paved for the smooth succession of Ayatollah Montazeri to the Imamate.

The observers who predicted the imminent demise of the regime of the mollahs in Iran have constantly underestimated not only the political astuteness of the ruling ayatollahs, but also their resolve, determination, and sense of historical mission. More seriously, they have underestimated the fact that the ruling clerical elite sees its fate as irrevocably tied to the destiny of the Islamic Republic. Unlike the Shah and his political and military elite, they have nowhere to go outside of Iran and are committed to defend the regime and to fight for it to the last man.

Divisive factors within the militant clerics have been kept at bay by absorbing their energy in achieving the establishment of Islamic theocracy through institution building and through ideological control of existing institutions. Since 1979, militant clerics have manned the Islamic Revolutionary Courts, which have meted out summary justice to "the enemies of God" and "the corrupters on earth," and have devoted themselves to the enforcement of Islamic morals with increasing firmness. They have also created and manned the Political-Ideological Bureaus of the various branches of the armed forces and governmental agencies. In addition to over ten thousand clerics who have joined various revolutionary bodies, a few hundred have entered service of the Judiciary in order to Islamicize this core institution of the Islamic theocracy. Many more are serving as *imam jom'eh*s and are engaged in making the Friday congregational prayer into the basis for another core institution of theocracy with an increasingly coordinated and centralized structure. It is to this sustained enterprise of institution building that we turn next.

Distinctive Institutions of the Islamic Republic of Iran

In August 1982, Ayatollah Khamene'i, the third president of the Islamic Republic, declared that "the Revolution cannot be lasting unless it is embodied in a framework of legal order." This declaration foreshadowed a speech by Khomeini in which he urged the judges to enforce the Sacred Law and throw away any law inherited from the previous regime that they had reason to believe was contrary to the Sacred Law of Islam. There was immediate orchestrated support for the Imam's pronouncements on the full Islamicization of the legal system, and procedural measures to realize this goal were announced. Almost a year later, in July 1983, President Khamene'i made the following declaration: "Political-Ideological Bureaus [attached to the army, police, and governmental agencies] are an essential and vital foundation of the regime of the Islamic Republic." These two declarations by President Khamene'i capture the ruling clerical elite's objective after the period of violent political struggle with armed opposition groups. The objective can be described as the construction of an Islamic theocratic state with a codified legal system and with total ideological control over all institutions and all spheres of life. This objective has required the creation of a set of new Islamic institutions and the ideological control of the existing and non-Islamic institutions.

During the first six months of the revolution, 248 military men, including 38 generals and 103 officers were executed. In addition, 128 men, including only 3 generals, were sentenced to imprisonment or internal exile. Furthermore, the army, police, and gendarmerie were purged by the retirement of 302 officers, of whom 210 were generals.[16]

The extensive purges of the state bureaucracy under Raja'i, first as prime minister and then as president, installed an ideologically committed (*maktabi*) cadre of Islamic activists in charge of the technical and managerial personnel of the state. This group has constituted a lay second stratum of the Islamic Republic in charge of the administrative and technical affairs of the state. They typically share a common social background (lower-middle-class and bazaari families) and a shared career pattern (the first generation in their families to obtain higher education and leave the traditional bourgeois occupations). The data supplied on twenty-four of the victims of the June 1981 explosion who belonged to this second stratum—a representative random sample *par excellence*—give us interesting information on its composition and orientation. This information, summarized in Table 15 (see the Appendix), shows them to be typically young (three-quarters of them under 40), with one-half having graduated from Iranian universities and one-third from foreign universities, mostly in technical fields or medicine. This lay second stratum, currently headed by the prime minister, Musavi, has faithfully put the state at the service of the clerical rulers of Iran. As lay servants of the ayatollahs, they know their place. Once he had organized his cabinet in the spring of 1980, Prime Minister Raja'i wasted no time in declaring, at the Seminar on Cultural Revolution in Tabriz: "If our Fundamental Law had only the single item of *velayat-e faqih*, it would have been worth giving many more times as much of our blood for it."[17] Toward the end of December 1984, in response to the Imam's order to strengthen the clerically controlled judiciary, Prime Minister Musavi declared: "The servant government (*dawlat-e khedmatgozar*) and the officials in charge of the administration of the country will execute the Imam's decree (*farman*) as a religiously incumbent duty."[18]

Much more extensive than the initial purge of the armed forces was the "Islamicizing" ideological purges that began in late September 1979 and had resulted in the discharge of some 12,000 military personnel, the majority of whom were officers, by the time of the Iraqi invasion a year later. Two groups of army officers reacted to these purges by conspiring to organize coups d'etat during the summer of 1980. The first was uncovered in June 1980 and the second, more serious conspiracy, the Nawzheh plot, in July. Together, they implicated at least 600 officers over 100 of whom were executed.[19] In 1982, according to one report, some 70 officers were executed for involvement in Qotbzadeh's plot to overthrow Khomeini.[20] The last reported instance of a military plot occurred on May 27, when five high-ranking officers of the air force were arrested for conspiracy to bomb Khomeini's residence.[21] By the end of 1983, all these attempts had been foiled, and the clerical control of the army was secure.

Last but not least among the twentieth-century political institutions taken over by the ayatollahs was the Majles. The Majles, the central organ of the

Constitution of 1906, was retained in the clericalist constitution and duly elected in March 1980. Its first important act was to change its name from the "National" (*melli*) Consultative Assembly to the Islamic Consultative Assembly (*Majles-e shuraye eslami*). The Islamic Revolution appropriated the major achievement of the Constitutional Revolution and subordinated it to clerical sovereignty as it had the major achievement of the Pahlavi era— the modernized state and army. By the time the First Islamic Majles completed its term in 1984, the amateur revolutionaries elected as its deputies had become expert parliamentarians. The speaker, Hashemi-Rafsanjani, and other parlimentary leaders were reelected in 1984, and the Majles began its second session with impressive vigor and efficiency, demonstrating its independence by not approving several members of the Cabinet presented to it by Prime Minister Musavi in August 1984. It has since distinguished itself by the openness of its debates and the extensiveness of its legislation.

In addition to taking over the existing governmental structure and the armed forces, the revolution itself gave birth to a number of very important organizations and centers of power. Immediately after the revolution Islamic revolutionary committees (*komitehs*) sprung up from the traditional networks of the clerics and craftsmen, retailers and youths of the city quarters. The committees soon constituted a power structure rivaling the government and a crucial basis for the militant clerics in their power struggle with Bazargan and Bani-Sadr. After the ousting of these liberal nationalists and the direct clerical takeover of the state, the ruling clerical elite understandably sought to undermine the autonomy of the committees and to coordinate their activities with those of other organs. The revolutionary committees were put under the supervision of the Ministry of the Interior in July 1982, and clerical control over them was preserved while they were integrated into the governmental system. At the time, there was some friction between the revolutionary committees, which served as an internal security force, and the volunteers of the Mobilization Unit of the Corps of the Guardians of Islamic Revolution.[22] The minister of the interior put a commission in charge of working out a satisfactory division between the functions of the committees and other organs such as the police and the Guards. Late in May 1983, a nationwide gathering of the heads of the committees was organized to further discuss problems of rationalization of organization and coordination of functions with other forces of law enforcement. By 1984, security matters were clearly left to the Revolutionary Guards and the committees assumed charge of smuggling offenses and dealing with the drug traffickers.

In May 1979, the Corps of the Guardians of Islamic Revolution (*Sepah-e Pasdaran-e Enqelab-e Eslami*) was formed to counter the power of the army and, according to later statements by its founders, to strengthen the hand of the IRP in its bid for total power. In subsequent years it was greatly strengthened and homogenized through the discharge of non-Islamic elements, and it played a major role in the war with the Kurdish insurgents. Its charter was ratified by the Majles in September 1982. Internally, the Revolutionary Guards proved their effectiveness by extirpating the Mojahedin and other armed opposition groups. Friction between the Guards and the army increased in the second half of 1984 with the first intimations that the ayatollahs intended to incorporate the regular army into the Corps of the Guardians of the Islamic Revolution. Brigadier General Zahir-Nezhad, who had served Khomeini faithfully as the chief of staff, opposed the takeover of the middle ranks of the army by the Revolutionary Guards, who were insisting on a large number of political promotions all the way up to the rank of colonel, and resigned in October 1984. He was replaced by a colonel who accepted these demands. In September 1985, Khomeini finally sounded the death knell of the regular army by issuing a decree for the creation of ground, air, and naval branches of the Corps of the Guardians of the Islamic Revolution to replicate those of the regular army. By the summer of 1987, the naval branch of the Corps had made itself conspicuous in the waters of the Persian Gulf but little information is available on its air force. Although the logic of the Islamic revolutionary ideological state requires the absorption of the armed forces by the three branches of the Corps, the clerical elite of Iran seem too astute to press this logic to its conclusion and may well find in the duality of the present military organization and the rivalry between the regular army and the Guards the key to neutralization of future challenges from the armed personnel and thus to perpetuation of their rule.

The Courts of the Islamic Revolution, like the revolutionary committees, had been set up, under clerical judges, immediately after the fall of the monarchy in February 1979. With the installation of theocratic government and the onset of the drive for the integration and rationalization of the power structure in 1982 to 1983, there was an attempt to integrate the revolutionary courts into the Ministry of Justice, and a scheme for that purpose was drawn up by the Supreme Judiciary Council and approved by the Majles on May 1, 1983. The attempt was overruled by Khomeini, who was persuaded by the contrary argument of the revolutionary prosecutor general that the integration of the Revolutionary Courts into the Ministry of Justice in its present state would "jeopardize the reputation and the future of the revolution," as the courts of the Ministry of Justice did not have the stamina to stand up to terrorists and armed opposition groups. If the integration took place, according to critics of the move, the country would perish.

The primary argument that impressed Khomeini was undoubtedly that the swift and usually deadly justice of the Revolutionary Courts was necessary to the survival of Islamic theocracy, not only by extirpating counterrevolution, but also because of the increasing reliance on the function of its Special Branch for the Affairs of Guilds and Trade in combating "economic terrorism," that is, profiteering and hoarding. There was, however, a secondary but important consideration: the Courts of the Islamic Revolution were Islamic whereas those of the Ministry of Justice, which still had to operate under the legal codes of the previous regime and *ad hoc* instructions from the Supreme Judiciary Council, were not. To minimize divergences from the Sacred Law of Islam, the Shah's modernized Family Protection Law had been repealed and the Special Civil Courts (*dadgahha-ye medani-ye khass*) had been set up to deal with cases in the sensitive area of family law. Nevertheless, the Islamicization of the legal system, begun under Ayatollah Beheshti in 1979, had not yet produced concrete results. This was due to enormous difficulty of codifying the Shi'ite Sacred Law, on the one hand, and the acute shortage of qualified religious jurists on the other, as well as the reluctance of the Shi'ite *'ulama* to become judges.[23] Nevertheless, a group of jurists under the direction of the Supreme Judiciary Council were assiduously at work, with a sense of historical mission as the first codifiers of the Sacred Law in Shi'ite history. They revised many of the provisions of the commercial code of the previous regime and drew up an Islamic penal code, which became law in the summer of 1982. Given the historically unprecedented nature of the enterprise, the Supreme Judiciary Council suggested that all Islamic laws be provisionally enforced for a period of five years. Having been reassured by this law-making activity that there was enough codified Islamic law to avoid chaos, Khomeini made a pronouncement on the suspension of all existing non-Islamic laws in August 1982. Since then the work of codification has continued, and there has been a sustained effort to attract and train at least one thousand religious judges to correct the current overwhelming presence of secular judges. The Supreme Judiciary Council regularly interviews and appoints young clerical applicants with a modicum of religiolegal training, and a judiciary college has been set up to train judges for the newly Islamicized judiciary system.[24]

By 1984 Khomeini must have been sufficiently impressed by these Islamicizing efforts to allow the Supreme Judiciary Council to proceed with integration of the Revolutionary Courts into a unified judiciary system. On January 22, 1984, Revolutionary Prosecutor General Musavi Tabrizi, who had opposed the merger, had to announce the integration of the Revolutionary Courts into the Ministry of Justice and then submitted his resignation. Early in February 1984, the president of the Supreme Judiciary Council, Ayatollah Musavi Ardabili, expressed his satisfaction with the absorption of the Revolutionary Courts in the Ministry of Justice and stated, with unusual precision, that the judiciary of Iran had become 80 percent Islamic: some 500 clerics had been attracted to judiciary service, and efforts were being made to overcome the further need for 500 qualified persons.[25]

One of Khomeini's earliest acts in the Islamic revolution was the revival of the Friday congregational prayer and its full utilization as a political platform. He appointed prayer leaders, *imam jom'eh*s, in all large and small towns. The *imam jom'eh* of the town, who is usually also Khomeini's representative in the same town or region, leads the congregational prayer on Fridays and delivers a political sermon. The political nature of the Friday congregational prayer is clearly brought out by its description in the Iranian media as "the devotional-political prayer" and the "enemy-smashing and unity-generating" gathering of "the lovers of God."

Ever since the demotion of Shari'at-madari and the full-scale promotion of the *velayat-e faqih,* there had been suggestions, especially by Ayatollah Montazeri, for organizing the *imam jom'eh*s into a centralized national agency. In the second half of 1983, concrete steps were taken toward this end. A scheme published in September 1983 envisioned a centralized headquarters for the *imam jom'eh*s in Qom, with a hierarchical structure corresponding to the administrative division of the country into province, city, city quarter, and the rural hinterland of the city. The organization was to be used for the propagation of the *velayat-e faqih* and of Islam and for strengthening the link between "the cleric and the layman." Mosques had already become centers for the distribution of rationed goods and had been collecting information on families living in the area around them. According to the new scheme, the mosques, under the supervision of the *imam jom'eh*s, were also to keep files on every household in their area, distribute essential foodstuffs, and gradually absorb all local groups so as to eventually replace the revolutionary committees.[26] Whether the scheme was officially sanctioned and scheduled for implementation is not known. But it is certainly indicative of the direction of clerical policy.

From the autumn of 1983 onward, for about a year, Friday sermons of the *imam jom'eh*s of large and small towns were extensively covered in the daily *Ettala'at*. Predictably, these sermons sought to perpetuate clerical rule by preaching the ideas of the *velayat-e faqih* doctrine. Less obvious perhaps were the prominence of foreign policy issues, especially themes of struggle against American imperialism and exportation of the revolution. One can only speculate on the effects of such sermons in remote towns, but it is plausible to assume that they have made for unprecedented politicization of the youth and for their concern with international politics, which can explain the continual supply of volunteers from such towns for the Mobilization Corps and the army.

In his speech to a Seminar of the Councils of Provincial Imam Jom'ehs in December 1983, Ayatollah Montazeri went so far as to envision, admittedly in the distant future, the removal of the dual division of authority between the governor and *imam jom'eh* in each province (*ostan*), which he considered uncharacteristic of a true Islamic theocracy.[27] Meanwhile, following a decree issued by Khomeini a month earlier, the members of the Central Secretariat of the *Imam Jom'eh*s were appointed in September 1984 and met with Ayatollah Montazeri to receive instructions on holding regular meetings with government officials in charge of local and regional administration. In October, President Khamene'i duly referred to the *imam jom'eh*s as "the great pillars of the Revolution, the speaking tongue of the Leader and the strong arm of general mobilization."[28]

Khamene'i was even more emphatic about the mobilization role of the *imam jom'eh*s in the republic when he addressed the members of the Secretariat of the *imam jom'eh*s of Azerbarjan as the Secretary General of the Islamic Republic Party on December 22, 1984. Ayatollah Beheshti and the other founders of the IRP had foreseen a major role for it in an Islamic one-party state. However this idea was not shared by many clerics, including Ayatollah Musavi Ardabili, who felt constrained to resign his membership in the party's political bureau upon assuming the chairmanship of the Supreme Judiciary Council. Furthermore, the performance of the IRP in mobilizing support for the regime did not prove to be too impressive. By 1984, it was becoming clear that the IRP would only play a secondary role in the Islamic regime. It was decided that its role in political mobilization would henceforth be ancillary to that of the *imam jom'ehs*. Thus, it was to the *imam jom'ehs*, in March and April 1984, that Khomeini entrusted the task of mobilizing the people for participation in the elections for the Second Majles. By the end of 1984, the IRP's local network was in effect put under the broad supervision of the *imam jom'ehs* as the Islamic Republic's "strong arm of general mobilization." Late in summer 1985, it was decided to reduce the activities of the IRP and close some of its branches. On June 1, 1987, Khomeini ordered the complete suspension of the activities of the IRP in response to a request by President Khamene'i and Majles Speaker Hashemi-Rafsanjani. In their request, Khamene'i and Hashemi-Rafsanjani, surviving founders of the IRP, had pointed out that the party had achieved its purpose of establishing the *velayat-e faqih* and the distinctive institutions of the Islamic Republic of Iran; its activities would henceforth encourage party politics (*tahazzob*) and have a divisive effect on the community. The suspension of the Islamic Republican Party underscores the unique identity of religious and political community in the Islamic Republic of Iran. By thus rejecting the party in favor of the mosque (controlled by the *imam jom'eh*) as the organ of communal unity and mobilization, the clerical rulers of Iran have once more demonstrated their remarkable determination not to imitate foreign—in this case, East European—models of political organization.

In addition to these core Islamic institutions that distinguish the Islamic Republic of Iran from other contemporary political regimes, a host of other organizations operating in social, economic, and charitable fields has come into being as a result of the revolution. These organizations were set up by special decrees issued by Khomeini. As they are at best secondary features of the Islamic political order, only some of the more important ones need to be mentioned: The Foundation for the Disinherited (*Bonyad-e Mostaz'afin*) and the Committee to Aid Imam Khomeini were set up early in March 1979, followed by the Housing Foundation (*Bonyad-e Maskan*) in April and the Jihad for Reconstruction (*Jehad-e Sazandeqi*) for rural reconstruction in June 1979. Subsequent organizations include the Foundation of 15 Khordad (June 5, the date of the 1963 uprising), the Commission (*setad*) for Economic Mobilization, and the Commission for the Reconstruction and Renovation of the War Zones.

The Islamic theocratic state has all along been conceived of as a totalitarian state with full control over the moral attitudes and political opinions of all its citizens. The clearest manifestation of theocratic totalitarianism have been the ruthless persecution of the Baha'i religious minority—a community of over a quarter of a million, some two hundred of whom have been executed and the rest forced to convert or subjected to the most horrendous disabilities. But it is not by any means confined to sectarian intolerance. A plethora of institutions for the strict enforcement of morals—conceived of as enjoining the good and forbidding the evil according to the Sacred Law—and an elaborate machinery of indoctrination and intelligence have come into being, mostly since 1981. Khomeini's desire for an intelligence service consisting of 36 million persons, that is, the entire nation, should not be taken lightly, nor should his instructions to students in September 1982 to closely observe their teachers and classmates and report any deviant behavior to authorities.

Let us begin with the organizations in charge of the enforcement of morals and indoctrination. The Revolutionary Courts are in charge of the enforcement of morals. They mete out summary justice to the offenders who are arrested by the committees, Revolutionary Guards and vigilante groups regularly patroling the cities. Sentences passed for offenses such as drinking, improper attire for women, unlawful sexual intercourse, and homosexuality are speedily carried out by a specially created judiciary police. A particular terror to secularized women is the vigilante group, the Sisters of Zaynab, who, completely covered in black veils, patrol the streets in special cars and pounce on unsuspecting offenders guilty of the slightest improper exposure. The Ministry of Islamic Guidance is the chief governmental agency in charge of Islamic propaganda.

In July 1984, the Vigilante Patrols for Combatting the Forbidden (*Gashtha-ye Mobarezeh ba Monkarat*) began operating under the direction of the central bureau of the revolutionary committee. In accordance to a law passed by the Majles on May 17, 1983, a Ministry of Intelligence was established, headed by a cleric with the highest qualification in religious jurisprudence (*ejtehad*). In due course, the judge of the Revolutionary Court for the Army, Mohammad Rayshahri, became the republic's first minister of intelligence. The Supreme Council of Islamic Propaganda (*tabliqat*) has also been set up. In addition, Islamic societies (*shuraha-ye eslami*) have been established in all organizations and enterprises of consequence to act as watchdogs for Islamic conformity. Islamic societies of ministries and governmental departments (i.e., of the Teachers of Iran, the armed forces, and factories) are of particular importance. There have been nationwide conventions of Islamic societies, and there is a Commission (*setad*) for the Coordination of the Islamic Societies of Governmental Departments and Agencies. The Bureau for Islamic Propaganda and the Council (*shura*) for the Coordination of Islamic Propaganda should also be mentioned. Finally, the national radio and television network, the Islamic Voice and Vision, is under clerical control.

While the effect of Islamic propaganda on the adult population is not easy to assess, the regime's success in indoctrinating school children and the young appears to have been considerable. As we have seen, non-Islamic school teachers and high school students were extensively purged in 1980 to 1981. In August 1982, just before the beginning of the 1982 to 1983 school year, the Minister of Education boasted that 70,000 of the existing teaching cadre had been familiarized with Islamic ideology. They were to be joined by 18,000 newly trained persons. The textbooks and syllabi have been carefully revised and their contents made Islamic. The clerical rulers of Iran knew what they were doing when they set out to captivate the minds of the school children. In October 1983, it was remarked, in connection with an educational seminar of biology high school teachers, that clerical supervision over the writing of textbooks was a "vital necessity." This explains why the young are so much more enthusiastic about the regime than their elders and are enthusiastic about enrolling for the army, the Guards, and the Mobilization Corps if they are male.

Political-ideological bureaus are attached to batallions, brigades, and divisions in all branches of the armed forces and to the police, the gendarmerie, and the ministries. Their function is to assure the ideological commitment to, and knowledge of, Islam on the part of the officers and government employees, who in addition to proper Islamic behavior, have to pass examination in Islamic ideology. The personnel of the police and the gendarmerie also have to take classes and pass an examination in Islamic ideology. Clerics in charge of ideological guidance are attached to all army units, including platoons and companies, and constitute a factor of considerable importance in the ideological control of the army. To indicate the extent of the penetration of the armed forces by the Islamic ideological commissars, it is enough to point out that by the summer of 1984, some 12,000 clerics had been sent to the Iraqi front and over 700 of them have been killed.[29]

With the onset of the campaign for *velayat-e faqih,* the recruitment criteria of the armed forces and government became explicitly discriminatory. Except for technical agencies and technical positions requiring expertise and qualifications in scarce supply, the applicants are compelled to subscribe to the theocratic principle of *velayat-e faqih.* This is true not only of prospective army cadets and employees of the Ministries of Foreign Affairs and Commerce, but also of applicants for the position of bilingual male typist at the National Oil Company.

An important procedure for maintaining Islamic ideological conformity among government officials, officers of the internal security forces, and the personnel of revolutionary corps and organizations is the holding of Congregational Prayers of Unity (*namaz-e vahdat*) in which they are expected to participate.

Paradoxically, the one armed force with which ideological problems have surfaced is the Corps of Revolutionary Guards. This paradox can be explained by the fact that the Guards, considering themselves the pillar of the Islamic Republic, are not susceptible to the mixture of clerical persuasion and intimidation that can be applied to the officers of the army and internal security forces. On some rare occasions, as in Isfahan (where the problem is greatly exacerbated by the rivalry between the *imam jom'eh* and the influential octogenarian Ayatollah Khademi, who controls most of the city quarters and the bazaar), there have even been instances of breakdowns of discipline and fighting among the Guards. But the problem was best illustrated in a speech by Ayatollah Mahdavi-Kani in June 1983, in which he accused a "line" among the Guards of insulting the jurists of the Council of Guardians for supporting feudal landlords and capitalists and of having said that the theological "professors of Qom are like fortune-tellers." The commander of the Guards retorted by maintaining that Ayatollah Mahdavi-Kani was an arch sower of dissent. The clash brings out the antipathy of the Guards for the more conservative ayatollahs—and conversely their sympathy for the clerics of the "Line of Imam"—and points to the possibility of a divided ruling clerical elite, which haunts Khomeini, and which he is determined to avoid at all costs.

Last, but no less important, there has been a remarkable development in counterrevolutionary intelligence. Some of the intelligence and security organizations inherited from the previous regime, such as the Protection and Security Unit attached to the prime minister's office, were modified and retained, and by a law passed in September 1983, advisory councils were set up at the province and town levels to coordinate internal security under the control of the Ministry of the Interior. But the distinctive and most effective intelligence and counterinsurgency organizations are recent creations of the Islamic revolution.

In a September 29, 1983, speech on the first anniversary of the formation of the counterrevolutionary patrol groups, the Vengeance of God (*Sar Allah*) and the Qari'a (*Calamity/the Hour of Last Judgment*), President Khamene'i hailed their intelligence successes. Referring to the fact that in the early days of the revolution a number of malicious and ignorant people chanted slogans saying that we had no need for intelligence services, Khamene'i added: "In such an atmosphere, world intelligence organizations like the KGB and the CIA imported what they wanted."[30] Following the creation of the Vengeance of God, this was no longer the case. About one month earlier, the speaker of the Majles, Hashemi-Rafsanjani had congratulated the Vengeance of God and had unabashedly boasted that "the revelations of the repentants from the armed groups of combatants [with God] is the important and distinctive fruit of the Islamic Revolution."[31] In January 1984, President Khamene'i emphasized that "intelligence work is as important as [being present] on the front."[32]

The patrol group, the Vengeance of God, was formed at the height of the violent clashes with the Mojahedin and other armed opposition groups when the Intelligence Unit of the Corps of Revolutionary Guards seemed inadequate for dealing with these threats. The patrol groups regularly circulate in the streets of towns, arrest and interrogate suspects, use repentants who are permanently or temporarily released from jail to identify individuals associated with opposition organizations, and search suspected houses. They have been effective both in discovering organized underground networks and in creating an atmosphere of terror in towns by their constant patrolling. With the formation of the Vengeance of God, the regime became more efficient in repressing organized opposition. Reporting on the massive arrest of the members of the Tudeh Party in April and May 1983, the commander of the Guards, Reza'i, could boast of his Corps' experience and expertise and contrasted the efficient clampdown on the Tudeh to the bunglings of the Guards in connection with the Mojahedin and the supporters of Bani-Sadr a year or two earlier.

The success of Vengeance of God in the cities has inspired the formation of the Army of God (*Jund Allah*) patrol group in the gendarmerie, to be used for counterinsurgency and armed smuggling in the countryside. In January 1984, the Patrol Group Helpers of God (*Ansar Allah*) began operating in the cities to combat profiteering and hoarding.

The Revolutionary Guards and the various and more specialized vigilante and patrol corps set up by the ruling ayatollahs have thus established an effective and continuous reign of terror that sustains clerical domination over Iran and makes organized opposition and protest impossible. By the end of 1984, the development of this intelligence system, more far-reaching and far more prying than the Shah's, had completed the transition from temporal to theocratic absolutism. The turban had replaced the crown.

Continuities Between Pre- and Post-Revolutionary Iran

The most striking and paradoxical element of continuity between pre- and post-revolutionary Iran is the continuing growth of the state. By 1984, as Table 10 shows, the number of persons directly employed in the state bureaucracy (the ministries), excluding the armed forces, had reached 1,770,000 (see Appendix). Of these, over 70,000 were employed by the revolutionary organizations.[33] In addition, the industrial private sector was largely nationalized, employing over 1000 persons so that, by 1983, 96 percent of industrial enterprises were controlled by the state. If we add to the above-mentioned figure, the 68 percent of the industrial labor force or some 370,000 persons employed in the public sector,[34] we obtain the figure of 2,140,000, which is already 863,000 larger than the number employed by the state under the Shah. Furthermore, as we have seen with regard to the organization of the congregational prayer leaders, the bureaucratic structure has now been extended from the state to the hitherto amorphous hierocracy itself. The same is undoubtedly true of the bureaucratization of the seminaries in Qom and elsewhere.

This striking continuity of the pre-revolutionary trend in the growth of the state confirms Tocqueville's insight that revolutions make the states they intended to destroy all the stronger. It also supports Weber's contention that the modern bureaucratic state is indestructible. This outcome is especially paradoxical in Iran in view of the intentions of the leader of the Islamic revolution. There can be little doubt that Khomeini's original design was for a withered state that would be deprived of judiciary authority and would remain permanently subservient to the hierocracy. For this reason, he did not allow Ayatollah Beheshti to run for president against Bani-Sadr and opposed the occupation of the highest offices of the state by clerics until the explosion that killed Beheshti and many others at the end of June 1981.[35] One may go further and state that if there ever was a revolution whose leader wanted to destroy the modernized bureaucratic state and army and go back to the good old days, it was the Iranian revolution. The fact that the state has survived and grown as an unintended consequence of the Islamic revolution shows the limits to the intended teleology of the revolution as prefigured in its ideology and embodied in the distinctive institutions of the Islamic Republic of Iran surveyed in the previous section.

Other elements of continuity between pre- and post-revolutionary Iran are not paradoxical because they are not contrary to the intentions of the leaders of the Islamic revolution. The most important of these is the continuation of the process of national integration and the deepening of the penetration of the central government into the rural periphery. The political and ideological penetration of the villages and tribal areas has greatly increased, especially as a result of the war with Iraq. In addition to the increased presence of the agents of the revolutionary organizations, especially the Islamic Rural Councils, described in April 1984 as "the executive arm of the state and the propagator of the culture of the Islamic Revolution in villages,"[36] Jihad for Construction had setup some 37,000 ideology classes and 28,000 classes for teaching Arabic and the Koran, and had distributed large quantities of books, magazines, film, tapes, and nearly three million posters in rural areas by 1984.[37] National economic integration, too, has proceeded through central economic planning and the activities of the Jihad for Construction. The latter organization claims to have built nearly 30,000 kilometers of roads and 12,000 bridges, brought electricity to over 5,000 and water to over 6,000 villages, and sent 67,000 medical groups to villages by March 1985.[38] By 1984, there were 27,431 kilometers of rural roads in Iran, twice the pre-revolution figure.[39] In the same period, the Ministry of Education has reportedly built 19,000 new schools in villages.[40] Finally, the national integration of the villages is enhanced by the increasing penetration of the rural periphery by the mollahs. In July 1984, the Organization for Islamic Propaganda claimed to have sent over 225,000 missionaries to village during the five preceding years.[41]

Like the Pahlavi regime, the Islamic Republic has fiercely suppressed all ethnic separatist or autonomist movements, most notably the Kurdish movement, and has sought to dominate and integrate the ethnic minorities. As the separatist ethnic minorities, the Kurds, the Turkmans, and the Baluch, are mostly Sunni, the major channel for their integration has been through the Sunni *imam jom'ehs* who are regularly convened for national conferences and receive suitable instructions from the Shi'ite ruling clerical elite in Tehran.

III

REFLECTIONS ON
THE ISLAMIC REVOLUTION

The best lack all conviction, while the worst
Are full of passionate intensity.
"The Second Coming," W. B. Yeats

9

The Revolutionary
Transformation of Shi'ism

Khomeini led a political revolution in order to preserve Shi'ism and restore it to its former glory. He was not the first revolutionary whose aim was the restoration of the good tradition. Nor was he an exception in transforming the tradition he sought to restore by means of revolution. In fact, he has been instrumental in a far-reaching transformation of the Shi'ite tradition that can best be characterized as a revolution *in* Shi'ism. The two central features of this revolution are the radical transformation of the Shi'ite theory of authority and of the Shi'ite law. Let us examine these separately.

Transformation of the Shi'ite Theory of Authority

There has been remarkable consensus among the Shi'ite jurists throughout the centuries regarding the interpretation of the "authority verse" of the Koran (IV: 59). The major Shi'ite Koran commentaries, al-Tusi's (d. 1067) *Tibyan,* al-Tabrisi's (d. 1153 or 1158) *Majma' al-Bayan* and Ardabili, the Moqaddas's (d. 1585) *Zubdat al-Bayan,* assert that "those in authority" (*ulu'l-amr*) are neither the secular rulers (*amirs*) nor the *'ulama*—neither of whom is immune from error and sin—but rather the infallible (*ma'sum*) Imams, 'Ali and his eleven descendants.[1] The Shi'ite consensus on the interpretation of the authority verse continued to hold until our time, until the onset of Khomeini's formulation of a new Shi'ite political theory. It is worth quoting the most influential contemporary Koran commentary, Tabataba'i's *Tafsir al-Mizan* (written in the 1950s and 1960s) on verse IV: 59:

> The conclusion is that it would not make sense to take the verse *"ulu'l-amr minkum"* to refer to "the people of binding and loosening" as a social sodality, in whatever sense. Therefore "those in authority" must refer to the individuals from the *umma* who are infallible (*ma'sum*), whose recognition depends on the explicit designation of God or his Messenger, and to whom obedience is incumbent. All this corresponds only to what have been related from the Imams of the House of the Prophet, may peace be upon them, as "those in authority."
>
> As for the assertion that "those in authority" are the rightly-guided

caliphs, the lords of the swords (*amirs*), or the '*ulama* who are followed in their sayings and views, *it can be completely refuted* in two ways: firstly, the verse indicates the infallibility of "those in authority" and there is undoubtedly no one who is infallible in any of the above categories (except for 'Ali, may peace be upon him, according to Shi'ite belief); secondly, there is no evidence to support any of these three opinions.[2]

The Shi'ite notion of authority implied in the above interpretations of the authority verse was not confined to Koran commentaries but also informed Shi'ite jurisprudence. For instance, Shaykh Mortaza Ansari (d. 1865), the most important jurist of the second half of the nineteenth century, in his discussion of authority, sought first, "to demonstrate how absurd it is to reason that because the Imams should be obeyed in all temporal and spiritual matters, the *faqihs* are also entitled to such obedience; and second . . . that in principle no individual, except the Prophet and the [infallible] Imam, has the authority to exert *wilaya* [Arabic variant of *velayat*] over others."[3] As was pointed out in Chapter 1, the *Osuli* movement finally established the religious and juristic authority of the '*ulama* on behalf of the Hidden Imam. As a consequence, certain aspects of *wilaya* were in fact transferred to the jurists. A number of highly specific functions of the Imam covered in the medieval treatises in jurisprudence, such as Mohaqqiq al-Hilli's (d. 1277) *al-Mukhtasar al-Nafi'* under the rubric of *Wila' al-imama*,[4] were now said to devolve, during the Occultation, upon the Shi'ite jurists by virtue of their collective office of "general vicegerency" (*niyabat 'amma*). However, as late as the 1960s, the *wilaya* transferred from the Imam to the jurists had the highly specific and well-delineated traditional connotations:

> The *velayat* of the fully qualified *faqih*, according to indubitable evidence, is the authority over the affairs of those minors who have no specific parents, and over the insane, so that he may manage their affairs according to expediency, and also authority over the wife of a person who has disappeared as regards maintenance and divorce . . . and the supervision of those *awqaf* which are without a specific administrator, and the upholding of the *hudud* and judgeship and ruling according to the Sacred Law (*hokumat*) and resolution of hostilities and investigation of claims and upholding of rights, and the like.[5]

In his bid to overthrow the Shah from exile in early 1970, Khomeini took a bold step by asserting that the *velayat-e faqih* went beyond these specific types of authority and included a general right to rule. The *velayat-e faqih* thus assumed the meaning of the sovereignty of the jurist. Khomeini extended the arguments of the early *Osuli* jurists, which were designed to establish the *legal* authority of the '*ulama* on the basis of a number of Traditions from the Prophet and the Imams, to eliminate the duality of religio-legal and temporal authority altogether. As was pointed out earlier, Khomeini argued that during the Occultation of the Imam, his right to rule devolves upon the qualified '*ulama*.[6] This formulation still preserved the Shi'ite juristic pluralism, as *velayat* was presented as the collective prerogative of all Shi'ite jurists or at least all the *maraji'-e taqlid*. About a year or two later, Khomeini attempted to reduce this juristic pluralism to a unitary theocratic leadership to be installed by an Islamic revolution. Having reaffirmed that the *'ulama* "possess with respect to government all that the Prophet and the Imams possessed," Khomeini maintains that:

> *wilaya* falls to the just jurist. Undertaking a government and laying the foundation of the Islamic state (*al-dawlat al-islamiyya*) is a duty collectively incumbent on just jurists.
>
> *If one such succeeds in forming a government it is incumbent on the others to follow him. If the task is not possible except by their uniting, they must unite to undertake it.* If that were not possible at all, their status would not lapse, though they would be excused from the founding of a government.[7]

In less than a decade, Khomeini's theory was embodied in the Constitution of the Islamic Republic of Iran. On the basis of a revolutionary reinterpretation of *velayat-e amr* and an equally revolutionary reinterpretation of Imamate as the principle of continuous (*mostamarr*, i.e., uninterrupted by the Occultation of the twelfth Imam) theocratic leadership, the ruling jurist is identified as the *valiy-e amr* and his supreme office is interchangeably defined as *imamate* and leadership (*rahbari*). The constitution defines the Islamic Republic of Iran as "an order based on the belief in:

1. the one God (there is no god but God) and the restriction of sovereignty and legislation to Him, and the necessity of submission to His command. . . .
5. Imamate and continuous leadership, and its fundamental role in the perpetuation of the Islamic revolution." (Article 2)

Article 5 asserts that during the Occultation, *"velayat-e amr* and the Imamate of the *umma* is upon the just and pious . . . jurist." A commentator on the constitution unabashedly declares: "the *ulu'l-amr* refers equally to the Imam and the Deputy (*na'eb*) of the Imam, and the Deputy of the Imam is the jurist who is installed in this position with the necessary conditions."[8] It is important to note that the terms *velayat-e amr* and *vali-ye amr* are deliberately chosen to imply derivation from the Koranic *ulu'l-amr*. This implied derivation is specious as the words *vali* and *ulu* are in fact etymologically unconnected.

The clerical ideologues have sought to link the Islamic revolution irrevocably to the establishment of supreme clerical leadership. In the words of the late Ayatollah Motahhari, "the analysis of this revolution is not separable from the analysis of the leadership of the revolution." Or, as another Islamic ideologue has remarked more recently:

> This revolution is the integrations of religion and politics, or better put, it is the refutation of the colonialist idea of "separation of religion from politics." In other words, this revolution is the acceptance of the servitude of God for deliverance from the servitude of non-God, and the

acceptance of the sovereignty of God for salvation from the domination of *taghut,* a sovereignty which is realized through obedience to the Prophet and the Infallibles and obedience to the learned and just *'ulama.*[9]

Incidentally, it is interesting to note that no mention is made of the principle of consultation (*shura*) or democracy as a defining characteristic of the Islamic republic. The principle of *shura* makes its appearance only in Article 7. Madani's commentary explains the subsidiary role of consultation. The principle of consultation is accepted but as a subsidiary to the principle of Imamate. "Islamic consultation is only possible when Imamate is dominant. In other words, consultation is at the service of Imamate." The Koranic verse III: 153 (*wa shawirhum fi'l-amr,* etc.) is said to imply that the actual decision maker is the Prophet who was also the Imam. The commentator adds that the advocates of the *shura* during the drafting of the constitution either did not firmly believe in Islam or were contaminated by syncretic thinking, and were trying "to link the *shura* to the principle of national sovereignty."[10]

As we have seen, Khomeini's theory of theocratic government extends the Shi'ite norm of juristic authority as elaborated in the nineteenth century into a new sphere previously not covered by it: government. From the Shi'ite point of view, this theory is open to two forceful objections. The first is that the mandate or authority of the Shi'ite *'ulama* during the Occultation of the twelfth Imam cannot be extended from the religiolegal to the political sphere. The second objection is that the mandate in question refers to the *collective* authority of *all* Shi'ite jurists and cannot be restricted to that of a single supreme jurist nor, by extension, to a supreme council of three or five jurists. The above doctrinal objections to *velayat-e faqih* were voiced by a number of *maraji'-e taqlid* and other ayatollahs.[11]

As we have seen, the dissident ayatollahs were decisively suppressed in the spring of 1982. After the demotion of Grand Ayatollah Shari'at-madari in April 1982, propaganda on the doctrine of *velayat-e faqih* was stepped up in the government-controlled media. One interesting means of promoting the theory was the publication of the wills of the "martyrs" of the war. Throughout May and June 1982 (and subsequently), the newspapers would regularly publish the martyrs' profession of faith in *velayat-e faqih* and their praise for the Imam and the militant clerics. Statements to the effect that obedience to the *'ulama* as "those in authority" is incumbent upon the believer as a religious duty were often excerpted from the will and made into headlines in bold letters.

It should be evident that Khomeini's attempt to subordinate juristic pluralism in the form of the voluntary submission of the Shi'ite believers to the Grand Ayatollahs as *maraji'-e taqlid* has been at the expense of the latter.[12] The relationship between the interpretation of the new supreme leadership as *velayat-e amr* and the old positions of *marja'-e taqlid* remains a thorny theoretical issue. Khomeini himself could not put forward any juristic argument and justified his position on the purely pragmatic grounds of the necessity of maintenance of order in society. Recent discussions have not gone beyond Khomeini's pragmatic justification of the superiority of one *faqih* over the others. Madani, for instance, conceives of the relationship between the supreme leadership and the *marja'iyyat* as one between the general and the particular: the supreme leader has to be a *marja'-e taqlid* but not every *marja'-e taqlid* can undertake the supreme leadership. Furthermore,

> The maintenance of order in society necessitates that when the Leader or the Leadership Council is accepted, all should obey a single authority in social and general problems of the country within the framework of the Islamic Constitution. Such obedience is implied in the title of *"vali-ye amr"* and "Imamate of the *umma*" and applies to all members of society without exception, and in this respect the *mojtahed* and the non-*mojtahed*, the *marja'* and the non-*marja'* are in an equal situation.[13]

Despite this accommodation, however, the future of the institution of *marja'iyyat* is in question. Doubts have been raised as to the legitimacy of individual as distinct from collective *ejtehad* now that, for the first time in the history of Shi'ism, an Islamic order has been created. One need only draw out the implications of this typical passage to understand that the institution of *marja'iyyat* has a dark future:

> With the establishment of Islamic government *marja'iyyat*, in practice and officially, took the form of leadership and rule over society; and the *velayat-e faqih*, which in past history had almost never been applied from the position of government, and had always been realized in a defective and incomplete manner, with this revolution reached perfection in practice and occupied its true station.[14]

The *maraji'* themselves are under no illusion about the future of the institution of *marja'iyyat*. It is hardly surprising that Ayatollah Sadeq Ruhani felt obliged to make the following declaration on the occasion of the election of Ayatollah Montazeri as the future *faqih* in November 1985: "My duty is to say that I see Islam in danger, that the *marja'iyyat* is in danger."[15]

The militant *'ulama* who followed Khomeini in the 1960s and 1970s sought to defend and revitalize the Shi'ite tradition through a political revolution. To secure the leadership of this political revolution for themselves, they have revolutionized the Shi'ite political ethos whose distinctive mark had been the secularity of temporal rule and desacralization of political order.[16] To establish and propagate their new conception of authority, the clerical rulers of Iran have incessantly insisted on the sacred character of all authority and thus the thoroughgoing sacralization of the political order. Here are four examples. In a lecture on the newly established Islamic order, the late Ayatollah Motahhari emphatically maintained that in Islamic government, authority is sacred (*moqaddas*). This is so because the offices of government and judgeship devolves upon the *'ulama*.[17] Ayatollah Rabbani Amlashi, temporary *imam jom'eh* of Tehran, would accordingly tell his congregation:

> Obedience to the *velayat-e faqih* is an incumbent duty (*vajeb*). In the Islamic Republic obedience to the law is incumbent like the daily prayer and fasting, and disobeying it is like disobeying the Islamic Sacred Law.[18]

Ayatollah Meshkini, the *imam jom'eh* of Qom, takes a step in a different direction to sacralize politics:

> Political activity is an incumbent (*shar'i*) duty. Today, one of the most important acts of devotion (*'ebadat*) is political activity because without politics our religiosity (*diyanat*) will not last.

Finally Ayatollah Mo'men, member of the Council of Guardians, takes a further step to sacralize *all* authority, legal and political:

> The legitimacy and legality of whatever is done and whatever institutions exist is due to the fact that they are buttressed by the *velayat-e faqih*. As the *velayt-e faqih* is at the head of all affairs and the main guarantor of the current laws of the country, it is the *divinely-ordained duty of all the people* to follow every law which is passed and given to the Islamic government for execution. . . . Disobeying such a law is forbidden (*haram*) as drinking wine is forbidden by Islam.[19]

On occasion, the sacralization of politics even necessitates going beyond the requirement of the Constitution of the Islamic Republic. For instance, according to the Constitution, participation in the elections are voluntary. However, as Madani points out in his commentary, people are usually enjoined by the Imam and the religious authorities to participate in the elections as Muslims fulfilling a religiously incumbent duty. However, "even though this matter becomes incumbent according to the Sacred Law, non-participation is [punishable] not materially but spiritually!"[20]

The month of January 1988 witnessed the culmination of the revolution in Shi'ism. On January 6, the Imam reprimanded President Khamene'i for a misunderstanding and representing of the Mandate of the Jurist in his declaration that the authority of the Islamic government can be exercised only within the framework of the commandments of the Sacred Law. A government in the form of the God-given, absolute mandate (*velayat-e motlaq*) was "the most important of the divine commandments and has priority over all derivative divine commandments . . . [it is] one of the primary commandments of Islam and has priority over all derivative commandments, even over prayer, fasting, and pilgrimage to Mecca." Five days later, in another letter that set the tone for a chorus of affirmations and "clarifications" by the ruling clerical elite, Khomeini referred to the president as a brother who supported the Absolute Mandate of the Jurist. On January 22, a chastened President Khamene'i thus propounded the principles of the new theocratic absolutism:

> The commandments of the ruling jurist (*vali-ye faqih*) are primarily commandments and are like the commandments of God. . . . The Regulations of the Islamic Republic are Islamic regulations, and obedience to them is incumbent . . . [they are all] governmental ordinances (*ahkam-e hokumati*) of the ruling jurist . . . in reality, it is because of

the legitimacy of the mandate [of the jurist] that they all acquire legitimacy. . . .

The Mandate of the Jurist is like the soul in the body of the regime. I will go further and say that the validity of the Constitution . . . is due to its acceptance and confirmation by the ruling jurist. Otherwise, what right do fifty or sixty or a hundred experts have . . . ? What right do the majority of people have to ratify a Constitution and make it binding on all the people? The person who has the right to establish the Constitution for society is the ruling jurist. . . .

It is the ruling jurist who creates the order of the Islamic Republic . . . and requires obedience to it. Opposing this order then becomes forbidden as one of the cardinal sins, and combating the opponents of this order should become an incumbent religious duty. . . .[21]

The paradox of the actual insignificance of the political ethics in the Shi'ite Sacred Law (the paucity of political provisions that reflect the age-old secularity of political authority and order in Shi'ism) against the claim that Islam is a total way of life and a total ideology, which is above all political and activistic has struck some observers. The clerical rulers of Iran sought to resolve this paradox by drawing on the distinction between the "primary commandments" (*ahkam awwaliyya*) and "secondary commandments" (*ahkam thanawiyya*). The first were said to derive from the sources of the *shari'a*, the second from expediency and the necessity of maintaining order and avoiding chaos. Both categories were said to be binding on the believer as a religious obligation. Thus, for the first time in Shi'ite history, sacrality was claimed for a category of secondary commandments, which were derived *not* from the juristic competence of the Shi'ite *'ulama but from their alleged right to rule.*[22] In January 1988, the charade of the "primary/secondary" distinction was finally given up and all governmental ordinances (*ahkam-e hokumati*) were said to belong to the category of primary commandments and to be immediately incumbent. All pretense has finally been set aside. The Absolute Mandate of the Jurist (*velayat-e motlaqeh-ye faqih*) has replaced the Mandate of the Jurist in the official terminology of the rule of God in Iran.

Thus sacralized, the new Shi'ite political order demands unconditional obedience from the individual as an incumbent religious duty. In Khomeini's words:

The preservation of the Islamic Republic is a divine duty which is above all other duties. It is even more important than preserving the Imam of Age (*Imam-e 'asr*), because even the Imam of Age will sacrifice himself for Islam. All the prophets from the birth of the universe to the present were sent to strive for preservation of God's word . . . Islam is a divine endowment. . . . Its preservation is an inexcusable individual obligation (*vajeb-e 'eyni*).[23]

Transformation of the Shi'ite Law

The legal history of Islam in general and of Shi'ism in particular is marked by the dualism of judiciary organization. From the early 'Abbasid period, public law and the administration of criminal justice remained the province of the ruler and his tribunals, while the science of jurisprudence and the Sacred Law (*shari'a*) developed as a "jurists' law" rather than a "judges' law."[24] Its being a jurists' law reinforced the infusion of Islamic law with ethical considerations and its relative indifference to the administration of justice. The medieval legal literature abounds with expressions of distaste on the part of the jurists for the office of judge.[25] The moralistic antipathy of the pious jurists to the administration of justice militated against any *procedural* rationalization of Islamic law,[26] the result being the informality that Weber singled out and used for the designation of his category of *"qadi*-justice." The administration of Islamic law, *qadi*-justice, was marked by the absence of procedural formalism. A perceptive French observer of seventeenth-century Iran was struck by the informality of the procedure at the *qadi's* home, and by the lack of coordination between judges and absence of hierarchical rationality in the judiciary organization.[27]

The procedural informality of *qadi*-justice went hand in hand with the dubious status of written documents. The decisive factor in establishing evidence in Islamic law is the oral testimony of a witness. At best, documents constituted only corroborative evidence. Their value as evidence derived from the moral probity of the individual who testified to their authenticity.[28] This concern with "normative witnessing" has been singled out as the most striking characteristic of the Islamic judiciary procedure, thus emphasizing the *personal* character of the Islamic administration of justice.[29]

The duality of judiciary organization—with the ruler's tribunals administering public law and criminal justice according to custom (*'orf*), and the religious courts enforcing provisions of the Sacred Law in personal and family law and transactions—moral idealism in jurisprudence and informality and personalism in the administration of justice were thus the distinctive features of the legal system of Shi'ite Iran in the nineteenth century. Paradoxically, the revolutionary break with this traditional pattern of the past came with Khomeini's attempt to revive and restore the Shi'ite tradition in 1979.

After the revolution of February 1979, the triumphant Shi'ite heirocracy inherited the political and hierarchical judiciary organization of the Iranian nation-state, as formally rationalized by seven decades of Western-inspired modernization. The declared aim of Ayatollah Khomeini had been to transform the Pahlavi state into a theocracy and to Islamicize its judiciary system. I suspect that before embarking on this project Khomeini and his clerical followers did not realize that the attainment of these goals would entail a legal revolution in Shi'ism. But embark on their project they did, and the legal revolution they initiated is in full swing.

As is well known, Max Weber saw the modern state as the typical societal organization of rational-legal domination. The true Islamicization of the modern state into a Shi'ite theocracy would require a drastic transformation of the Shi'ite Sacred Law. From being a jurists' law it was to be transformed into the law of the state. Law finding as the typical activity of the Islamic jurists was to be replaced by legislation and codification. Shi'ite law was to be extended to cover public law fully. It was also to cover criminal justice. Its penal provisions, unenforced for a millennium, were to become fully operative. Procedurally, Shi'ite law was to be enforced through the modified mechanism of the inherited judiciary organization modeled on the West European civil law systems. All this meant that the moral idealism of Shi'ite law had to give way, at least partially, to practical realism, its procedural informality to a formally rationalized bureaucratic court system, and its often unpractical personalism to more impersonal and efficient procedures involving much greater reliance on written documents and impersonal forms of evidence.

Once the historic decision to take over the state and implement theocracy within its legal framework was taken by Khomeini and the militant clerics, it was no longer possible to recreate the religious court alongside the civil courts. The entire modernized judiciary apparatus had to be taken over and Islamicized. The commitment to do so was declared in the Constitution of 1979. Article 156 included in the functions of the judiciary the revival of Islamic market regulations (*hesba*) and "the enacting of the penalties and provisions of the Islamic penal code." Article 157 established the Supreme Judiciary Council to reform the judiciary organization where necessary, to prepare laws suitable for the Islamic Republic, and to appoint just and worthy judges. It was to consist of the president of the Supreme Court, the prosecutor general of Iran, and three *mojtahed* judges, as per article 158. Article 162 further specified that the president of the Supreme Judiciary Council and the prosecutor general also had to be *mojtaheds*. Article 170 required the nullification of government regulations and administrative orders that were at any time found to contravene Islamic laws and regulations or to be in excess of the authority of the judiciary branch. A Court of Administrative Justice would decide on cases in this regard.

Article 163 is semantically interesting: "The attributes and qualifications of judges will be determined by law (*qunun*) according to the criteria of jurisprudence (*feqh*)." It reflects the revolutionary intent of the triumphant Shi'ite hierocracy to replace traditional law-finding by modern legislation—legislation which, as indicated in articles 156 and 157, was to include codification, notably of penal provisions of the Sacred Law. Since 1980, the clerical rulers of Iran have embarked on a historically unprecedented comprehensive program of codification of the Shi'ite Sacred Law. As the civil and commercial codes of the Pahlavi period had already taken many provisions of the Shi'ite Sacred Law into account, they only needed minor revisions, except for the Family Protection Law of 1975, which was repealed because it diverged from Shi'ite law by strengthening women's rights. The penal code, however, was altogether a different matter.

The Law of Punishments and Talion (*qanun-e hudud va qesas*) of October 1982 revives and codifies the atavistic penal provisions of the Sacred Law, which had generally remained in abeyance since the takeover of criminal justice administration by rulers' tribunals very early in Islamic history. Penalties such as amputation of the limbs, stoning, and crucifixion are [re]introduced and specified in gruesome detail, and severe punishments, including death, are decreed for adultery, homosexuality, and other sexual offenses. Punishment is made the private right of "those entitled to the blood" (*awliya'-e dam*), that is, the victim's heirs, who have the option of executing the criminal personally.[30] (This option has since been exercised several times.) An interesting novel feature of the law (articles 198–200) is the definition of the armed opponents of "Islamic government" as "fighters [of God and His Prophet]" (allusion to Koran 9:108) and "corruptors on earth" (allusion to Koran 2:10, 57, etc.). Article 201 states that the proof of fighting God and spreading corruption on earth can be established by confession, once, or by "the testimony of only two just men." The penalties are execution, hanging, cutting of the right hand and the left foot, exile, and crucifixion (articles 202–211).[31] The Code of Indemnities (*diyat*), establishing blood money, wergild, and indemnities in archaic measurements for a long range of bodily injuries, was subsequently raified by the Council of Guardians in December 1982.[32]

The radical departure from the Shi'ite tradition in the enactment of these codes is implicitly acknowledged by the provision that they be experimentally enforced for a period of five years. A more radical transformation of the Shi'ite legal tradition is in fact brought about by the unacknowledged appropriation of European legal material contained in the revised civil and commercial codes. But perhaps the most revolutionary transformation of the Shi'ite tradition concerns the fundamental attitude toward the administration of justice.

The Shi'ite jurists' pious reluctance to accept judgeship was in no small measure due to the severity of the Islamic penal prescriptions and the consequent fear of miscarriage of justice. Reflecting the typical Shi'ite traditional attitude to judging, an authoritative statement of the Shi'ite doctrine in the 1930s, citing a Tradition from the first Imam, 'Ali, that "the *qadi* is on the edge of hell" and another from the Prophet: "He who is appointed a *qadi* has been slaughtered without a knife," states that the awareness of the dangers of judging made the great Shi'ite teachers and dignitaries of the past refrain from it and often made them advise governments on peace and resolution of conflict. "Therefore, we too, as usual follow our pious forebears and act in the same manner."[33] Having radically diverged from this traditional attitude, Khomeini, Montazeri, and the other ruling clerics have constantly extolled the virtues of judging and its incumbency upon the clerical community. However, radical as this transformation of attitudes is, it does not represent change in Shi'ite jurisprudence as judgeship has always theoretically been the collective (*kefa'i*) moral obligation of the 'ulama. Where a corresponding radical change in Shi'ite jurisprudence has occurred is in connection with the procedural rules for establishing evidence and issuing a verdict. This change has had the consequence of reducing the importance of "normative witnessing" and allowing the judge to rule on the basis of evidence produced by modern methods of criminal investigation.

In Islamic law generally, there appears to be little room for the discretion of the judge in determining the facts of the case. The formal means for establishing proof (*bayyina, hojja*) are the testimony of witnesses, oath, and confession.[34] In Shafi'i law, the judge is allowed more discretion and can rule on the basis of his own knowledge (*'ilm*) when he is personally certain about the facts of the case. Even here, however, such personal knowledge is not sufficient as regards criminal punishments, referred to as "the rights of God" (*hoquq Allah*), where one of the above-mentioned formal legal proofs is necessary.[35] In Shi'ite jurisprudence, the principle of "judging according to one's own knowledge (*'ilm*) has been a matter of controversy. It has been generally accepted as an ancillary to the main formal means of establishing proof, that is, weighing of contradictory testimonies, rectification of error or mendacity of witnesses, determination of contempt of court.[36] However, beyond the general acceptance of this ancillary role, there has been controversy as to whether the principle is valid as regards "the rights of God" or "the rights of people" (*hoquq al-nas*) or both or neither.[37] Even Sangelaji, who favors the general validity of the principle in all legal spheres, ends his discussion with a revealing question:

> If the judge can act according to his own knowledge, and if a case is referred to him whose facts he [personally] knows, is it or is it not incumbent [*vajeb*] upon him to deal with that case? According to the apparent meaning of the received evidence, if there is no other judge, investigation is incumbent upon him, otherwise it is permissible for him to refuse that case.[38]

Khomeini transformed the substance of this ancillary principle, making it an independent means for establishing evidence and issuing a verdict; the Penal Code of 1982 transformed the moral idealism of the previous discussions of the topic into positive law. Already in the mid-1960s, in his manual of practical jurisprudence, Khomeini had ruled:

> The judge is permitted to rule [issue a verdict] on the basis of his own knowledge [*'ilm*] without needing [formal] proof or confession or oath, be it concerning "the rights of God" or "the rights of people." Furthermore, he is not permitted to rule according to formal proof if it contradicts his knowledge or to administer an oath upon a person who is a liar in his opinion.

Nevertheless, the moralism typical of Shi'ite jurisprudence is retained in the ensuing sentence: "Yes, it is permissible for him not to accept to serve as judge if he has such an opinion provided he is not the only one available."[39] An authorized commentary on the manual, published after the revolution, diverges from the traditional jurisprudential classifications by stating that there are two means of establishing evidence: (1) knowledge of the judge (*'ilm-e qadi*)' and (2) proofs according to the Sacred Law, consisting in turn of confession, proof (*bayyina*) through testimony, and oath.[40] The commentator adds:

> If the judge acquires knowledge and certainty concerning the facts of the case through the reliable modern means and instruments such as fingerprinting, medical examination, the infra-red camera which photographs the past and other instruments for detection of crime, this knowledge is more valid for him than . . . proofs of the Sacred Law such as confession, witnesses and oath which are of the next [i.e., lower] order.[41]

Khomeini's rulings in jurisprudence were translated into positive law in the Code of Punishments and Talion of 1982:

> Article 27—The ways to prove murder in the court are as follows: 1—Confession, 2—Testimony [of witnesses], 3—Oath, 4—Knowledge of the judge. Article 120—The administrator of the Sacred Law (*hakem shar'*) can execute the divine penalty on the basis of his own knowledge, be it concerning "the right of God" or "the right of people." It is necessary that he mention the basis of his knowledge. Concerning a "right of God," the process is not dependent on any person's petition, but concerning a "right of people," the execution of the penalty is dependent on the petition of the person entitled to that right.[42]

The president of the Supreme Judiciary Council has been constantly complaining of the shortage of qualified *mojtahed*s to serve as judges although the seminaries of Qom have been producing jurists as fast as possible. Meanwhile, the categorical affirmation of the knowledge of the judge as an independent means of establishing evidence and the corresponding emphasis on the discretion of the judge have laid the legal foundations for the complete takeover of the bureaucratic machinery for the administration of criminal justice by the religious jurists.

10

Significance of the
Islamic Revolution:
A Comparative Perspective

Iran's Islamic Revolution is significant in world history with regard both to its causes and its consequences. These will be considered separately in this final chapter.

Causes and Preconditions of the Islamic Revolution

Collapse of the Monarchy

The emphasis of recent scholarship on the role of the state, its repressive capacity, and its ability to weather serious crises has brought out the fact that revolutions often owe their success more to the internal breakdown and paralysis of the state than to the power of revolutionary groups.[1] It has been argued that the decisive factor in the occurrence of a revolution is the fragility of the political system.[2] Centralization of monarchical states reduces the degree of pluralism in society and increases its political fragility. Among the political regimes of the modern world, monarchies are especially fragile and vulnerable to revolution. They have the property of focusing discontent on a single person. Tocqueville, who considered that the hatred of the ancien régime dominated all other passions throughout the French Revolution, also showed how the hatred of the ancien régime became fatally focused on a single person, the king: "To see in him the common enemy was the passionate agreement that grew."[3] The same can be said about the Shah, the demand for whose ouster was the one common factor that brought together almost all disparate sections of Iranian society. Furthermore, the same property of the monarchical system in Iran goes a long way toward explaining the meteoric rise of Khomeini as antimonarch and the Shah's counterimage.

Fragility is also characteristic of what we might call neo-patrimonial states. In these states, in contrast to the ideal type of the absolutist state in which the king is only the first servant of the state, government is extremely personal and the chief executive encourages divisions within the army and the political elite in order to rule. Such neo-patrimonial states are particularly prone to collapse and ensuing revolution once the ruler breaks down.[4] The Mexican (set in motion by the death of Porfirio Díaz in 1911), the Cuban, and the Nicaraguan revolutions can be cited in support of this proposition. The Shah's regime combined the weaknesses of the neo-patrimonial state with the old vulnerabilities of monarchy.[5] He had painstakingly constructed the state machine around his person, and there can be no doubt that the collapse of the man preceded the collapse of the machine. This collapse was evident in the Shah's monumental wavering and indecision—for example, he could not make up his mind to appoint a prime minister from the liberal, nationalist opposition until it was far too late, in his inconsistent combination of rewards and threats, and in his highly inhibited use of force.

The neo-patrimonial character of his state notwithstanding, the Shah did have a disciplined and well-equipped army and police force. However, he simply refused to use them effectively to repress the revolutionary movement. The Shah pretended to be using the army. He declared martial law in some cities in the late summer and installed a military government in November of 1978. But, as we have seen, after the Black Friday massacre of September 8, he muffled the army, to the outrage of his generals. This is reflected in the low casualties: about three thousand killed in the whole of Iran over five months of revolutionary turmoil. On December 21, 1978, even General Azhari, the parlor general who the Shah had made prime minister, complained to the American ambassador of the demoralization of the army due to orders forbidding them to fire except in the air, no matter how badly abused or pressed. "You must know this and you must tell it to your government. This country is lost because the king cannot make up his mind."[6]

Unlike the czar's army in 1917, the Shah's remained by and large intact and loyal until he departed on January 16, 1979. As we have seen, the strain of confrontation with the people did not seriously affect the morale and discipline of the armed forces before the announcement of his departure. It was only *after* the Shah's departure that the process of disintegration of the army under political pressure seriously set in. It is not being asserted that the use of the army for massive repression would have prevented the revolution. We will never know what would have happened if the Shah had ordered the army to be brutally repressive in October and November 1978 while it was unaffected by the revolutionary turmoil. Our analysis in Chapter 6 suggest that the army could only have been used effectively by the Shah himself, and not by the generals without him; it may or may not have disintegrated or split. We cannot engage in counterfactual history. The fact remains, however, that the army had not disintegrated even by January 16, 1979. And the opposition knew it.

The army officers had a strong sense of professional identity, but no attachment to any solidary social group, nor to any organized interests. Furthermore, the Shah had chosen his top generals carefully to assure they could not act in concert against him, and he had succeeded. The generals could have acted under him, but he did not let them. They could not act against him, but they could not act for themselves or for any other group either. Some of them finally made a deal with the clerical opposition in desperation. Tilly has correctly emphasized the importance of coalitions linking the revolutionary challengers to the military.[7] Although the term "coalition" may be too strong, the agreement worked out by Bazargan and Beheshti with a number of the generals, with active encouragement by Khomeini and through the mediation of the American ambassador, was of crucial importance in bringing about a split in the army and its consequent neutralization in February 1979.

The Shah's regime collapsed despite the fact that his army was intact, despite the fact that there was no defeat in war, and let me add, despite the fact that the state faced no financial crisis and no peasant insurrections. Where does all this leave the usual generalizations about revolutions? Mostly in the pits. War has time and again been considered the midwife of revolution, and peasant insurrections have been considered indispensable by many currently fashionable theories.[8] The inferences we can draw from the case of Iran are as follows: financial and fiscal crises, or for that matter the extractive capacity of the state and heavy taxation, are not necessary for the occurrence of revolution. Furthermore, it is possible for the societal structure of domination to collapse without the participation of the peasantry, and a major war and/or defeat of the army is not necessarily a precondition of revolution. The revolution in Iran clearly shows that a political order may collapse without any of these conditions. I may add that already with the Cuban Revolution, there had been an instance of a revolution without rebellion of the peasantry and without a major defeat in the war. Skocpol, whose theory of revolution puts a great deal of emphasis on both these allegedly necessary conditions, cavalierly dismisses Cuba in one half of a footnote.[9] Furthermore, she does not face the theoretical consequences of the absence of these factors in an article subsequently written about the Iranian Revolution. She is rightly determined to bring the state into the picture but does so in an unsatisfactory way, largely by deploying a new pet phrase, "the rentier state." The basic idea is misleading in that the rentier state—a state based on "rent" in the form of oil revenue rather than on taxation—was in fact no other than the state created by Reza Shah from the early 1920s to 1941 when the revenue received by the state from the Anglo-Iranian Oil Company was in fact small—some 10 to 15 percent of government revenue—and miniscule compared to the oil revenue in the 1970s. Skocpol musters a modicum of other plausible but *ad hoc* subsidiary themes to account for the Iranian Revolution. However, she never faces the problem of reconciling the Iranian Revolution with her theoretical schema of 1979.[10] The truth of the matter is this: the collapse of the political order in Iran demonstrates that there is more to a system of authority than coercion. There is also the normative factor of legitimacy. What brought the Pahlavi regime down was not the disintegration of its army but rather its loss of legitimacy and a massive nationwide campaign of civil disobedience. The structure of authority in Iran crumbled from within and at the center. What crippled it was not any peasant rebellion on the periphery—and are such insurrections on the periphery really crippling as Skocpol alleges?—but the fact that the manpower of the state bureaucracy, including its various economic organizations, refused to recognize its master as legitimate and refused to obey him.

However, there is one generalization borne out by the revolution in Iran.[11] The Shah was seriously compromised by his close and subservient association with the United States; and the American military and economic presence and the presence of a large European work force acted as a major stimulus to mass mobilization. This anti-foreign motive in challenging the legitimacy of the societal structure of domination finds parallels in the English, French, Russian, Chinese, and Cuban revolutions and in East European fascism.

The State, the Hierocracy, and Civil Society in Shi'ite Iran

It would be a mistake to equate the societal structure of domination with the state. For Weber, its major components were the state *and* the church. He defined the two institutions of legitimate authority analogously and took care to analyze the relationship between the church and civil society when appropriate.[12] This point is substantively crucial because the unique feature of Iran's Islamic Revolution is that it is simultaneously a crucial stage in the hierocracy-state conflict and a modern political revolution. It is a composite of two phenomena whose counterparts in Western history are separated by centuries. The absolutist states of Europe had already won the protracted contest with the Roman Catholic Church before the coming of the early modern European revolutions.[13] In the history of Iran the analogous contest between the state and the hierocracy occurred much later. Shi'ism was declared the state religion of Iran in 1501, but the hierocracy remained heteronomous and subordinate to the state for a long time. Only at the end of the eighteenth and beginning of the nineteenth century did the Shi'ite hierocracy consolidate its power and autonomy. The curtailment of the power of the hierocracy and the appropriation of many of its prerogatives and functions by the state took place in this century. However, the Shi'ite religious authorities were and remained doctrinally and institutionally independent of the state. They retained their autonomous religious authority, and they retained their control over appreciable resources, independent of the state bureaucracy.[14]

The Western revolutions were against state *and* church. The church had been anglicanized in England, gallicanized in France, and disestablished by Peter the Great in Russia; in all instances, it was part and parcel of the monarchical regime. In the Islamic Revolution in Iran, the entire beleaguered Shi'ite hierocracy rose against the state. (Incidentally, this was partly due to the Shah's fateful ineptitude in not splitting the Shi'ite hierocracy in time; there is now evidence that some of the grand ayatollahs were ready for a compromise from the summer of 1978 onward, and the split in fact did occur after the revolution.)

For analytical reasons, too, it is important to conceive of the societal structure of domination in more inclusive terms. Revolutionary situations occur because of the disintegration of central authority. With the disintegration of the authority of the state, other elements of the societal structure of domination, other forms of legitimate authority, assume greater importance. Corporations and individuals with legitimate authority in other spheres of life can extend their authority to the political sphere and assume positions of leadership. In such situations, they emerge as natural leaders of the people. The hierocracy and men of religion can use their traditional authority in this fashion and have often done so, for instance, in Spanish history.[15] Many of the high-ranking members of the Shi'ite hierocracy thus led the popular opposition to the monarch during the Constitutional Revolution in 1905 and 1906. Similarly, in 1978, many groups and individuals who wanted the Shah out but had no interest whatsoever in a theocracy accepted Ayatollah Khomeini's leadership. When central authority disintegrated during the Constitutional Revolution, both the hierocracy and the tribal forces of the periphery stepped into the central political arena. With tribes eliminated as structures of military power, only the hierocracy had the capacity to act when the authority of the state disintegrated in 1978.

The centralization of the state necessitates the concentration of economic, coercive, and symbolic resources. It entails encroachments upon local and provincial privileges and fiscal and constitutional immunities; and it entails the dispossession of certain privileged social groups. It thus sets in motion an intense and continuous political struggle. The reaction of privileged groups and of autonomous centers of power against the expansion and centralization of the state is a major source of revolutions.[16] This is the case with most if not all of the early modern European revolutions: the revolt of the Comuneros of the cities of Castile against Charles V in 1520, the revolt of the Netherlands in reaction to the centralizing policies of Philip II in the 1560s, the French Civil War of the sixteenth century, the revolt of the Catalans after Olivares had consigned their constitutions "to the devil," and of Portugal in 1640, the early phase of the English Revolution,[17] the Fronde and the aristocratic pre-revolution of 1787 to 1788 in France.[18] In all these cases, estates and corporations whose autonomy and inherited privileges were threatened by the state reacted, and they usually found men of religion as their allies. The dispossessed or debt-ridden nobility of the Netherlands, for instance, found allies in Calvinist preachers and iconoclasts.[19] In Iran of the 1970s, the chief dispossessed solidary group capable of reaction and the preachers were one and the same group.

Three privileged social groups were the major victims of the centralization of the state under the Pahlavis. The first consisted of the tribal chiefs. As we have seen, Reza Shah's pacification campaigns of the 1920 and early 1930 broke the power of the tribal chiefs and eliminated many of them. Many tribal chiefs survived, and some even entered the Pahlavi political elite, but did so as individuals and without an autonomous power base.

The second group was the landowning class. They not only survived Reza Shah's state building but were in fact helped by its legal reform.

Mohammad Reza Shah's land reform in 1962–63 partially dispossessed the big landlords but completely liquidated them *as a class*. Once the Majles was dissolved, the "feudal" landowning class had no autonomous institutional basis and could not react against its complete political and partial economic dispossession by the state. The *'ulama* constituted the third group. Under Reza Shah, the Shi'ite hierocracy came under fierce attack by the centralizing Pahlavi state. The state deprived it of all its judiciary functions, eliminated its prebendal, fiscal, and social privileges, and greatly reduced its control over education and religious endowments. In the face of Reza Shah's determination and severity, the hierocracy did not react in any significant fashion. As we have seen, the relations between the hierocracy and the monarchy had improved after the fall of Reza Shah, especially in the late 1940s and 1950s when the monarchy was weak and the hierocracy alarmed by the threat of Communism. The state resumed its aggressive posture in the 1960s and 1970s, this time encroaching upon the religious sphere in the strict sense. In contrast to the landowning class, the partially dispossessed Shi'ite clerical estate did have an autonomous institutional basis. It could react to the expansion of the state, and eventually did so.[20]

In a political struggle set in motion by the centralization and modernization of the state, dispossessed social groups that retain an institutional basis for reacting against the expanding state nevertheless need to create *coalitions* with other social groups and classes if they are to succeed. In the early 1960s, badly coordinated attempts were made to forge a coalition by elements from the hierocracy, the landlords, and the tribal chiefs, but the separate uprisings of Khomeini's followers and the Qashqa'i and Boyr Ahmad tribes of Fars in 1963 were ruthlessly suppressed.[21] In 1978, an effective coalition did come into being and did carry out a revolution.

Thanks to the unifying hatred of the Shah, the revolutionary coalition of 1978 came to include the bulk of Iran's urban population. The Islamic Revolution was an urban revolution and the peasantry did not play a role in it. Nor did the industrial working class. All the other segments of the urban population actively opposed the Shah and accepted Khomeini's revolutionary leadership. The militant clerics' two most important coalition partners consisted of the new middle class—government employees, school teachers, the intelligentsia, and white-collar workers from the service sector—and the traditional bourgeoisie of the bazaar.

The coalition between the Shi'ite clerics and the new middle class was highly unstable. It rested on fraudulent silence on the part of the former and on wishful thinking on the part of the latter. It did not last long. Having ejected the Shah, Khomeini wasted no time in liquidating the Westernized intelligentsia.

The coalition between the revolutionary clerics and the traditional bourgeoisie, on the other hand, rested on more tangible grievances on both sides and on a more solid historical basis. It proved more enduring. It was the most recent instance of the alliance of the mosque and the bazaar, which resembles, among other instances, the alliance of the urban bourgeoisie and the church in the eleventh and twelfth centuries in Western Christendom. It was forged again in the late 1970s under the immediate impact of the Shah's destruction of the seminaries in Mashhad and his massive antiprofiteering campaign against the bazaar merchants and retailers.

Why did the new middle class lose? After all, it could have gone the other way, as it did in the case of Nasser's temporary coalition with the Muslim Brothers, who had very wide popular support and were in some ways much better organized than the mollahs. In twentieth-century Iran, the centralizing state had atomized society to a considerable degree. It had detached the tribal chiefs and dissolved the landowning class, and it had created an intelligentsia, a bureaucratic class, a body of army officers and, last, an industrial/entrepreneurial group, all of whom were unattached to any solidary social community, be it a tribe, an estate, or a corporation. However, in partial contradistinction to eighteenth century France, the atomization of Iranian society was not complete; three elements of the old civil society escaped it: (1) the Shi'ite clerical estate, (2) the bazaar and traditional bourgeoisie, and (3) urban communities in certain older city quarters that were dominated by the previous group. To these one should add: (4) new urban communities created by chained migration from rural areas and small towns into the larger cities. Given this background, it is not surprising that the atomized new middle class proved to be the proverbial Marxian sack of potatoes while the other solidary social groups in the coalition were capable of remarkably concerted political action and soon took over.[22]

The Shah had kept the new middle class under constant supervision of the secret police and had not allowed it to form associations or gain any political experience. Furthermore, the ability of the new middle class to act was seriously impaired by the isolation of the army officers from the rest of its elements. The political representatives of the new middle class could not easily form a coalition with the army, because it was too closely identified with the Shah and his regime. They therefore decided to form a coalition with the Shi'ite hierocracy.

According to Tilly, contenders who are in danger of losing their place in a polity are especially prone to "reactive" collective action. He rightly observes that for centuries the principal form of collective action followed the reactionary pattern, that is, it was reactive and communal. However, thanks to social evolution, this is no longer the case, and collective action has become predominantly "proactive" in modern times.[23] This conceptual distinction seems of dubious value as a whole set of revolutions analyzed in this chapter are both reactive and proactive. In reality, collective action typified as reactive by Tilly does not lose its importance after the middle of the nineteenth century, and it usually continues to draw on communal traditional solidarities. Whenever these communal solidarities are class solidarities, they pertain not to rising but to *declining* or threatened social classes. What we are alerted to by the Islamic Revolution in Iran is the undeniable importance of reactive action in the revolutionary movements of the last two centuries, including those taken by Marx to be revolutions of rising classes.

Fascinating evidence for the importance of reactive action and traditional communal solidarities in revolutionary movements has recently come to light as regards the very groups who inspired Marx with a theory of revolution that has distorted our understanding of the phenomenon for over a century. The myth of the middle class in the English and the French Revolutions has long been exploded, notable by Hexter and Cobban. Trevor-Roper's characterization of the English Revolution as the declining, "mere gentry's" revolution of despair contains an element of truth but also much exaggeration.[24] On the other hand, we now know that the revolutionaries of 1789 were *not* the capitalist bourgeoisie,[25] and that the revolutionaries of the first decades of the nineteenth century in England and of the Continent in 1848 were *not* the industrial working class. The English revolutionary working class of the early decades of the nineteenth century in fact turns out to have consisted of the artisans and craftsmen who were threatened by capitalist industrialization and were holding onto the memory of the golden age of the community of small producers based on mutual ties and cooperation.[26] A recent study of these "reactionary radicals," as one observer calls them, concludes that "commitment to traditional cultural values and immediate communal relations are crucial to many radical movements." Communal relations are seen to be important resources for mobilization as they enable traditional communities to remain mobilized for a long time and in the face of considerable privation.[27] Shopkeepers and artisans predominate in the French insurrections of the 1830s.[28] The same group of artisans reacting against industrial capitalism and proletarization, who drew their standards and idiom of protest from the past, turn out to be the backbone of the 1848 revolutions in France and Germany. In France, the journeymen's brotherhoods that perpetuated traditional corporate consciousness and solidarities of the *ancien régime* constituted the leading revolutionary element in 1848. In Germany, the artisan elements were prominent in the revolutionary movement of 1848, while the proletariat were the most quiescent of all social groups.[29]

"Reactionary radicals," concludes Calhoun "have seldom, if ever, been able to gain supremacy in revolutions. But at the same time, revolutions worthy of the name have never been made without them."[30] With the Islamic Revolution, a group of reactionary radicals under the leadership of the custodians of the Shi'ite tradition have at last gained supremacy in this theoretically most interesting of modern revolutions.

Let me also mention some movements Marx did *not* study. Generally speaking, the Islamic Revolution has this in common with peasant rebellions: it draws on corporate solidarities and communal and kin ties and consequently has many conservative and defensive features.[31] In Mexico, there was the massive peasant rebellion of 1810 led by Father Hidalgo and Father Morelos, both parish priests.[32] In the 1830s in Spain, we have the Carlists, whose aim has been described as the "restoration of 'monkish democracy'." The clergy led the Basque and Aragonese prosperous yeomanry in rising to defend their local autonomy and their *fueros* against the centralizing policy of the Bourbon government.[33] Moving to the present century, there was the revolt of Zapata in defense of the local autonomy of traditional agrarian communities against the expanding *haciendas* in Mexico. Thanks to the devout Zapatistas (laws of 1915 and 1917) and to President Cárdenas (1934–40), the Mexican revolution established the security of the *ejido,* community-owned, inalienable individual or communal holdings in the villages. It should be added that the outcome of the Mexican revolution would have been much less secularist and more conservative if the Cristero movement, organized by priests and lay Catholics in reaction to the anticlerical policies of central government—with the motto "Viva Christ the King"—had succeeded in 1927 to 1928.[34]

Turning to fascism, the pernicious idea that it was a movement of the petty bourgeoisie has finally been laid to rest.[35] The petty bourgeoisie was somewhat overrepresented in most fascist movements, and it is undoubtedly overrepresented in the Islamic movement in Iran. But it is overrepresented in all sorts of radical movements. We find the "little people" (*menu peuple*), in the religious riots in sixteenth century France on either side.[36] We find them among the stormers of the Bastille,[37] and as we have just seen, we find them among the nineteenth century radicals who, for E. P. Thompson, *made* the English working class. Recent studies clearly show that fascist parties were supported by elements from *all* social groups, but especially the dislocated, the dispossessed, and the declassed. What is more to the point, and not disputed, is that the *leadership* of the fascist movements came disproportionately from the déclassé and dispossessed: from demobilized army officers, displaced or unemployed bureaucrats—especially those dislocated by the redrawing of national boundaries, and the occasional dispossessed aristocrat. Nor did the Nazis fail to tap the traditional communal solidarities of the Protestant countryside.[38]

The Islamic movement in Iran and European fascism are similar in that they were led by dispossessed elements. But there are two important differences. The fascist leaders were a heterogeneous group whereas Khomeini's militant clerics formed a homogeneous solidary group. Secondly, the fascist leaders did not have exclusive control over any cultural assets and had to get their ideas where they could find them. The Shi'ite hierocracy were the custodians of a rich religious tradition. The consequences of these differences will become apparent presently.

Integrative Social Movements as Reactions to Social Dislocation

We can now turn to the social dislocation and moral disturbance consequent upon rapid social change as preconditions of revolution. Let us begin at the most superficial level.

The conspicuous consumption of high society and the abundance of the *nouveaux riches* gave plenty of cause for an acute sense of relative deprivation to the new middle class. To this was added at times the discomfort of absolute deprivation that resulted from an acute housing shortage, aggravated by the influx of a sizable foreign workforce and American advisers. In this context, it would be valid to speak of the widespread discontent of 1977 and 1978 as a confirmation of Davies's *J*-curve of continuous rising expectations and sudden frustration. However, it was argued in Chapter 6 that the problem was more deep rooted. What underlay the widespread desire for revolutionary change was fundamental disorientation and anomie, more than a superficial and short-run frustration of material expectation. As Durkheim pointed out, crises of prosperity generate disorientation by disturbing the collective normative order.[39] There can be no doubt about the tremendous confusion and disorder created by the massive inflow of petrodollars, as there can be little doubt about similarly caused current confusions in Nigeria and Mexico. The consequent sense of moral disorder and desire for the reaffirmation of absolute standards should not be minimized. There was widespread cultural malaise, ranging from general confusion and disorientation on the part of the *nouveaux riches* to sharply focused and intense rejection of foreign and antireligious cultural influences on the part of the mollahs and the bazaar.

Among the causes and preconditions of modern revolutions, national integration must be treated as analytically distinct from state formation, though concretely the two may go hand in hand and the latter may significantly contribute to the former, as was the case in twentieth century Iran. The political dimension of national integration consists in increased political awareness and mobilization that manifests itself in demands for inclusion in political society—citizenship—and for increased political participation. The striving for incorporation into the national political society, a distinct phenomenon of modern public politics, can be viewed in comparative perspective as a special case of the more general need for integration into societal community.

The socialist and fascist mass movements were part and parcel of the extraordinary wave of mass political mobilization and national integration that swept through Europe in the early decades of the twentieth century.[40] They acted recognizably as vehicles for the integration of the recently mobilized masses into societal community. However, one should not forget that religious movements have often performed the same function in a parallel manner.

Political mobilization comes about as a result of basic social change, which also entails considerable social dislocation. Social change dislodges a large number of persons from the strata into which they were born into new social and occupational groupings. These persons yearn for and demand inclusion into new forms of societal community. Religious movements and sects are the age-old channels for the integration of dislocated individuals into new communities, political movements and parties, the new channels. The Islamic Revolution demonstrates that the old and the new can combine.

Urbanization and the expansion of higher education in the fifteen years preceding the revolution are the two dimensions of rapid social change most relevant to our problem. Both these factors contributed significantly to the rise of the Islamic movement. Thousands of religious associations spontaneously came into being in cities and in universities and acted as the mechanism of social integration for a significant proportion of the migrants into the cities and of the first-generation university students. The Shah's parallel deliberate attempt to integrate these same groups into his one-party political system proved a fiasco.

There is nothing new about dislocated, uprooted men and women finding new moorings in religious associations, sects, and revivalist movements. In England, for instance, many masterless men became sectaries in the sixteenth and seventeenth centuries.[41] Already in the 1570s, we encounter the Presbyterian classes attended by laymen, but in the 1620s and 1630s, Puritan lectureships took root in towns to an astonishing degree, to the dismay of the Anglican Church. Laymen became patrons and paymasters of Puritan lecturers, and congregations clustering around the lecturers became "models for ideological party organization."[42] The story bears a striking similarity to the growth of lay religious associations in Iran in the 1960s and especially 1970s, where the mollahs preached—at first in person but later, when demand outstripped supply, through cassette players—to avid audiences of urbanites. We find an even closer parallel in the rise of Methodism. In the eighteenth and early nineteenth centuries, migrants into the new industrial towns of England flocked to the assemblies of the Methodist preachers. Viewed from the perspective of the need for integration into the societal community, the sociological cogency of Halévy's famous thesis becomes clear: the Methodist Revival integrated the recently urbanized masses into societal community and thus prevented a political revolution in England.[43]

Fascism, too, acted as the vehicle of integration of rural to urban migrants into societal community. In Germany, for instance, "many of the new urbanites failed to complete their cultural adjustment to city life and instead remained curiously vulnerable to the agrarian romanticism of *volkisch* ideologues."[44] One-half of the top Nazi party leaders were born in large villages.[45]

Literacy and Puritanism went hand in hand. The same is true of the growth of Islamic scripturalism. Islamic fundamentalism spread in Iranian universities, as Puritanism had spread in Oxford and Cambridge.[46] Many of the Islamic activists of the 1970s, who currently form the lay second stratum of the Islamic regime, discovered "the true Islam" in university associations, as Cromwell had been reborn in Cambridge. It is important to note that fascism spread in European universities in a parallel fashion. In Eastern Europe in particular, university students and young activists constituted the core of the fascist parties and their leaders. Rumanian fascism is in this respect of particular interest. Its leaders, Codreanu and Mota, in the early 1920s founded university associations for Christian reform and national revival in the universities of Iasi and Cluj respectively.[47]

The *combination* of higher education and social dislocation is of particular importance in explaining the politicization of integrative movements. The key to the social composition of Islamic and university activists of the 1970s is that they came to big city universities from small towns, or they were the first generation from their traditional lower-middle class background to attend a university, or both.[48] These young men crucially contributed to the revolutionary politicization of the Islamic revival of the 1960s and 1970s just as the educated country gentlemen in England had contributed to the revolutionary politicization of Puritanism. The parallelism with Rumanian fascism is even more striking. As the last Iron Guard leader, Sima, put it, "in 1926–27, our universities were flooded by a big wave of young people of peasant origin . . . who brought with them a robust national conscious-ness and were thus destroying the last strongholds of foreign spirit in our universities."[49] According to Weber "legionary leadership came from the provincial, only just urbanized intelligentsia: sons or grandsons of peasants, school-teachers, and priests."[50]

Max Weber once remarked that with the advent of modern mass politics, the conditions of clerical domination itself changes. "Hierocracy has no choice but to establish a party organization and to use demagogic means, just like all other parties."[51] Rapid urbanization and the Shah's failure to integrate uprooted elements, especially the socially mobile, newly educated elements, into his political system offered Khomeini and the cornered Shi'ite hierocracy an unparalleled opportunity for creating a politicized revolutionary mass move-ment. Using the organizational network of the lay religious associations and Islamic university students, the mollahs periodically organized massive anti-Shah demonstrations and closures of the bazaar, which amounted to a gen-eral strike of unparalleled duration. Perhaps it could even have brought down stronger regimes, but this we will never know. What is certain is that the clerically led general strike did bring down the fragile Pahlavi regime.

The Political and Moral Motives of the Supporters of Revolution and the Minor Significance of Class Interest

National integration engenders a widespread political motive for gaining power by inclusion in political society. This political motive is particularly strong for those social groups that are excluded from political society but can aspire to gain control of it through revolution. These political motives for supporting a revolution can become effective if the revolution offers the prospect of citizen-ship to the general population, and of rule to the aspiring social groups. In ad-dition, the political motive to retain or recover political power and institutional assets threatened or expropriated by the centralizing state can result in initiating or supporting a revolution. The moral motive for supporting a revolution may stem, negatively, from the moral condemnation of a regime because it is unjust, because it is servile to foreign powers, and/or because it is instrumental in spreading an alien culture and undermining authentic traditional cultural and religious values. The moral condemnation of the regime as unjust may in turn be due to its perception as tyrannical and/or it may be due to a sense of relative deprivation. Positively, the moral motive for supporting the revolution may re-sult from the acceptance of the modern myth of revolution as a redemptive col-lective act. Finally, class interest can act as a motive for supporting the revolution if the economic interests of a class, so defined by virtue of their position in the mode or system of production, are either protected or furthered by doing so. Within this schema, let us examine the motives that can be plausibly attributed to the social groups who supported the revolution against the Shah.

Political and moral motives are closely intertwined as regards the Shi'ite hierocracy. The primary material interest of the clerical leaders was to regain the prerogatives and functions they had lost as a result of the centralization and modernization of the state. This statement applies to the leading clerical militants who came from traditional urban backgrounds, were in their forties or fifties at the time of the revolution, and had a keen awareness of the dispossessions of the Shi'ite hierocracy by the Pahlavi state. The younger militant clerics, by contrast, were heavily drawn from humbler rural and small town backgrounds and saw all avenues of upward social mobility blocked for their profession under the Pahlavis.[52] An Islamic government would guarantee them rapid social ascent and full incorporation into the political system.

Both the clerical leaders and the militant seminarians were morally indignant at the spread of immorality, libertinism, and alien culture by the Pahlavi regime. In a significant statement, Khomeini's son identifies the conservative members of the Shi'ite hierocracy who supported the revolution against the Shah as those whose motivation was exclusively moral.[53]

The political and moral motives are also entwined as regards the intensely politicized lay Islamic activists. These first-generation provincial and lower-middle-class university students and graduates, mostly in applied sciences and engineering, also saw themselves barred from the westernized high society and high government positions. They were moved by the desire to remove these barriers to their upward social mobility. It would be absurd to attribute any class interest to this young "petty bourgeois" group other than the desire to gain power and entry into the political system, to move up on the social ladder, and to put an end to a cultural climate they found alien and deeply resented.

The motives of the new middle class were both political and moral. Many of the members of this socially mobilized class—and let us not forget the recently mobilized middle-class women who figured prominently in the anti-Shah demonstrations—wanted inclusion in the political society. They considered the Pahlavi regime tyrannical and unjust and accepted the myth of revolution. It should be noted, however, that the potency of the political myth of revolution made the new middle class, especially the middle-class women, suicidally join the Islamic revolutionary movement against their class interests.[54]

The traditional bourgeoisie—the merchants of the bazaar, the petty bourgeoisie of distributive trades, and the craftsmen of the bazaar guilds—was the one social group for which class interest was the primary motive for overthrowing the Shah. This class felt threatened by the developmental economic policies of the state, which among other things, excluded them from easy access to credit, and by the encroachment of the modern sector of the economy on its territory in the form of competing machine-made goods and new distributive networks of supermarkets and chain stores. To this motivating class interest was added a sense of relative deprivation caused by the tremendous gains made by industrialists connected with the Pahlavi court and considerable moral indignation caused by the disregard of Islam and traditional values under foreign cultural influence.

Teleology of the Islamic Revolution

Integrative Revolution, Moral Rigorism, and the
Search for Cultural Authenticity

As an integrative revolution, the Islamic Revolution in Iran has greatly enlarged the political society and increased political participation, especially in small towns. In doing so it has accommodated the political motive of the urban masses of incorporation into the political society. It has also accommodated the political motive of two of the most politicized social groups that aspired to rule: the clerics and the lay Islamic intelligentsia. The *'ulama,* who had a strong motive to regain their lost power, have done so. It is remarkable that intellectuals, lay and clerical, constituted over two thirds of the representatives elected to the First Majles in 1980. (See Appendix, Table 16.)

The fact that integrative social movements are reactions to social dislocation and normative disorder explains the salience of their search for cultural authenticity and their moral rigorism. "Fascism was a revolution, but one which thought of itself in cultural, not economic terms."[55] The same is true of the Islamic Revolution, which saw itself even more emphatically in those terms, even when not explicitly so, as in the Islamic cultural revolution against Westernism and (Eastern) atheistic communism that was inaugurated with the closure of the universities in April 1980. More fundamentally, Iran's secular judiciary system has been systematically Islamicized, the Shi'ite Sacred Law codified for the first time in history, and Islamic morals, including coverage of women, strictly enforced by especially created official vigilante corps.

Disoriented and dislocated individuals and groups cannot be successfully integrated into a societal community without the creation or "revitalization" of a moral order.[56] Walzer emphasizes that Puritanism was primarily a "response to the *dis*order of the transition period." Ranulf correctly underscored the moral rigorism of Nazism and compared it to Puritanism.[57] The intense and repressive moralism of the Islamic revolutionaries in reaction to the moral laxity and disorder of Pahlavi Iran is strictly parallel to Puritan moralism in reaction to the moral laxity and sensuality of the Renaissance culture and the Nazi moralism in reaction to the decadence of Weimer culture. Furthermore, the parochial rejection of cosmopolitanism is a common feature of the Islamic Revolution, Nazism, and especially of Eastern European fascism.[58] The vehement rejection of cultural Westernism in favor of revitalized Christianity in Rumania and Hungary finds a close parallel in

Khomeini's more systematic and successful determination to extirpate Western cultural pollution by establishing an Islamic moral order.

In remarkable contrast to the efficacy of political and moral motives in the teleology of the Islamic Revolution stands the inefficacy of class interest. As shown in Table 16 (see appendix), the bazaar, which had a salient class interest in overthrowing the Pahlavi regime, is minimally represented in the Majles. Furthermore, it has been fighting a losing battle against the statist economic policies of the Islamic government, which we have not had the space to discuss. It should also be noted that the one social class that has benefited from the policies of the Islamic government more than others,[59] the peasantry, did *not* participate in the revolution.

The Revolutionary Ideology and Its Adoption by Latecomers

Let us first consider the confused teleology of the revolutions of early modern Europe. They were made by men for whom restoration was the key word, and who "were obsessed by *renovation*—by the desire to return to an old order of society." These revolutions were marked by an absence of *ideology,* and by a corporate or national constitutionalism, "which was mainly the preserve of the dominant social and vocational groups."[60] As for the English Revolution, it is true that "with the nature, source, and grounds of political legitimacy all up for grabs, there was almost inevitably a great effusion of claims to legitimacy on all sorts of grounds, old and new."[61] Nevertheless, two elements predominate in the teleology of the English Revolution: parliamentarianism and Puritanism and its offshoots.

If the French Revolution remedied one thing for all the subsequent ones, it is the absence of ideology. It gave birth to Jacobinism as the classic form of modern revolutionary ideology. The ideas of constitutional representation and national sovereignty were coupled at the beginning. However, as the revolution progressed, the source of legitimacy drifted from representation of estates to the symbolic embodiment of the will of the people. The claim to embody the will of the nation as a single homogeneous entity could only be made through the manipulation of the maximalist language of consensus. Presumed embodiments of the will of the people became the sole and sufficient basis of legitimacy. During the period of Jacobin ascendency, revolutionary legitimacy triumphed, and with its triumph, revolutionary ideology "filled the entire sphere of power" and "became coextensive with government itself."[62] The distillation of the Jacobin experiment was the modern political myth of revolution. Revolutionary legitimacy became an autonomous and self-sufficient category.

In the nineteenth century, revolutions became "milestones in humanity's inexorable march toward true freedom and true universality."[63] With Leninism, this conception of revolution was coupled with the Jacobin myth and became the justification for the seizure of power by revolutionaries who proclaim themselves in charge of realizing the next stage of sociohistorical development.[64] With the consolidation of Marxism-Leninism in Russia, Leninist revolutionary ideology "obtained control over the interpretation of world history."[65] It is this control that is challenged by the fascist and the Islamic revolutionaries, while upholding, like the Bolsheviks, the myth of revolution as an act of redemption and liberation of oppressed masses and nations.

Both fascism and the Islamic revolutionary movement are *latecomers* to the modern international political scene. As such, they share a number of essential features. The foremost of these is the appropriation of the legitimatory political myth of revolution. The Italian Fascists boasted of their "revolutionary intransigence" and the Nazis contrasted their revolution, the revolution of the German *Volk,* to the "subhuman revolution" of 1789.[66] Iran's revolutionaries similarly take great pride in the historic mission of the Islamic revolution.

"Economics was indeed one of the least important fascist considerations."[67] The same is very much true of the Islamic revolution. Khomeini, responding to complaints about the state of the economy, once remarked that the Islamic Revolution was not made so the Persian melon would be cheap. (In fact, the Persian melon has not become cheaper; and per capita income in 1985 was about 30 percent below the prerevolution level.) Furthermore, like the European fascists, the Islamic militants aim at integrating all classes, including the working class, into a national community, albeit as just a first step toward their eventual integration into a universal Islamic community of believers. The fascists substituted "nation" for "class" and developed the conception of "the proletarian nations." Class conflict was thus replaced by the conflict *between* nations, rich and poor. With the Islamic revolutionaries in Iran, we have an identical transposition of the theme of exploitation of one class by another into the exploitation of the disinherited (*mustaz'af*) nations by the imperialist ones.[68]

"The fact that fascism is a latecomer," writes Linz, "helps to explain, in part, the essential anti-character of its ideology and appeal." Furthermore, "it is paradoxical that for each rejection there was also an incorporation of elements of what they rejected."[69] Like fascism, the Islamic revolutionary movement has offered a new synthesis of the political creeds it has violently attacked. Like the fascists, the Islamic militants are against liberal democracy because they dislike its secularity and consider it a foreign model that provides avenues for free expression of alien influences and ideas. Similarly, both groups are anti-bourgeois, resenting the international cosmopolitan orientation in the new middle class. Both movements are anti-Marxist, that is, anti-communist, while appropriating the ideas and certainly the slogans of social justice and equality. However, the Islamic revolutionary movement has the considerable advantage over fascism of combining this anti-character with strong traditionalism. Here we can see the consequence of the fact that the dispossessed leaders of the Islamic Revolution were not a heterogeneous but a homogeneous solidary group and, furthermore, one that guarded the Shi'ite religious tradition. In contrast to both the Nazi "Revolution of Nihilism" and the striking lack of reference to Japan's own intellectual tradition in the writings of the leaders of the fascist New Order Movement of the late 1930s,[70] the Islamic Revolution combines the rejection of other alien political ideologies with a vigorous affirmation of the Islamic religious and cultural tradition. I have therefore characterized it as "revolutionary traditionalism."[71]

In addition to their anti-character and other incidental features, fascism and the Islamic revolutionary movements both have a distinct constitutive core. Racism and anti-Semitism were the most obnoxious features of European fascism, but as Mosse and others have convincingly shown, not their core component. The constitutive core of fascism that transcended the European variant and continues to live in a variety of forms as a vigorous ideological force in the Third World is the combination of nationalism and socialism. As George Valois put it in 1925, "nationalism + socialism = fascism." The marriage of nationalism and socialism was inevitable after World War I.[72] This fact by far transcends the particular conditions of any dispossessed stratum, any European country, or for that matter, of inter-war Europe. It was arrived at by different fascist leaders in different European countries, and it has been arrived at independently by many Third World ideologues since 1945.

An enduring feature of fascist ideology has been its insistence on the reality of the nation and the artificiality of class. To the emotionally unattractive idea of perpetual class struggle, the French fascist thinker Marcel Dèat contrasts the appeal of belonging to a community untainted by divisive conflict and fragmentation: "The total man in the total society, with no clashes, no protrations, no anarchy."[73] The Arab nationalist thinkers sought to utilize the appeal of belonging to a community by similarly replacing class by nation. The advocates of Islamic ideology only needed to take one step further and to replace nation by the *umma,* the Muslim community of believers.

In the same way, the emergence of an Islamic revolutionary ideology has been in the cards since the fascist era, irrespective of the plight of the dispossessed Shi'ite clerical estate in Iran. The latter did have the advantage of institutional autonomy and of independence in the exercise of religious authority, something the Sunni Islamic ideologues like Rashid Rida could only dream of. But the clerical estate was exceedingly slow in creating a consistent and comprehensive ideology in order to defend itself against the state. In fact, as we have seen, the Islamic fundamentalist *ideology,* which condemned the secular state as an earthly idol claiming the majesty that is God's alone, was developed elsewhere by the Muslim Brotherhood and by publicists and journalists, such as Mawdudi in Indo-Pakistan and Qutb in Egypt. When Khomeini finally rose against the Shah, he imported this internationally current Islamic ideology as a free good. The categories of this fundamentalist ideology were combined with the specifically Shi'ite clericalist theory of the mandate of the jurist; its influence became more pronounced after the elimination of the moderates and Islamic modernists in 1980 to 1981 and is easily noticeable in the speeches of the political elite of the Islamic Republic of Iran. By 1982, Ayatollah Safi had no difficulty whatsover in combining the advantages of the ideologies of Mawdudi and Qutb with the clericalist ideas of Khomeini: The government of the jurist on behalf of the Hidden

Imam is the true government of God on Earth, vowed to the implementation of His Law. All other political regimes are ungodly orders, regimes of Ignorance, and of *taghut*. The Islamic revolution will continue until the overthrow of all these regimes.[74]

The Old and the New in Revolutionary Traditionalism and the Teleological Irrelevance of Progress

I hope enough has been said to demonstrate that the Islamic Revolution underlies the importance of reactive and reactionary elements in all revolutions. Both Nazi ideologues and historians of Nazism, notably Bracher, have insisted that it contained revolutionary and reactionary elements.[75] Mannheim showed that the ideology of proletarian revolution, too, incorporated the romantic, reactionary critique of the Enlightenment.[76]

The Islamic Revolution constitutes a wry comment on the debate among historians as to whether the early modern European revolutions were conservative or modern, reactionary or progressive. It also demonstrates that revolutionaries often act in defense of traditional values. Baechler is right when he notes, "contrary to appearances and accepted belief, conservative revolutions are supported less by the elite than by the people."[77] Not surprisingly, some important teleological elements in the clerically led popular movements such as Carlism and the Cristero movement[78] find resonance in the Islamic movement in Iran: repudiation of foreign and cosmopolitan influences and values, and vehement opposition to anticlerical policies of modernizing governments and, of course, to atheism. Marx's famous idea that the French revolutionaries parodied the Roman republicans because they had not yet developed a political language of their own should not automatically be generalized. The revolutionaries who draw on traditional imagery can vary greatly regarding their knowledge of and professional identification with tradition. The ayatollahs were the official custodians of the Shi'ite tradition and knew the methodology of Shi'ite jurisprudence. In the past six years they have proven this by their sustained efforts to Islamicize Iran's judiciary system, by institutionalizing very substantial political functions for the Friday prayer leaders, and by presiding over the strict enforcement of Islamic morals.

The Islamic revolutionary traditionalism does have its modern trappings. The Constitution of the Islamic Republic pays lip service to equality and especially social justice, and it guarantees freedom of the press, of the expression of political opinion, of political gatherings and groups—provided that they are not contrary to the interests of Islam. Finally, there is another modern element that is more than a trapping: the Majles. The constitutionalism of the early modern European revolutions was little more than the idealization of customary practice and was closely linked to the estates' aim of preservation of local liberties. In Iran, constitutionalism was imported as a panacea in 1906. Nevertheless, seventy years is a long time, and even the mollahs had used the constitutionalist ideology when opposing the Shah. The Majles is consequently an enduring feature of the Islamic regime. Its legislation, however, is rigorously supervised by the clerical jurists of the Council of Guardians. In addition, both the ruling clerics and the second stratum of the regime, consisting of lay Islamic activists, have a keen interest in technology. They love broadcasting and being televised and interviewed by the press, and they love organizing seminars and congresses and using modern-sounding phrases such as "political-ideological" bureaus.

In addition to the vitality of parliamentary legislation and effectiveness of parliamentary supervision of the executive branch of government, other features of the Islamic revolutionary ideology as a latecomer have been translated into the institutions and policies of the Islamic Republic of Iran. Perhaps the most telling of these is the commitment to rural development and improvement of the lot of the peasantry. The ideological derivation of the rural developmental policies of the Islamic government is clearly established by the fact that the peasantry did not participate in the Islamic Revolution.

When the notions of revolution and progress are linked, as has been done persistently since the nineteenth century, a line can be conveniently drawn between revolution and counterrevolution. The evidence offered in this book makes it impossible to draw such a line between revolution and counterrevolution. It has been pointed out that all revolutions contain counterrevolutionary elements. The reverse is also true: all counterrevolutions must incorporate revolutionary innovations in order to restore what they consider the traditional order. This is clearly the case with Islamic revolutionary traditionalism in Iran. As has been shown in Chapter 9, it has in fact brought about a revolution *in* Shi'ism.[79]

The Teleological Relevance of Religion

Comparative evidence not only requires that we sever the conceptual link between revolution and progress but also suggests that we may do better if we link revolution and religion. Religion was an important factor not only in the Puritan Revolution but in all early modern European revolutions except the Fronde.[80] Walzer is right when he says the Puritan Marian exiles of the 1550s were the forerunners of modern revolutionary ideologues.[81] But the same is true of the clerics of the Catholic League thirty years later.[82] In 1640 the Puritan preachers were calling the House of Commons God's chosen instrument for rebuilding Zion.[83] But their Catholic counterparts were equally engaged in revolutionary activity in Catalonia in the same year, as seen in the commander of the Spanish king's forces in Rossello complaining of the sedition and licentiousness of the clergy:

> In the confessional and the pulpit they spend their entire time rousing the people and offering the rebels encouragement and advice, inducing the ignorant to believe that rebellion will win them the kingdom of heaven.[84]

There are striking parallels between the Puritan Revolution and the Islamic Revolution. For Cromwell as Moses we have Khomeini as Abraham and Moses all in one; for the Puritan saints we have the militant mollahs;

and for the fast sermons of 1642 to 1649[85] we have the gatherings at forty-day intervals in 1978 to 1979 to commemorate the "martyrs" and, since the revolution, the Friday sermons at congregational prayers. However, there are important differences that affect the teleologies of these respective revolutions. There were strong anarchic elements in Puritanism, especially Independency, as the true Church within the corrupt church. Such anarchic innerworldly millenarian precepts of the Independents militated against their acceptance of a presbyterian national church government. They could also lead in the direction of the Levellers' conception of man as a rational being in the image of God and hence to natural rights. The corporate solidarism of the militant Shi'ite *'ulama* contrasts strongly with the factionalism of the Puritan saints, as their methodologically grounded legalism contrasts with the saints' millenarian idea of Christ as the Lawgiver. Although Shi'ite millenarianism played an important role in the Islamic revolution, it did not have any of the divisive and anarchic consequences of Puritan millenarianism because the clerics were firmly in control of its interpretation and, in fact, partly derived their legal/juristic authority from it. Finally, the revolutionary Shi'ite clericalist theory of the mandate of the jurist is in sharp contrast to the idea of congregational representation—note Presbyterianism.[86]

The situation seems different as regards modern revolutions, and it is. But let us see how. Tocqueville knew that the French Revolution had produced a new religion. The French Revolution aimed at "nothing short of a regeneration of the whole human race. . . . It developed into a species of religion, if a singularly imperfect one, since it was without a God, without a ritual or promise of a future life. Nevertheless, this strange religion has, like Islam, overrun the whole world with its apostles, militants and martyrs."[87] The terms "secular religion" and "political religion" have aptly been used to describe communism and fascism.[88] Modern revolutions *do* require political religions. The crucial issue is this: Is there any necessary incompatibility between religion and political religion?

The Bolshevik Revolution was militantly atheistic. But before we draw any conclusions, let us think of its totally imported ideology, and of the exceedingly narrow social base of its political elite. What about the French Revolution? Tocqueville did not see any incompatibility between Christianity and the political religion of the revolution. Anticlericalism and the campaign against religion stemmed from the identification of the Church with the ancien régime and not from any widespread anti-Christian sentiment.[89] What about fascist revolution? European fascism was often associated with anticlericalism. But this association is neither general nor fundamental. The Nazis glorified the mythical pre-Christian German tradition and were antireligious. The same is true of other fascist movements in Western and Northern Europe. At the other end of the spectrum, however, the Rumanian, the Hungarian, the Slovak, and the Croatian fascist movements were emphatically Christian and aimed at establishing Christian corporatist states.[90]

The Slovak Republic established by Father Hlinka's People's Party, presided over by Father Tiso, and the Ustasha movement in Croatia offer interesting parallels as regards clerical leadership and participation.[91] But the most illuminating parallel is between Shi'ite revolutionary traditionalism and the Rumanian Iron Guards, the Legion of Archangel Michael. Both movements are characterized by extraordinary cults of suffering, sacrifice, and martyrdom. Priests figured prominently in the legionary movement, side by side with university students. Legionary meetings were invariably preceded by church service, and their demonstrations were usually led by priests carrying icons and religious flags. The integral Christianity of the Legionaries differentiated them from the Nazis and the Italian Fascists. This they knew. As one of their leading intellectuals explained, "Fascism worships the state, Nazism the race and the nation. Our movement strives not merely to fulfill the destiny of the Rumanian people—we want to fulfill it along the road to salvation." The ultimate goal of the nation, Codreanu and others emphasized, was "resurrection in Christ."[92]

Finally, we must consider Brazilian Integralism, the most important fascist movement in Latin America. Its founder, Plinio Salgado met Mussolini in 1930. The meeting made a deep impression on him; he certainly saw no incompatibility between the fascist political religion and Catholicism. He returned to Brazil to "Catholicize" Italian fascism. Taking advantage of an extensive network of lay religious associations, which had been brought into existence by Cardinal Leme, he founded the Brazilian Integralist Action with the aim of creating a corporatist, integralist state. Integralism appealed to Catholic intellectuals by its promise of a "spiritual revolution," and of an integral state, which comes from Christ, is inspired in Christ, acts for Christ, and goes toward Christ." Salgado accordingly criticized the "dangerous pagan tendency of Hitlerism" and lamented the lack of a Christian basis in Nazi ideology.[93]

I trust few would find the statement that the myth of revolution is a modern form of millenarianism objectionable. Russian communism was the secular millenarianism of the Third Rome, and Nazism was the secular millenarianism of the Third Reich, "the Thousand Year Reich of national freedom and social justice."[94] As was the case with religion and political religion, political and religious millenarianism are by no means mutually exclusive. The religious chiliastic element may appear predominant, as in the Taiping Rebellion, which aimed at establishing the Heavenly Kingdom of Great Peace,[95] or it may play an important subsidiary role as in the Puritan Revolution in England and the Islamic Revolution in Iran.

In the Puritan Revolution we encounter two forms of millenarianism: the milder, more innerworldly millenarianism of the Independent divines, and the better known, activist millenarianism of the Fifth Monarchy men. There can be no doubt that revolutionary political millenarianism played a crucial role in the motivation of the Iranian intelligentsia and other groups. But in addition, the Shi'ite doctrine contains an important millenarian tenet: the belief in the appearance of the Twelfth Imam as the Mahdi to redeem the world. This belief was most convenient for Khomeini's revolutionary purpose, as it had been for the founder of the Safavid Empire in 1501.[96]

Conclusion

The success of the Islamic revolutionary ideology is the novel and teleologically distinct mark of the Islamic Revolution in Iran. The Islamic revolutionary ideology is a powerful response to the contemporary politicized quest for authenticity. It has been constructed through the unacknowledged appropriation of *all* the technical advantages of the Western ideological movements and political religions with the added or rather the emphatically retained promise of otherworldly salvation. In a sense it has considerable ideological advantage over Nazism and communism, both of which clashed with religion. Rather than creating a new substitute for religion, as did the Communists and the Nazis, the Islamic militants have fortified an already vigorous religion with the ideological armor necessary for battle in the arena of mass politics. In doing so, they have made their distinct contributions to world history.

Appendix

Table 1. State Revenues in the Beginning of the Eighteenth Century

State	Estimated Population (in millions)	Tax Revenue (in millions of livres tournois)	State Taxes Per Capita (in livres tournois)
France	21	500–570	24–27
Spain	6	110 [about ¼ from the American Empire]	18 [14 with adjustment for empire]
England	5	70	14
Iran	6–10	36	3.6–6.0

Sources: T.M.: 185–86; Ardant 1975: 200; Cambridge Economic History of Europe, IV (1967): 65–67 for populations, 458 for exchange rates.

Table 2. Safavid Central Bureaucracy in 1722

Functions	Functionaries	Scribes
Grand Vazir and Vazir of Isfahan	2	?
Sadrs, in charge of religious courts and colleges, and administrators of endowments	5	?
Divanbegi, in charge of secular courts, and his subordinate the Ghassalbashi	2	?
Mollabashi	1	?
Constable of the Realm	1	—
King's Secretariat	6	10
State Secretariat	1	28
Taxation and Finance	18	51
Army Administration and Finance	12	17
TOTAL	48	(over) 106

Sources: T.M. and Dastur al-Muluk.

Note: Excluded from the list are the deputies of the Grand Vazir, the supervisors of various royal workshops and the proctor of government stores. All the employees under these functionaries are also excluded. Also excluded are the functionaries in charge of the administration of the capital, Isfahan, a large number of functionaries in charge of different sections of the king's household, various masters of ceremonies, the poet laureate, the royal astrologer and physicians.

The Vazir of Isfahan is included as he was also in charge of the reception and entertainment of the diplomatic envoys. In three instances of slight variation between the two sources, the figure for the number of scribes is taken from T.M.

Table 3. Composition of Iranian Population in the Nineteenth Century

Period	Urban	Rural Agrarian	Tribal Nomadic
1812	1/6	—	—
1850–60	1/5	under 1/2	over 1/3
1900	1/5	over 1/2	1/4

Sources: Issawi 1971: 26–28; Gilbar 1976: 145, 149; Amirahmadi 1982: 91; Sheil 1856: 393–402.

Note: The estimate for the mid-century tribal and nomadic population is on the basis of the numbers attached to Sheil's listing of the tribes (Issawi 1971: 28) in preference to his statement that the tribes constituted possibly half of the population (Sheil 1856: 393). The estimates of urban population around 1900 are just over a fifth, or some 21 percent.

Table 4. Qajar State Revenue from Taxation in Cash and Kind in the Nineteenth Century

Year	Revenue (in millions of *tumans*)	Revenue (in millions of £ sterling)
[1722]	[0.785]	[2.1]
1800–9	2 or 3	2 or 3
1833–34	2.46	1.23
1851–52	3.2	1.6
1867–68	4.9	1.97
1876	4.9	1.9
1888–89	5.5	1.65
1900	7.2	1.4

Sources: Malcolm 1820: 336, 343; Fraser 1834: 296; Sheil 1856: 388–89; Curzon 1892, 2: 481; Jamalzadeh 1916–17: 118; Issawi 1971: 336–37, 361; Keddie 1969: 8.

Note: The figure for 1833–34, as reported in 1836 (Issawi 1971: 361) is corroborated by Fraser's estimate. Sheil's figure, presumably for 1851–52, is corroborated by the figure of 3.37 million *tumans* given by Jamalzadeh for 1853–54. The figure for 1876 represents the average between Jamalzadeh's figure (5.07 m.) and Issawi's (4.75 m.). The figure for 1900 was given in pounds in the source and has been converted to *tumans* at the current exchange rate.

Table 5. Composition of the Annual Expenditure of the Iranian Government in Thousands of *Tumans* and Percentage of Total

Purpose/Year	1867–68	%	1884–85	%	1888–89	%
Army	1725	47	1900	40	1810	43
Administration	725	20	900	19	1050	25
Salaries of princes, ministers and government employees	(725)	(20)	(—)	(—)	(—)	(—)
Foreign Office	(—)	(—)	(750)	(16)	(100)	(2)
Other departments/employees	(—)	(—)	(150)	(3)	(150)	(4)
Pensions of government servants, etc.	(—)	(—)	(—)	(—)	(800)	(19)
Shah's court	500	13.5	900	19	500	12
Pensions and allowances	250	7	750	16	560	13
Clerics	(250)	(7)	(—)	(—)	(—)	(—)
Princes and nobles	(—)	(—)	(—)	(—)	(500)	(12)
Subsidy to Qajar tribe	(—)	(—)	(—)	(—)	(60)	(1)
Colleges	(—)	(—)	30	0.6	40	1
Other	500	13.5	270	6	263	6
TOTAL	3700	101	4750	100.6	4223	100

Sources: Curzon 1892, 1: 482–83; Issawi 1971: 32, 337.

Note: The figure of 150 under "Other departments/employees" for 1888–89 is taken from Curzon (1: 483) and comprises "Revenue collectors, writers, accountants, secretaries in various administrations, priesthood and clergy."

Table 6. Composition of the Revenues of the Iranian Government in Thousands of Current *Tumans*

Year	Revenue	%	Land Tax and Revenue from *Khaleseh*	%	Customs	%	Mint, Passport, Post Telegraph, Stamp, Tolls, Opium, Royalties, & Other	%
1888–89	5,537	100	4,618	83	800	14.5	119	2
1911 and 1912 (average)	9,939	100	2,294	23	4,751	48	2,894	29
1924–25	23,750	100	5,128	22	10,258	43	8,364*	35

Sources: Curzon 1982, 2: 481; Jamalzadeh 1916–17: 145; Mochaver 1937: 292.

* Of which 1,736 (7% of total revenue) from oil royalty.

Table 7. Selected Items Estimates of Government Expenditure in 1911

	In 1000 *tumans*	As % of total
Ministry of War	3,732	27
Pensions	3,000	22
Court	675	5
Welfare (Ministry of Education, Endowments, and Public Welfare)	308	2.3
Development (Ministry of Agriculture, Trade, and Industry)	28	0.2
TOTAL	13,619	100

Sources: The budget for 1328 (prepared by Sani' al-Dawleh and presented after his assassination): Issawi 1971: 368; and Shuster 1912: 308 for pensions.

Table 8. The Iranian Army in the Twentieth Century

Year	Size of the Army (in thousands)	Population (in thousands)	Armed Men (per 1000 persons)
1920	23	11,470*	2.0
1922	29	11,780*	2.5
1941	184	14,760*	12.5
1966	145*	25,789	5.5
1976	390*	33,709	11.5

Sources: Iranian Army, n.d.; *Salnameh,* 1353: 67; *National Census of Population and Housing,* 1976, vol. 186: xiv, 67.

* Estimates. The figure for those serving in the army are not given directly in the Iranian government statistics in the 1960s and 1970s but are included in the category of "Government Employees-Non-classified Occupations!"

Table 9. The Tribal Population of the Iran in Millions

Year	Total Population	Tribal and Nomadic Population	Tribal Population as % of Total
1900	10*	2.47*	25
1932	13*	1*	8
After 1941	15–16*	2*	12–13
1966	25.8	0.71	3
1976	33.7	0.35	1**

Sources: Bharier 1971: 26–27, 31; *National Census of Population and Housing,* 1976, vol. 186: xiv.

* Estimates.
** See n. 63, p. 230.

Table 10. Government Employees in Thousands (excluding the armed forces)

Year	Number	Total No. of Household in the Population	Per 1000 Households
1956 (1335)	[203]*	3,986	[51]*
1972 (1351)	856	5,905**	145
1976 (1355)	1,277	6,712	190
1983 (1362)	1,770	8,377**	211

Sources: Bill 1972: 63–65; *Salnameh,* 1353: 69; *National Census of Population and Housing,* 1956, and 1976, vol. 186: 67; *Salnameh,* 1363: 89.

* These figures are based on the internal statistics of the Ministries and very probably under-represent the total number of employees which are more accurately represented by the national statistical figures for the later years.
** Estimates.

Table 11. Urban Population of Iran

Year	The Urban Population of Iran	
	In Thousands	As % of Total Population
1956 (1335)	5,954	31.4
1966 (1345)	9,794	38.0
1971 (1350)	12,398	41.3
1976 (1355)	15,854	47.0

Sources: The Iranian national censuses of 1956, 1966, and 1976; *Salnameh,* 1353/1976.

Table 12. Internal Migration in Iran

Year	Total Population (in thousands)	Migrants	Migrants as % of Total Population
1956 (1335)	18,955	2,081	11.0
1966 (1345)	25,079	3,224	12.9
1972 (1351)	29,526	4,275	14.5
1976 (1355)	33,709	5,056	15.0

Sources: Same as those for Table 10.

Table 13. Enrollment in Institutions of Higher Learning

Academic Year	Number of Students
1970–71	74,708
1975–76	151,905
1978–79	175,675

Source: Salnameh, 1361/1982–83: 117.

Table 14. The Iranian Labor Force and (Inferred) Class Structure, 1956–77 (in thousands)

	1956	As % of Labor Force	1977	As % of Labor Force	Change from 1956 to 1977 (as % of Labor Force
Predominantly new middle-class sectors:	(852)	(14)	(1,907)	(18)	(+4)
Government services	248	4	780	7.4	+3.4
Banking and other services	582	9.6	1,040	9.8	+0.2
Utilities	12	0.2	65	0.6	+0.4
Oil*	(10)	(0.2)	(22)	(0.2)	(0)
Predominantly traditional bourgeois sectors:	(563)	(9.3)	(1,005)	(9.5)	(+0.2)
Commerce	355	5.9	725	6.8	+0.9
Transport and communications**	208	3.4	280	2.6	−0.8
Predominantly working-class sectors:	(1,167)	(19.2)	(3,513)	(33.1)	(+13.9)
Construction	336	5.5	980	9.2	3.7
Mining and manufacturing, including					
Industrial workforce***	—	—	(700)	(6.6)	—
Employed in handicraft and small workshops	—	—	(1,780)	(16.8)	—
Oil*	(15)	(0.2)	(33)	(0.3)	(+0.1)
Predominantly peasant sector:	(3,326)	(54.8)	(3,800)	35.8)	(−19)
Agriculture	3,326	54.8	3,800	35.8	−19
Wholly unemployed	158	0.3	375	0.4	+0.1

Source: Halliday 1979: 176–83.

* Figures for the employees in the oil industry are split on the basis of the assumption of the constancy of middle/working-class ratio as inferred from Halliday 1979: 179.

** This sector includes a large number of taxi, bus, and truck drivers. Though new occupations dating from the development of interurban bus networks in the latter part of Reza Shah's reign and the large-scale importation and later assembling of motorcars in Iran after World War II, drivers and operators of these transportation networks maintained their ties with the popular city quarters, and to some extent with the bazaar, and developed a distinct popular ethos.

*** As estimated by Halliday 1979: 182.

Table 15. Profile of the 24 Lay IRP Members, Belonging to the Second Stratum of the Islamic Republic of Iran, Who Perished in the Explosion at the IRP Headquarters in June 1981* (N = 24 for all calculations)

	Number	Percentage
Age at the time of assassination		
24–30 years	3	12.5
31–40 years	15	62.5
41–44 years	5	21
Not known	1	4
Educational attainment		
Bachelor's degree	9	37.5
Master's degree	7	29
Doctorate	5	21
Not specified [presumably high school]	3	12.5
Last university attended		
Iranian	13	54
Foreign	8	33.3
None	3	12.5
Field of specialization		
Medicine	2	8
Technical/Engineering	6	25
Economics, [business] administration and accounting	10	42
Other	3	12.5
None	3	12.5
Special features of career		
Prominence in organizing Islamic student associations in Iran and abroad	7	29
Self-taught knowledge of religious sciences	3	12.5
Political activism in the 1960s	1	4
None specified	13	54

Source: Special Supplement to *Ettela'at,* 7 Tir 1361 (June 28, 1982).

* Other "martyrs" of the explosion commemorated in the special anniversary supplement to *Ettela'at* can be grouped as follows: 15 clerical Majles deputies of whom 3 had doctorates; 13 lay Majles deputies of whom 3 had doctorates; 2 other clerics.

Table 16. Distribution of the Members of the First Islamic Consultative Assembly (Majles) in 1980

	Percentage
Status at the time of election (N = 216*)	
Clerical	47
Lay (of whom 34% had some formal religious education)	53
Place of birth (N = 214**)	
Cities†	20
Small towns††	62
Villages	16
Holy cities of Iraq	2
Occupation (N = 289***)	
Teachers and students at institutions of secular learning	39
Teachers and seminarians at institutions of religious learning	31
Professionals	17
Agriculture	4
Government employment	3.5
Trade (the bazaar)	1.5
Other	4
Father's occupation (N = 228****)	
Agricultural	30
Clerical	28
Trade and crafts	25
Government employment	4.5
Worker	4.5
Other	8

Source: Majles-e Shura-ye Eslami 1983: 118–89, 204–5.

* Excludes three members who resigned to take up positions in government.
** Two cases could not be identified.
*** More than one occupation computed in 73 cases.
**** More than one occupation computed in 12 cases.
† The ten largest cities, with populations ranging from 187,000 to 4.5 million in 1976.
†† Other towns, with populations ranging from 5,000 to 187,000 in 1976.

Notes

Abbreviations

E'temad (1)   E'temad al-Saltaneh, Mohammad Hasan (1970/1349) *Sadr al-Tavarikh*, Tehran: Vahid.

E'temad (2)   E'temad al-Saltaneh, Mohammad Hasan (1971/1350) *Ruznameh-ye Khaterat-e E'temad al-Saltaneh*, Afshar, I., ed., Tehran: Amir Kabir.

Hedayat (1)   Hedayat, Mokhber al-Saltaneh, M.-Q. (1965/1344) *Khaterat va Khatarat*, Tehran: Zavvar.

Hedayat (2)   Hedayat, Mokhber al-Saltaneh, M.-Q. (1984/1363) *Gozaresh-e Iran. Qajar va Mashrutiyyat*, Tehran: Entesharat-e Noqreh.

Kalantar   Eqbal Ashtiyani, A., ed. (1946/1325) [Mohammad, Kalantar of Fars] *Ruznameh-ye Mirza Mohammad Kalantar-e Fars*, Tehran: Sherkat-e Chap.

M.M.   *Mozakerat-e Majles-e Shura-ye Melli*, Tehran: Ruznameh-ye Rasmi-ye Keshvar-e Shahanshahi-ye Iran. 1946 onward.

Rostam   Moshiri, M., ed. (1969/1348) [Mohammad Hashem Rostam al-Hokama'] *Rostam al-Tavarikh*, Tehran.

Salnameh   *Salnameh-ye Amari-ye Keshvar* (Statistical Yearbook), Tehran: Sazeman-e Barnameh va Budjeh.

T.M.   Minorski, V. ed. (1943) *Tadhkirat al-Muluk. A Manual of Safavid Administration* (*circa 1137/1725*), London: Luzac and Co.

Yahya   Dawlatabadi, Yahya (n.d.) *Hayat-e Yahya*, Tehran: Ebn Sina. 4 vols.

Introduction

1. Sick 1985: 164–65.
2. The term was coined to describe the unceremonious undoing of the provisional prime minister, Mehdi Bazargan, by the clerical elite. See Chapter 7.
3. That is, a community based on shared values as distinct from primordial and local ties.
4. On the other hand, I have used the terms "cleric" and "clerical" despite their specifically Christian connotation. For the justification of the attribution of clericalism to the Shi'ite *'ulama*, see Arjomand 1984a: 138.

Chapter 1

1. All the above themes are treated at greater length in Arjomand 1984a.
2. Weber 1978: 1160.

222 *Notes*

3. See Keddie 1972; Amanat 1988.
4. Yahya, 1: 50–54, 134.
5. T.M.: 32, 33, 155; Persian text: 11. The alienation of the Qizilbash commander by the last Safavid monarch, Soltan-Hosayn, must be considered one of the causes of the military weakness of central government that made it vulnerable to the Afghan invasion. See Röhrborn 1966: 37.
6. Savory 1985: 891. In his search for military technology in the 1590s, Shah 'Abbas first sent an envoy to Russia and received the czar's envoys twice. In the last years of the decade, however, he opted for the English and received a military mission headed by Sir Robert Sherley (Yekrangiyan 1957: 9–10).
7. Savory 1980: 228.
8. Röhrborn 1966: 33, ch. 3.
9. T.M.: 175–82; Tapu Defteri, no. 904.
10. Barker 1944: 5.
11. Lewis 1973: 208.
12. Of the European states in that period, the less integrated and centralized Austrian monarchy, with a population nearly as large as that of France and a state revenue one-fifth as small—i.e., about three times the revenue of the Iranian state—comes closest to Safavid Iran from the fiscal point of view (Ardant 1975: 200).
13. Pages cited in Barker 1944: 7.
14. Lambton 1953: 121, 124–26.
15. Röhrborn 1966: ch. 1.
16. Lambton 1977: 108.
17. Kalantar: 17, 71–86; Rostam: 246–49, 186, 436–46.
18. Issawi 1977: 162; Perry 1979: 238.
19. Fasa'i: 61.
20. Lambton 1953: 131.
21. Kalantar: 24.
22. Kalantar: 26.
23. Perry 1979: 237.
24. Kalantar: 42–43, 79–80.
25. Kalantar: 20–21, 44.
26. Kalantar: 41.
27. Fasa'i: 21.
28. Kalantar: 84–85; Rostam: 447–49.
29. Fasa'i: 48.
30. Fasa'i: 44. A decade or so later, the same attitude is reported by Malcolm who observes that the word *shahri* (citizen) was "used in Persia as a term of contempt, to signify unwarlike [*sic*], the soldiers being all men of wandering tribes." Cited in Oberling 1974: 151.
31. Fasa'i: 64; emphasis added. In a letter to the governor of Georgia, Aqa Mohammad claimed sovereignty over Georgia *because Shah Isma'il the Safavid* had ruled over it. (Fasa'i: 66). On his claim to Safavid descent, see Rostam: 51 ff.
32. Bakhash 1971: 139. In the small size and circumscribed power of his bureaucracy, the first Qajar monarch followed the precedent of Nader Shah Afshar and Karim Khan Zand (Perry 1979: 218).
33. Lambton, EI, 3; 1104.
34. E'temad (1): 71, 145; Bakhash 1971: 140–43; Meredith 1971: 63–65. As late

as 1885, an Austrian administrator would write: "A Ministry in Persia consists of the minister and some scribes, without any determinate place of office, or any of the apparatus that appears indispensable to Europeans. The bureau is set up at whatever spot the minister happens to be, whether in his house, or in an ante-room, or a court of the Royal Palace, or perchance in the street or in a coffeehouse. A swarm of scribes buzzes after the chief on all his marches, each bearing with him in his pocket the necessary writing apparatus and documents" (cited by Bakhash).

35. Nader Shah, whose fiscal ruthlessness we have noted, is reported to have collected 2,950,000 *tumans* or 3.48 million pounds sterling in taxes annually (Pigoulevskaya et al. 1970, 2: 641; Issawi 1971: 387 for the exchange rate). However, the government revenue under the more humane and just rule of Karim Khan Zand can be estimated at about one-half million *tumans* or 500,000–600,000 pounds sterling. This estimate is arrived at on the basis of Perry's figures, using populations of districts with unspecified tax assessments as weights, the assumption being that they paid the same amount of tax per capita (Perry 1979: 230–31; source, Rostam).

36. Issawi 1977: 388–89, n. 46.

37. Amirahmadi 1982: 210.

38. At the height of Amir Nezam's centralizing reforms, in 1850–51, the remittance from the province of Fars was 168,000 *tumans* out of the total revenue of 400,000 (Adamiyyat 1969: 272).

39. Eqbal 1976: 177–78.

40. Eqbal 1976: 196.

41. In the mid-nineteenth century, Sheil estimated that "the ryots pay double the amount of their assessments" (Sheil 1856: 388–89). In 1881–82, an Austrian, Baron von Teufenstein, paid 25,000 francs for the office of the tax collector of Saveh and cleared 80,000 francs (Curzon 1892, 1: 442).

42. Sheikholeslami 1971.

43. This rate of rotation throughout the century seems about the same for Shiraz, the provincial capital of Fars (Fasa'i: 422–23), as for the less important city of Kashan (Zarrabi 1956: 404–7).

44. Eqbal 1976: 195.

45. Adamiyyat and Nateq 1977: 388–89.

46. Cited in Amanat 1983: xxii. See also Eqbal 1976: 181–93; Lambton 1953: 139, 148.

47. Lambton 1953: 131–32 and private communication.

48. In the tax assessment drawn up for the city of Kashan in 1817–18, all the revenue in kind and over one-half of the revenue in cash came from the royal domains. About one-fourth of the revenue came from customs and tax on silk and one-eighth for other agricultural land (Zarrabi 1956: 416).

49. Aqasi's other measures to increase the power of central government, most notably recovery of overdue taxes, were more difficult to implement and aroused the intense hostility of the notables (Amanat 1983: xxi–xxiii).

50. Lambton 1953: 152.

51. Lambton 1953: 152; Bakhash 1983.

52. Demorgny 1915: 31, 123–24.

53. Amirahmadi 1982: 152.

54. Lambton 1979: 1104, also 1107–8.

55. Sheil 1856: 384–85; Lorenz 1974: 189.

56. Curzon 1892, 1: 581.
57. Amirahmadi 1982: 236.
58. Sheil 1856: 385; Curzon 1892, 1: 392.
59. Eqbal 1961.
60. Curzon 1892, 1: 595. During a state visit to St. Petersburgh, Naser al-Din Shah took a liking to the uniform of the Cossack guard and invited a few Cossack officers to organize a regiment in Iran. The Iranian Cossack Brigade was commanded by Russian officers until 1920, and some 140 Russians continues to serve in it after the October 1917 revolution in Russia (Yekrangiyan 1957: 94–106).
61. Curzon 1892, 1: 416.
62. Zill al-Sultan 1907: 79; Sheikholeslami 1971: 106.
63. Rostam: 471. Rostam exaggerates a little. In reality, the number of royal progeny was more modest. The official court historian Sepehr lists the names of 158 wives, 60 sons, and 48 daughters (Sepehr, 2: 140–73).
64. Fraser 1838: 400–401.
65. Molkara: 169–70.
66. This occurred in April, 1885 (E'temad: 352; Bamdad 1968, 3: 15).
67. On the male favorites, the eunuch and other influential servants of the harem, see Bamdad 1968, 3: 5–7, 20–50, 219–23; Bamdad 1971, 5: 201–4.
68. Bamdad 1968, 3: 35–37.
69. E'temad (2): 534; Bamdad 1974, 6: 34–36.
70. E'temad (1): 298; Lambton 1970.
71. Curzon 1892, 1: 499–500.
72. Lambton 1970: 437–38; Adamiyyat and Nateq 1977: 422.
73. Bakhash 1978: 280–81.
74. Fraser 1834: 279.
75. As was the case with the execution of officials (see p. 30), the influence of the political culture of the "civilized" West affected these practices in the second half of the nineteenth century. Confiscation of the property of officials became less frequent, though it was still done in 1876 (E'temad (2): 15, 56). Sale and farming of offices, however, continued throughout the latter part of the century (e.g., Molkara: 168–69), albeit with uneasy conscience. In 1885, Naser al-Din Shah is reported to have become angry to hear the English newspapers had written that government offices are auctioned in Iran (E'temad: 352).
76. E'temad (1): 141; E'temad (2): 312. For the historic roots of the notion of *ne'mat* (benefit/benefaction) in Islamic culture, see Mottahedeh 1980.
77. Curzon 1892, 2: 484–85; Esfahanian 1975, 4: 100–101, 221–74; Adamiyyat and Nateq 1977: 428–29.
78. Bakhash 1978: 276.
79. The government expenditure figures reported for the province of Fars in 1851 show 12,000 *tumans* for public works or 3% of the total expenditure of 400,000 *tumans* (Adamiyyat 1969: 272).
80. Barker 1944: 6.
81. Barker 1944: 10, 20.
82. Barker 1944: 14–25.
83. Arjomand 1984: 98–100.
84. Rostam: 51–52, 57–63, 237–40; Arjomand 1988.
85. Garthwaite 1983: 54.

86. Meskub 1981: 104–6.
87. Rostam: 253.
88. Rostam: 309, 399. On the title of *vakil al-ra'aya,* see Perry 1978. Perry, however, overstates the extent of substitution of ra'aya for dawleh in Karim Khan's title.
89. E'temad (1): 137.
90. Adamiyyat 1969: 261.
91. E'temad (1): 261.
92. Bakhash 1978: 96.
93. E'temad (1): 297.
94. Sheikholeslami 1971: 107.
95. Bakhash 1978: 28–38.
96. Bakhash 1978: 134.
97. Donboli 1825–26: 132–43.
98. Donboli 1825–26: 138–39.
99. Algar 1969: 74.
100. Donboli 1825–26: 139.
101. Eqbal 1976: 208.
102. Lambton 1953: 164; Lorenz 1974: 188–89.
103. Sheil 1856: 384.
104. Adamiyyat 1969: 267–75; Lorenz 1974: 160–67.
105. Algar 1969: 131–33.
106. Adamiyyat 1969: 338–43.
107. Curzon 1892, 1: 590.
108. Bakhash 1978: 84–89. The specialized courts survived. However, there seems to have been no clear delimitation of the jurisdiction of the minister of commerce and the specialized court of commerce (Molkara: 169–71).
109. Bakhash 1978: 78–80, 96–103.
110. Oberling 1974: 63, Yekrangiyan 1957: 141–43.
111. Bakhash 1978: 104, 112.
112. Bakhash 1978: 150, 160–61.
113. Bakhash 1978: 279.
114. Bakhash 1978: 261–75, 287–92.
115. Bakhash 1983.
116. Issawi 1971: 370.

Chapter 2

1. Issawi 1971: 70, 335.
2. Amanat 1983: xiv–xix.
3. Issawi 1971: 22.
4. Abrahamian 1982: 58–59.
5. Lambton 1958: 46.
6. Abrahamian 1982: 78.
7. Browne 1910: 420, n. 13; Lambton 1958.
8. Tocqueville 1955: 176–77.
9. Browne 1910: 112–18; Keddie 1969: 246–47.
10. Gilbar 1977: 296.
11. Yahya 2: 22–23. The Constitutionalists on occasion showed considerable

astuteness in securing the government's rejection of particularistic clerical demands. See Malekzadeh 2: 69.

12. Cited in Gilbar 1977: 198, n. 68.
13. Browne 1910: 353.
14. Theqat al-Eslam 1976: 11–13.
15. Theqat al-Eslam 1976: 18–24.
16. Lambton 1965a: 651. Twenty-two people were killed in one bloody clash with the Cossacks in the summer of 1906. See Abrahamian 1982: 83. It is not at all uncommon for the first phase of revolutions, the phase preceding the emergence of multiple sovereignty, to be much less violent than the subsequent phases. See Rule and Tilly 1975.
17. M.M., I, Pt. 1: 131–39. Popular associations were permitted by Article 21 of the Supplementary Fundamental Law.
18. Malekzadeh, 3: 82.
19. Hedayat (1): 151.
20. Lambton 1963: 57–58.
21. Hedayat (2): 188; Lambton 1963: 62.
22. The assemblies are said to be "mostly destructive of the Majles," and like the agitation over the Supplementary Fundamental Law, "all obstacles to acting to attain the fundamental objective." (Hedayat [1]: 151) By December 1907, Sa'd al-Dowleh had defected from the constitutionalist camp. It is worth noting that in addition to him, the well-known Constitutionalists Moshir al-Dawleh and Mo'tamen al-Mulk also served as ministers under the restored autocracy, June 1908–July 1909. See Nava'i 1977: 66.
23. Yahya, 2: 202–3; Malekzadeh, 3: 149, 155, 238.
24. McDaniel 1974: 78.
25. Lambton 1963: 77–78.
26. Except for a brief flurry of activity in November–December 1911, before the closure of the Majles in response to Russian pressure. See Shuster 1912: 176–77.
27. Yahya, 3: 141; Bahar 1978: 8–9; Khal'atbari 1983: 144–49.
28. Lambton 1963: 88.
29. Nava'i 1977: 7–12.
30. Adamiyyat 1976: 433–46. The connection in the mind of the Constitutionalists between the establishment of the national bank and the realization of national economic sovereignty is quite clear. It was proposed by the prominent merchant Mo'in al-Tojjar Bushehri in response to the first request by the government for the approval of a foreign loan. The decree ordering its establishment, issued in January 1907, included the following points: only Iranian nationals could hold its shares; and it was given the concession to issue notes upon the expiration of the current concessions held by the (British) Imperial Bank, the concession to explore underground and underwater resources and the concession to build railways and roads (Hedayat [2]: 175–76).
31. Malekzadeh, 2: 18–19, 104–5, 142–73.
32. M.M. I, Pt. 1: esp. 26–27, 44–46.
33. Shuster 1912: 333–34.
34. Shuster 1912: 290.
35. Shuster 1912: 304. On the inflated army payroll and bogus military expenditure, see also Keddie 1969: 10.
36. McDaniel 1974: 76.

37. Shuster 1912: 308. The estimate of 100,000 seems correct. In 1907, the Financial Committee of the Majles considered about 100,000 cases or items of expenditure. See Nava'i 1978: 118.

38. Ardant 1975: 221.

39. An income tax was reintroduced in 1843, accounting for another 10% during remaining years of the 1840s. Mitchell and Dean 1962: 387; Mitchell 1981: 744.

40. Most notably merchants Mo'in al-Tojjar and Amin al-Zarb, *mojtahed* Behbehani, and bureaucrats Mokhber al-Saltaneh and Vothuq al-Dawleh. See Yahya, 2: 125; Nava'i 1978: 210.

41. The Majles approved sending a telegram of welcome to the *anjoman* of the port of Anzali, where Amin al-Soltan wanted to land from his trip to Europe, with a vote of 77 to 4, with 9 abstentions. See Nava'i 1978: 127.

42. *Dastur al-'Amal-e Shahaneh*, Tehran, 1907.

43. Keddie 1969: 155.

44. When the Shah received Mokhber al-Saltaneh and Sani' al-Dawleh the evening before the assassination of Amin al-Soltan, he was extremely agitated and asked if they had come to demand his abdication. See Hedayat (2): 194.

45. His assassin was presumed to have been a certain 'Abbas Aqa, member of a radical *anjoman* who was found dead not far from the scene of assassination. However, Mokhber al-Saltaneh believes Amin al-Soltan was assassinated by an agent of the Shah. See Hedayat (1): 157–58; Nava'i 1978: 121–35.

46. Safa'i 1966: 9–11.

47. Nava'i 1978: 117.

48. Safa'i 1966: 12–13.

49. M.M. I, Pt. 1: 11, 26–27. Taqizadeh became the chairman of the Committee for Financial Laws in February 1910. Vothuq al-Dawleh subsequently served as minister of finance from December 1909 to July 1910. See Nava'i 1977; Nava'i 1978: 117–18.

50. M.M. I, Pt. 1: 94–163.

51. Adamiyyat 1976: 446–50, exp. 449.

52. Hedayat (2): 200; Nava'i 1977: 30; Nava'i 1978: 110–19.

53. Hedayat (1): 159.

54. Sani' al-Dawleh 1907: 11. The resonance of the German writings on public finance in Sani' al-Dawleh's tract is clear. Von Stein held the opinion that "The State personifies . . . society and endows it with the power of volition and action." He also spoke of "the organic circuit in the inner most life of the state: tax potential creates the tax, the tax creates administration and administration in turn creates tax potential" (Musgrave and Peacock 1958: 19, 36). In the same vein, Wagner put forward his "law of increasing expansion of public, and particularly state, activities" in progressive countries" (Musgrave and Peacock 1958: 8).

55. Sani' al-Dawleh 1907: 17.

56. Adamiyyat 1976: 452–53.

57. Adamiyyat 1976: 454–55. He was accused of rebating the tax arrears of the Shah's reactionary Minister of Court, Amir Bahador-e Jang.

58. Malekzadeh 4: 239.

59. On May 18, 1911. See McDaniel 1974: 123; Yahya 3: 158.

60. Hedayat (1): 226; Yahya 2: 211; Shuster 1912: 33. The cost of its collection led to the repeal of the salt tax in June 1911.

61. Nava'i 1977: 188–91.
62. According to Mokhber al-Saltaneh, the assassins were instigated by the former prime minister, Sepahdar, and Sardar Mohyi, whose request for 20,000 *tumans* for his armed retinue Sani' al-Dawleh had turned down (Hedayat [1]: 222).
63. McDaniel 1974: 74; Adamiyyat 1976: 441.
64. McDaniel 1974: 90–92.
65. Shuster 1912: 49–50.
66. Shuster 1912: 306–12.
67. Shuster 1912: 304.
68. Shuster 1912: 70, 175, 190. Shuster envisioned a force of ten to twelve thousand eventually.
69. Shuster 1912, esp. 53, 67–68, 79, 142–46, 158–68, 210; McDaniel 1974: 190–93.
70. Shuster 1912: 115–16. In the early days of November 1911, Shuster sent an official with five treasury gendarmes to confiscate the property of Prince 'Ala' al-Dawleh, a notorious tax-evader. The prince immediately called the prime minister, Samsam al-Saltaneh, who was so touched by the plight of a fellow grandee threatened by fiscal reform as to exclaim in the Lori dialect of his tribe: "Has the world come to an end that 'Ala' al-Dawleh has to pay taxes? They want to confiscate his house. Tomorrow they will come to us" (cited in Yekrangiyan 1957: 150; see also Shuster 1912: 157–59).
71. McDaniel 1974: 197–98; Garthwaite 1983: 123.
72. Khal'atbari 1983: 58–106.
73. Garthwaite 1983: 116–19.
74. Browne 1910: 300; Hedayat (1): 239.
75. Hedayat (1): 214; Shuster 1912: 79.
76. Shuster 1912: 115. On this basic contradiction, see Klein 1980: 60–61, 67.
77. Hedayat (1): 142. It is instructive to note that, the government finances being what they were, the expenses for the gathering were paid by Mokhber al-Saltaneh who was the director of the military academy out of his own pocket.
78. Hedayat (2): 201.
79. About one tenth of the entire body of the Iranian officers of the brigade.
80. Yekrangiyan 1957: 130.
81. M.M., II: Session 208.
82. McDaniel 1974: 146–47, 175; Hedayat (2): 291; Yahya 3: 242.
83. Hedayat (1): 164–67; Yahya 2: 205, 213–16.
84. Hedayat (1): 165.
85. Hedayat (2): 208.
86. M.M., II: Session 246.
87. E.g., M.M., I, Pt. 1: 23, 151–52, 165. Bread remained on the agenda of the Second Majles also. Meanwhile, the municipal assemblies took great interest in the provision of foodstuff (Malekzadeh 3: 175).
88. M.M., II: Session 200.
89. Hedayat (1): 147–49.
90. McDaniel 1974: 67.
91. Kermani, 1: 321–27.
92. Malekzadeh 2: 176–77. For the English translation of the royal decree and its confirmation, see Lambton 1965a: 651.
93. Kermani, 2: 66–67.

94. Yahya 2: 108–9.
95. Kermani 2:85.
96. Kasravi 1: 224–26.
97. Kasravi 1: 281–85.
98. Rabino 1973: 12–13, 103–4.
99. Vijuyeh 1976: 14.
100. Kasravi 1: 370–75.
101. Yahya 2: 22, 24.
102. Yahya 2: 113–14.
103. Kasravi 1: 364, 375–76.
104. Rezvani: 168–73.
105. Rezvani: 185.
106. Rezvani: 171.
107. Rezvani: 170.
108. Rezvani: 199–200.
109. Rezvani: 169–70.
110. Rezvani: 200, 205.
111. Rezvani: 171.
112. Rezvani: 200.
113. Rezvani: 164; Yahya 2: 131. The majority of the deputies, according to Kasravi (Kasravi, 1: 569).
114. Najafi-Quchani: esp. 325, 460; Olfat. A copy of the Fundamental Law was first brought for endorsement to Hajj Mirza Hosayn, son of Mirza Khalil, then taken to Khorasani. Yazdi, seeing the seal of Khorasani and having heard that Hajj Mirza Hosayn had thought his sole endorsement sufficient, declined to endorse the document. Only then was it taken for endorsement to Mazandarani instead (Najafi-Quchani: 366).
115. Olfat. See also Najafi-Quchani: 460–62.
116. Kasravi 1: 362.
117. Lambton 1963: 50–51.
118. Kasravi 1: 505 ff.
119. See Chapter 4, pp. 79–80.
120. Yahya 2: 153.
121. Hedayat (1): 163; Kasravi 1: 526.
122. Vijuyeh 1976: 20–23.
123. Yahya 2: 260–61.
124. Tavakkoli 1949: 57.
125. Kermani 2: 166, 169–70.
126. Kermani 2: 183. Nuri was to repeat his *takfir* of the (Constitutionalist) Akhund in January 1909 (Kermani 2: 274).
127. Kermani 2: 241–42.
128. Zahir al-Dawleh 1972: 401–7.
129. Zahir al-Dawleh 1972: 414–15; Kermani 2: 265–70. He was 'Ali Aqa Yazdi, the father of the mastermind of the coup of February 1921, Sayyed Zia al-Din.
130. Kermani 2:266–67.
131. This is clearly indicated by the very drastic drop in the volume of cotton and tea traded with Britain and India, especially during the first six months of 1909. See Oberling 1974: 85, n. 7.

132. This was contradicted by Sayyed Mohammad Kazem Yazdi (Farid al-Mulk: 311). However, it is not difficult to imagine which of the conflicting injunctions the taxpayers would find more palatable.
133. Kermani 2: 306.
134. Kermani 2: 326–91.
135. Kermani 2: 404.
136. See, for instance, Browne: 407–8.
137. Kasravi 2: 10, 25–26; Fakhra'i, 1973: 101–4.
138. McDaniel 1974: 93–94.
139. In November 1909, as the most powerful member of Sepahdar's cabinet and the Minister of the Interior, Sardar As'ad appointed his brothers and cousins as provincial governors in the south. He also appointed the brother and enemy of the Qashqa'i chief, Sawlat al-Dawleh to the governorship of Shiraz (Garthwaite 1983: 119).
140. Safa'i 1966: 10.
141. Safa'i 1966; Oberling 1974: 96.
142. McDaniel 1974: 104–5.
143. McDaniel 1974: 137.
144. Garthwaite 1983: 122.
145. McDaniel 1974: 139, 142.
146. McDaniel 1974: 136, 140.
147. McDaniel 1974: 155, 158.
148. McDaniel 1974: 167.
149. McDaniel 1974: 158–59.
150. Malekzadeh 2: 26–28, 72–73, 170–72, 193–94.
151. Khalkhali 1907: 34, 43, 51–54.
152. Adamiyyat 1976: 259–67; Hairi 1977.
153. According to Mr. Badr al-Din Ketabi, Olfat on a number of occasions recalled the Akhund's words on the purpose of his proposed journey to Tehran in December 1911 before his unexpected death: "I am going to bring down the ass I have put on the roof" (*miravam an khari ra keh bala-ye bam bordeham pa'in biyavarem*). This is supported by Aqa Najafi-Quchani's account where the Akhund is described as a moderate constitutionalist as opposed to a revolutionary or a democrat (Najafi-Quchani: 458–59).
155. Yahya 2: 213–16; Hedayat (1): 164–67.
156. Zahir al-Dawleh 1972: 438; Yahya 3: 126–28.
157. Kasravi 2: 130; Hairi 1977: 115.
158. Garthwaite 1983: 114.
159. Yahya 3: 215–16.
160. Garthwaite 1983: 124.

Chapter 3

1. Naraqi 1976; Yahya, 4: 99–100, 116–22; Bahar 1978: 29–36.
2. Ivanov 1976: 32–52.
3. Yahya, 4: 123–49.
4. Yahya, 4: 221, 227–28.
5. Yahya, 4: 224.
6. Yekrangiyan 1957: 103–15.
7. Yahya, 4: 233–39, 245; Hedayat (2): 367.

8. Avery 1965: 254–55.
9. Yekrangiyan 1957: 171–75.
10. Hedayat (1): 331–34. Lahuti apparently had plans for a coup d'etat on the anniversary of the February 21, 1921, coup, but the agitation over the non-payment of salaries forced his hand. Yekrangiyan 1957: 340.
11. Hedayat (1): 349.
12. Iranian Army publications, n.d.
13. Yahya, 4: 252–57; Avery 1965: 258–59.
14. Hedayat (1): 350–52; Yahya, 4: 284–85.
15. Hedayat (1): 355, 358–60.
16. Bahar 1984: 27–63.
17. Yahya, 4: 343.
18. Being of humble origins, Reza Khan was not known by a family name. A certain Mirza Mahmud Khan Pahlavi, whose family name he fancied, was forced to resign [*sic*] this surname so that Reza Khan could assume it (Yahya, 4: 390).
19. Abrahamian 1982: 120, 132.
20. See Chapter 4, p. 81 and n. 24, p. 231.
21. Yahya, 4: 339. Mossadeq al-Saltaneh opposed the elevation of Reza Khan to kingship because, as a constitutional monarch, he would have to cease being the head of government and the country would thus be deprived of his considerable administrative ability. See Bahar 1984: esp. 351–55.
22. Yahya, 4 is particularly illuminating on this point.
23. Yahya, 4: 319; Hedayat (1): 372. There was some posthumous justice in that the plan for railway construction was introduced to the Majles in February 1927 by Sani' al-Dawleh's brother, Mokhber al-Saltaneh, now Reza Shah's prime minister.
24. Hedayat (1): 351, 361.
25. Oberling 1974: 151.
26. Ivanov 1976: 84–85; Tapper, 1983: 26–27.
27. Hedayat (1): 384–85.
28. Bavar, 1945: 73–75; Yekrangiyan 1957: 403–5; Oberling, 1974: 163–64.
29. Oberling 1974: 164.
30. Razmara 1941a: 99–101; Razmara 1941b: 8–25; Douglas 1951: 104–9.
31. Yekrangiyan 1957: 379–85.
32. Bavar 1945: 76; Oberling 1974: 154, 166–67.
33. Oberling 1974: 165.
34. Hedayat (1): 403. He was the son of the Constitutionalist Sardar As'ad, 'Ali-qoli Khan. Qavam al-Molk, the chief of the Khamseh tribal confederacy, also befriended by Reza Shah in accordance with his early divide and rule policy, was arrested with Sardar As'ad, but his life was spared. His son later married one of the Shah's daughters.
35. Hedayat (1): 386; Razmara 1941b: 24; Bavar 1945: 75–76; Yekrangiyan 1957: 388, 413.
36. Razmara 1941b: 27–35; Tapper 1983: 27–28.
37. Lambton, EI, 2: 656.
38. Avery 1965: 285.
39. Oberling 1974: 150, 153.
40. Yekrangiyan 1957: 391–92.
41. Oberling 1974: 156–62, esp. 162, n. 50.

42. Oberling, 1974: 164.
43. Vothuq: 35–36.
44. Hedayat (1): 371.
45. Bharier 1971: 62–78.
46. Barker 1944: 19, 61.
47. It should be noted, however, the National Bank was run by a cadre of seventy Germans including its director (Hedayat [1]: 390).
48. Elwell-Sutton 1941: 115–16.
49. Elwell-Sutton 1941: 78–79; Banani 1961: 59–60.
50. Greenfield 1934; Banani 1961: 68–80; Floor 1983: esp. 132.
51. Banani 1961: 80–84.
52. Reza Khan personally held the command of the northern brigade until he became Shah (Yekrangiyan 1957: 310). Then after, he frequently used army officers in management of his expanding estate in the north (n. 43 above).
53. Bharier 1971: 63–64.
54. Elwell-Sutton 1941: 108–11.
55. Bharier 1971: 194–96.
56. Sotoudeh 1936: 113–14.
57. Hedayat (1): 378. Tahmasbi had been Reza Shah's superior in the Cossack Brigade. The rumor spread that the Shah may have had a hand in his elimination. Either way, his being shot was no accident!
58. Abrahamian 1982: 144–45. In addition, some 150,000 adults were enrolled in the special adult literacy schools (Elwell-Sutton 1941: 143).
59. Hedayat (1): 371; Yahya, 4: 307.
60. Hedayat (1): 397–98.
61. Abrahamian 1982: 178.
62. Tapper 1983: 29.
63. The total number of tribal families in 1985 is given as 188,570. This would mean a total population of about 950,000 or over 2% of the population (*Kayhan Hava'i,* 29 January 1986).
64. Kanun-e Vokala' 1965.
65. Abrahamian 1982: 121.
66. Abrahamian 1982: 149–50.
67. Lambton 1953; Ivanov 1969: 71.
68. Razmara 1941b: 35; Lambton 1953: ch. 15.
69. Bharier 1971: 89–97.
70. Early in 1945 in connection with Millspaugh (Hedayat (1): 437).
71. Abrahamian 1982: 267–80.
72. Keddie 1981: 156.
73. Ashraf 1982: 1, 11–12; Hooglund 1982: 79, 81; Lambton 1984: 76–77. The destruction of the peasant-landlord relationship was completed in the 1960s, during the second and third phases of the reform, with the schemes for division of land between peasants and landlords. Though the redistributive effect of these phases was negligible, their sociopolitical effect in breaking the traditional links between peasants and landlords was profound.
74. Razavi and Vakil 1984: 28–35.
75. Kazemi 1980: 13.

Chapter 4

1. Weber 1978: 1163.
2. It is thus even possible to present Pope Gelasius, the author of the famous letter of 494 which contains the fundamental statement of the Christian theory of the two powers, as a monistic hierocrat. See Benson 1982: 19, 25, and 38, n. 7.
3. Tellenbach 1959: 153, emphasis added.
4. Ullmann 1949: 120.
5. Grignaschi 1966: 49; my translation.
6. Lambton 1974.
7. Algar 1969: 80.
8. Algar 1969: 129, n. 21.
9. The most striking recurrence of the millenarian strand in Twelver Shi'ism took place in 1844, one thousand lunar years after the occultation of the Twelfth Imam, when Sayyed 'Ali-Mohammad of Shiraz, the Bab, claimed to be the returning Mahdi. The Babi movement rapidly gathered momentum, and there were a series of Babi millenarian uprisings in the mid-nineteenth century. The *'ulama* dreaded this return of "extremism" (*gholovv*), whose success would have eliminated them and the orthodox Shi'ism they represented, and urged the government to suppress Babism. After the suppression of the uprisings and the persecution of the Babis, the movement split, and the majority of the Babis eventually followed Baha' Allah, who died in 1892, and became known as Baha'is.
10. Arjomand 1984a: 248, 323, n. 41.
11. Lambton 1965b; Keddie 1966.
12. In practice, the Qajar rulers did inherit from the Safavids the "caesaropapist" prerogative over a number of clerical appointments, most notably those of the *imam jom'ehs*.
13. In order to avoid confusing the reader, I have employed the phrase "the Shi'ite nation" even though the term usually used at the time was the Islamic nation (*mellat-e Islam*). This does not pose a problem as the Shi'a naturally equate Shi'ism with true Islam. In seeking to understand the situation by making comparisons with the West, we must not feel bound by contemporaneity and look to the period prior to the onset of the secularization of culture and the advent of democratic political theories. We do find a comparable situation in medieval Christianity where "each of the two powers the Church and the State can and must in case of necessity (*casualiter and per accidens*) assume for the weal of the whole body, functions which in themselves are not its proper functions" (Gierke 1958: 18).
14. Keddie 1969: 10, 157–64.
15. Browne 1910: 112.
16. Kermani 1: 94.
17. Kermani 1: 151.
18. Kermani 1: 225.
19. E.g., Kermani 1: 142–45, 353–57.
20. Kermani 1: 299–300, 310.
21. Arjomand 1981a: 177.
22. Behbehani was assassinated after a bitter confrontation with the Social Democrats in July 1910. His assassins were widely presumed to belong to that party.

23. *Yaddashtha-ye Khosraw Avval, Anushirvan-e Dadgar,* Tehran 1932/1310. On the title page we read: "This unique and most important historical document is offered to His Imperial Majesty by His Excellency Mr. Teymurtash, the Minister of the Sublime Court."
24. Hedayat (1): 352–68.
25. Bahar 1984: 62–63, 92–95.
26. The best graphic illustration of clerical support for Reza Khan is a photograph of his on a raised platform, surrounded by a dozen prominent clerics at the opening of the Constituent Assembly on 15 Azar 1304. The other delegates are standing at his feet (Safa'i 1975: 125).
27. Hairi 1977: 144–47.
28. Sadeqipur 1968: 72.
29. Cottam 1964: 54–55; Akhavi 1980: 74; Richard 1982.
30. Hedayat (1): 377–83.
31. Hedayat (1): 322, n. 1; Sharif-Razi 1973, 1: 264; 1974–75, 4: 621.
32. Hedayat (1): 404, 407, 411.
33. Abrahamian 1982: 152–53.
34. Akhavi 1980: 33, 40, 56–58.
35. Akhavi 1980: 132.
36. Fischer 1980: 89–94.
37. Eisenstadt 1978: 245–46.
38. Weber 1968: 1194; for comparative parallels.
39. Brenan 1943: 37, 43; Carr 1966: 45–48.
40. Chapter 5, pp. 93–94.
41. Sharif-Razi 1973, 1: 264–65.
42. Sharif-Razi 1973, 1: 265–69; Algar 1972: 242.
43. Floor 1980.
44. Bakhash 1984: ch. 2.
45. Algar 1972: 245.
46. Private interview in 1977.
47. Arjomand 1981b: 305, n. 26.
48. Makki 1945, 2: 244.
49. Arjomand 1981b: 298, n. 13.
50. Richard 1982.
51. See, for instance, the sardonic remarks of Ayatollah Motahhari, cited in Arjomand 1984b: 223.

Chapter 5

1. Koran, VI. 98.
2. Gellner 1981.
3. Arjomand 1986.
4. Kazemi 1980: 31.
5. I have documented this elsewhere. See Arjomand 1986: 96, Table 4.
6. The figure is actually for the Persian year 1352/March 1973 to March 1974. For the documentation, see Arjomand 1981b: 311–12.
7. Kazemi 1980: 63.
8. Arjomand 1981b: 313, n. 43.
9. For the references, see Arjomand 1981b: 313–14. On Shari'ati's thought, see Arjomand 1982.

10. Arjomand 1988.
11. Ruhani 1982a: 613, 681, 727–28, 861.
12. Raja'i 1983: 11.
13. Ruhani 1982a: 623, 727–28.
14. Ruhani 1982a: 702, 777–79.
15. Matahhari 1970–71.
16. See, for instance, the interview with Mohammad Hojjati Kermani in *Ettala'at*, 25 Ordibehesht, 1362.
17. Qutb 1973: 29, emphasis added.
18. See Yazdi, n.d., for the one-sided admiration of the Islamic student associations in the United States for the clerical leaders coupled with lack of information about the latter's positions and activities. See also, Bani-Sadr 1979.
19. Sharif-Razi 1974: 422–23 and 1979: 88–89, 98–99, 149, 162–65, 175–76, 224, 254–71, 308.
20. Bakhash 1984: 43.
21. Ruhani 1982a: 43–50, 553–54, 615, 642.
22. Khomeini 1971: 64.
23. Khomeini 1977: 94.
24. The speech of 14 Shawwal 1397.
25. Edwards 1927: 54.
26. In a mystical tract he wrote at the age of 29, he introduces himself as "Sayyed Ruhollah, son of the murdered scholar, Sayyed Mostafa." Khomeini 1981b: 18.
27. Ruhani 1982a: 20–25.
28. Khomeini 1981b: esp. 33 ff.; Khomeini 1981c: 365–77.
29. Arjomand 1984a: 74–77, 163, 269.
30. Ajami 1986: 120–21. On the appellation of Khalesi as Imam, see Arjomand 1988.
31. Shari'ati 1971: 11, n. 1.
32. Yazdi 1984: 46, 55.
33. Yazdi 1984: 155–61.
34. In his memoirs, Bani-Sadr claims credit for wrecking the agreement worked out in Tehran and persuading Khomeini to change his mind after Qotbzadeh had announced that Khomeini would receive Bakhtiar (Bani-Sadr 1982: 48–49). Whatever the truth of his claim, not much significance can be given to it as Khomeini's attitude to Bakhtiar as a person and to compromises of the sort proposed is amply clear from other evidence cited in this chapter.
35. Yazdi 1984: 162.
36. Sick 1985: 150; Yazdi 1984: 126–27.
37. *New York Times*, April 17, 18, and 19, 1979. In the following years, Khomeini destroyed Grand Ayatollah Shari'at-madari much more thoroughly and ruthlessly. In that instance, however, he was using not his charismatic authority and sheer force of personality but rather the formidable apparatus of repression he now controlled.
38. See Chapter 7, pp. 137–46.

Chapter 6

1. Sorel 1950: 122–23.
2. Bazargan 1984: 37–38.

3. Mawdudi 1947; Lerman 1981: 500.
4. Haddad 1983. The Islamic government has aptly acknowledged its ideological debt by issuing a stamp commemorating Sayyid Qutb.
5. Dunn 1972: 236.
6. Mandelstam 1983: 126.
7. Salnameh 1974: 91, 98.
8. Halliday 1979: 218.
9. Looney 1982: 261.
10. Abrahamian 1982: 498.
11. Bazargan 1984: 48–49.
12. The extent of participation of the recent migrants in the revolutionary movement is not clear. The serious clash between the government forces and the squatters of a shanty town that the government had begun to bulldoze in the latter part of August 1977 left some twelve persons dead and more than a hundred injured. It provoked attacks on several police stations by urban guerrillas and must be counted among the factors precipitating the fateful crisis of 1978 (Stempel 1981: 84). On the other hand, there is evidence that the squatters, as distinct from the better-established and better-housed recent migrants, did not play any significant role in the demonstrations of 1978–79 (Kazemi 1980: 92; Parsons 1984: 111).
13. Halliday 1979: 114.
14. Abrahamian's (1982: 510–18) assertions concerning the importance of the participation of the industrial working class in the revolutionary movement are quite misleading. Similar statements by Ashraf and Banuazizi (1985: 25, 33–34) reflect the same bias of the conventional wisdom and are in fact contradicted by the evidence they themselves offer. It is true that there was a wave of industrial strikes in September and October of 1978. However, the strikes in the private sector were quickly ended with wage settlements and the meeting of the workers' economic demands. The more persistent strikes were among the state-employed oil workers and workers in the government-owned tobacco factories. These perfectly fit the pattern of strikes among government employees discussed in pp. 115–16.
15. Hajj-Sayyed-Javadi 1979: 354.
16. Cottam 1986.
17. Bazargan 1984: 26, n. 2.
18. Brière et Blanchet 1979: 107.
19. Monnerot 1960: 17.
20. Graham 1979 for this and other economic details.
21. Davies 1962.
22. Parsons 1984: 44.
23. Stempel 1981: 81; Looney 1982: 140–67, 260–61.
24. Stempel 1981: 128–29, 148–49; Sick 1985: 31.
25. Stempel 1981: 88–89, 97–98.
26. Hajj-Sayyed-Javadi 1979: 436–86.
27. Abrahamian 1982: 510.
28. See Chapter 4, pp. 86–87.
29. Bazargan 1984: 24–25.
30. Yazdi 1984: 130–34.
31. *Ettela'at,* 21 to 23 Farvardin 1358. Strangely enough, the Shah cites Rabi'i's statement with tacit acceptance (Pahlavi 1980: 173).

32. In contrast to the first of that season's demonstrations, the one of September 4 in which the middle class could regard as the inauguration of Iran's Prague Spring, the victims of the massacre largely belonged to the religious party. The troops enforcing martial law fired on religious demonstrators as they were leaving a mosque, killing some 250 people.

33. The memoirs of General Qarabaghi, who was the minister of interior in Sharif-Emami's cabinet is replete with examples of absence of team spirit and consultation, high-handed decision making by the prime minister about which the cabinet members would learn through the media, contradictory measures, bungling, and inability to rule. See Qarabaghi: especially, 31–32, 72–78, 82, 85, 97, 119.

34. Qarabaghi: 62–64.

35. Stempel 1981: 119, 134; private interviews.

36. Parsons 1984: 83.

37. Qarabaghi: 48.

38. Yazdi 1984: 208–10.

39. Stempel 1981: 130–31; Parsons 1984: 97.

40. Qarabaghi: 69. On the frustration of the army under Sharif-Emami and the friction between him and General Oveisi, see Qarabaghi: 64, 67, 69.

41. Qarabaghi: 43, 75, 90.

42. Interview with General Hasan Tufanian on July 27, 1986. Incidentally, the administration of martial law did not entirely cease to be a state within the state after the installation of military government. See Qarabaghi: 91.

43. Qarabaghi: 67–69; Sick 1985: 75–79.

44. Stempel 1981: 133; Green 1982: 138.

45. Private interview. The first list published was a partial one and astutely omitted the names of the prime minister and the cabinet members. Only after the prime minister ordered an official enquiry by the Ministry of Justice into the matter, was the full list including his name and those of some of his ministers published! Later, when the official enquiry was completed, a number of prominent transmitters of money were put on a blacklist and forbidden to leave Iran. The most notable of these was the administrator of martial law, General Oveisi, who fled the country in the early days of January 1979. In this, we have one of the more hilarious examples of contradictions in government policies during the revolutionary crisis. See Qarabaghi: 92–94, 101.

46. Bazargan 1984: 45–46.

47. Yazdi 1984: 87; Huyser 1986: 167.

48. Sick 1985: 52–53.

49. In a secret four-hour meeting with the director of the SAVAK in late September 1978, Shari'at-madari, mentioning constant intimidation by Khomeini's militant followers, explained that he was constrained to oppose the Shah publicly in order not to lose the initiative to the militants but assured the Shah of his loyalty (Ruhani 1982b: 146–53).

50. Interview with Sir Anthony Parsons, and Parsons 1984: 92–93, 102.

51. Interview with the Shah's entourage. The failure is all the more glaring in view of the good connections between General Qarabaghi and Ayatollah Shari'at-madari who helped him, among other things, by securing the release of essential and perishable goods by striking customs officers after the Shah's departure. See Qarabaghi: 36–37, Huyser 1986: 162.

52. Stempel 1981: 121; Huyser 1986: 149.

53. Pahlavi 1980: 86.
54. Zonis 1983: 604; Parsons 1984: 27.
55. Pahlavi 1980: esp. Chapter 11; Sick 1985: 48.
56. Huyser 1986: 31.
57. Green 1982: 122–23.
58. Hobbes 1963: 4.
59. The figure given for the period January 1978 to February 1979 by Ashraf and Banuazizi (1985: 22) is 3,008.
60. Sick 1985: 56–57.
61. Parsons 1984: 84.
62. Stempel 1981: 143, 146; Parsons 1984: 109.
63. Sick 1985: 101.
64. The incident and the casualty figures were kept secret. According to Sick, 12 officers were killed by 2 enlisted men. Qarabaghi, who was in a position to know, cites a later newspaper report that put the casualty figure at 72 (Sick 1985: 111; Qarabaghi: 88).
65. *New York Times,* December 19, 1978; Sick 1985: 122; Qarabaghi: 85, 98 on desertions.
66. Parsons 1984: 122.
67. In his memoirs, Huyser is emphatic about these figures (Huyser 1986: 105, 160). According to Sick, however, he was reporting 500 to 1000 desertions per day in this period (Sick 1985: 140; confirmed in Brzezinski 1983: 384). The statements made in this chapter hold even if we accept the highest figure.
68. Sick 1985: 118.
69. Huyser 1986: 116.
70. Yazdi 1984: 138–39.
71. Yazdi 1984: 181–86; Sick 1985: 145–46. In his memoirs General Qarabaghi understandably seeks to minimize these contacts. Nevertheless, they are mentioned in all the opposition sources. See also Huyser 1986: esp. 209–10, 285.
72. Yazdi 1984: 154–55.
73. Yazdi 1984: 170–71; Qarabaghi: 315. The letter of resignation, however, had not been signed.
74. Qarabaghi: 120–21, 240, 329.
75. Huyser 1986: 206.
76. Qarabaghi: 136. Qarabaghi is deliberately vague about the period to which the figure pertains. I have assumed it pertains to the period after January 26, but in view of conflicting reports (see n. 67 above) it may also apply to mid-January.
77. Qarabaghi: 124–25, 155–56.
78. Qarabaghi: 121–22, 247, 281–82, 330.
79. Qarabaghi: 246–47.
80. Qarabaghi: 213.
81. Huyser 1986: 225, 241–42.
82. Stempler 1981: 168; Qarabaghi: 110–11, 370–71; Huyser 1986: 183, 217, 244.
83. Qarabaghi: 130–34, 266, 445.
84. Fardust appears to have supplied information to the Soviet Union as well. (Huyser 1986: 64, 203).
85. Qarabaghi: 355, 372.

86. *Le Monde,* February 8 and 9, 1979; Huyser 1986: 278. Qarabaghi does not mention this incident in his memoirs.
87. Brzezinski 1983: 379–80; interview with General Tufanian.
88. Qarabaghi: 207.
89. Sick 1985: 131.
90. Private interview.
91. Parsons 1984: 128.
92. Huyser 1986: 260.
93. The captured documents, published by Yazdi to demonstrate the existence of plans for a coup, only proves how far it was from the stage of execution and how logistically ill-prepared the generals were for carrying it out. See Yazdi 1984: 249–304, esp. 260–61, 266. Yazdi's inference on page 286 is particularly dubious.
94. Qarabaghi: 62, 106–7, 110, 199, 207; Huyser 1986: 95.
95. Sick 1985: 171.
96. Sick 1985: 152; Huyser 1986: 40, 53.
97. Huyser 1986: esp. 34–35, 46, 57–58.
98. Huyser 1986: 50, 68.
99. Huyser 1986: 88. Yet another contradictory aspect of the situation was brought out in discussions between Huyser and the Iranian generals on January 10, 1979, by the navy chief, Admiral Habibollahi who saw a conflict between support for Bakhtiar and the Shah's insistence on keeping control of the army (Huyser 1986: 56).
100. General Huyser told Qarabaghi to contact the opposition religious leaders on January 14, just on the day the generals were "distinctly upbeat" and beginning to feel they could count on U.S. support, thus nibbing their incipient confidence in the bud (Huyser 1986: 108–9).
101. This was especially so after the arrival on January 18 of Eric von Marbod for negotiations concerning military agreements and sensitive equipment. From that point onward, it must have seemed to the Iranian generals that Huyser was only there to safeguard American military interests (Huyser 1986: 240).
102. By February 8, 1979, the Iranian generals had so little confidence in the United States that they did not give General Gast a chance to speak at their meeting. Even that confidence evaporated on the fateful day of February 9 with a statement made by a political officer of the U.S. embassy that Bakhtiar was failing and the Bazargan/Khomeini axis rapidly gaining in strength (Huyser 1986: 278–81).
103. Huyser 1986: 95, 118, 144, 152, 157, 163, 176.
104. Huyser 1986: 168.
105. Huyser 1986: 146, 194, 205, 222.
106. Brzezinski 1983: 385.
107. Huyser 1986: 227–35.
108. Qarabaghi: 377–83.
109. Qarabaghi: 428–30.
110. Qarabaghi: 451; private interview.
111. Qarabaghi: 415–23, 430–35.
112. *Le Monde,* February 12, 13, and 14, 1979.
113. Qarabaghi: 455–69; Yazdi 1984: 194.

114. Qarabaghi: 39, 129, 169; Yazdi 1984: 206–23; private interviews.
115. Sick 1985: 60–61.
116. Vance 1983: 328–30.
117. Brzezinski 1983: 370.
118. Sick 1985: 101, 107–15.
119. Vance 1983: 328; Sick 1985: 348, n. 9.
120. Sullivan 1981: 222–24; Sick 1985: 128–36.
121. Vance 1983: 326–27. According to Vance, this is the main reason why Ambassador Sullivan was not instructed to contact the opposition.
122. Brzezinski 1983: 375; Sick 1985: 124–25.
123. Stempel 1981: 128–29. The date given for this event by Sick (p. 98) is November 18.
124. Bakhash 1984: 50.
125. Stempel 1981: 148–49.
126. Brière and Blanchet 1979: 105–6.
127. Vance 1983: 332–33; Sick 1985: 126.
128. Brzezinski 1983: 377, 383.
129. Huyser 1986: 18, 33, 37, 45, 58.
130. Sick 1985: 138.
131. Sick 1985: 140.
132. Qarabaghi: 186.
133. Yazdi 1984: 123–26; Sullivan 1981: 236–37.
134. Yazdi 1984: 142.
135. Brzezinski 1983: 388.
136. *Newsday*, February 7, 1979.
137. Hobbes 1914 [1651].
138. Cottam 1986: 82–83.
139. Brzezinski 1983: 375.
140. See p. 116.
141. Sick 1985: 118.

Chapter 7

1. Parsons 1984: 75–76.
2. Stempel 1981: 143.
3. Stempel 1981: 150.
4. Stempel 1981: 155.
5. Parsons 1984: 124.
6. Yazdi 1984: 114–15.
7. Huyser 1986: 198.
8. Yazdi 1984: 118–19.
9. Qarabaghi: 357–59.
10. Qarabaghi: 296, 298.
11. Abrahamian 1982: 526–27.
12. Bazargan 1984: 77.
13. Hourcade 1980: 30–31, 36–40, *New York Times,* March 22, 28; April 23, 24; May 31, 1979.
14. Bakhash 1984: 57.
15. Mahdavi-Kani told Bani-Sadr that forty to forty-five thousand members of the committees were purged (Bani-Sadr 1982: 99).

16. At the end of May 1979, Rahmatollah Moqaddam-Maraghe'i resigned his governorship of Eastern Azerbaijan in protest against the persistence of dual power. In an interview with the author on March 22, 1982, he stressed that the prestige of state that had displaced the Shah and the respect for its authority was tremendous, and that Bazargan could have eliminated the militiamen and the committees if he had been firm.
17. *New York Times,* April 26, 1979.
18. *New York Times,* May 7, 1979.
19. Khomeini 1981c: 268–74.
20. *New York Times Magazine,* October 28, 1979. He knows better now. See Bazargan 1984: esp. 141.
21. Hobsbawm 1959.
22. The former director of the Iran National Oil Company, dismissed for not displaying the proper subservience toward the mollahs, especially Khomeini's son-in-law.
23. Far from being rewarded for this service, Ahmad-zadeh, like Bani-Sadr, fell from the Ayatollahs' grace and his family suffered tragically.
24. Arani 1980.
25. Zabih 1982.
26. Ayatollah Beheshti had chosen Jalal al-Din Farsi, a prolific Islamic ideologue. Farsi, however, turned out to have been born in Afghanistan and not in Iran, as required by the Constitution. The IRP then had to settle for the lackluster Hosayn Habibi as their candidate.
27. *Le Monde,* February 19, 1980.
28. Khomeini 1981c: 291–92.
29. Khomeini 1981c: 291, 295–99; Bazargan 1984: 107, n. 1.
30. In an interview with the author on March 24, 1982, Bani-Sadr insisted that he was obliged to do so to forestall a plot to "Bazarganize" and remove him. Tapes recording the voice of the ideologue of the IRP, Dr. H. Ayat, do in fact indicate a Bazarganization plan for President Bani-Sadr. Nevertheless, Bani-Sadr's reconstructed account of the event is not convincing (Bani-Sadr 1982: 123–28).
31. Bakhash 1984: 122.
32. See further Chapter 9, p. 185.
33. Bakhash 1984: 111.
34. *Ettela'at,* 26 Mordad, 1359.
35. *Le Monde,* July 1, 1980; *Ettela'at,* 5 and 6, Mordad 1359.
36. Bakhash 1984: 112–13.
37. Arani 1981.

Chapter 8

1. It is interesting to note in this context that, according to a later statement by Khomeini's son, Ahmad, the vast majority within the Shi'ite hierocracy who neither liked nor disliked the Shah were forced to break their silence in 1976–77 because they were shocked by the "moral decadence" and "social filth" they held him responsible for (Abrahamian 1980: 26).
2. Khomeini 1981a, 1: 482, 502.
3. Algar 1972: 245; Khomeini 1981c: 202, 235.
4. Bakhash 1984: 32.
5. Bakhash 1984: 74–75.

6. Bakhash 1984: 78.
7. Arjomand 1985.
8. Arjomand 1984a: 182–83.
9. *Jomhuri-ye Eslami,* 31 Shahrivar 1361.
10. See Chapter 9, pp. 177–78.
11. The account given in this chapter generally relies on newspaper reports, which are cited specifically only for quotations or potentially contestable points. On the significance of the new terms, see Chapter 9.
12. The term was used repeatedly by Hashemi-Rafsanjani in June and July 1986.
13. *Kayhan Hava'i,* 19 Shahrivar 1365/10 September 1986.
14. Speech on October 6, 1983.
15. *Kayhan Hava'i,* 13 Azar 1364/4 December 1985; *Ettela'at,* 15 Azar 1364.
16. Rose 1984: 160–77.
17. Raja'i 1983: 169.
18. *Ettela'at,* 6 Day 1363.
19. Hickman 1982: 14–18; Rose 1984: 185–86.
20. *The Times,* August 17, 1982.
21. *International Herald Tribune,* May 27, 1983.
22. Interview with Ayatollah Mahdavi-Kani, Supervisor of the Revolutionary Committees, *Ettela'at,* 5 Mordad 1361.
23. *Ettela'at,* May 22 and 28, 1983.
24. The reluctance of men of piety to serve as judges is typical in the history of Shi'ism prior to its recent revolutionary politicization. This traditional antipathy to judgeship has been reinforced on the part of the qualified doctors of jurisprudence by the fact that religious courts have been abolished for more than two generations. As for the younger militant clerics, they find political and political-ideological work more exciting and infinitely less demanding.
25. *Ettela'at,* 15 and 17 Bahman, 1362.
26. *Mojahed,* No. 175, 5 Aban 1362.
27. *Kayhan Hava'i,* 30 Azar 1362/31 December 1983.
28. *Ettela'at,* 1, 9, and 22 Mehr 1363.
29. *Kayhan Hava'i,* 3 Ordibehesht 1363/2 May 1984; *Ettela'at,* 30 Khordad 1363.
30. *Foreign Broadcast Information Service,* 30 September 1983.
31. *Ettela'at,* 26 Mordad, 1362.
32. *Ettela'at,* 18 Day, 1362.
33. The exact figures for the year 1362 (March 1983–March 1984) are 1,768,488 and 72,319 respectively (*Salnameh,* 1363/1984–85: 89).
34. *Salnameh,* 1362/1983–84: 416, 434.
35. Bazargan 1984: 123, esp. n. 1.
36. *Ettela'at,* 20 Farvardin, 1363.
37. *Salnameh* 1362/1983–84: 722–23, 737.
38. *Kayhan Hava'i,* Special Issue, 23 Bahman 1364/12 February 1986.
39. *Salnameh,* 1363/1984–85: 599.
40. See n. 38 above.
41. *Ettela'at,* 30 Khordad 1363.

Chapter 9

1. Al-Tusi, *Tibyan,* 2: 236–37; al-Tabrisi, *Majama'al-Bayan,* 5: 202–3; Ardabili, *Zubdat al-Bayan:* 687.

2. Tabataba'i, n.d., 5: 398–99; emphasis added. The same traditional Shi'ite view is upheld by less important Koranic commentaries which appeared in the 1970s (Thaqafi 1978, 2: 70–73; Najafi 1979–80, 3: 277–81). Incidentally, the Persian translator of Sayyid Qutb's commentary added a footnote to point out that "in the commentaries of us Shi'ites "those in authority" refers to the twelve Imams, may peace be upon them" (Qutb 1956–57: 44, n. 1).

3. Enayat 1983: 162.

4. Al-Hilli 1964: 273.

5. Naseri 1972: 4–5.

6. Khomeini 1971; English tr., Khomeini 1981c: 1–166.

7. Calder, 1982: 14; emphasis added and translation modified.

8. Madani in *Sorush,* No. 176, 17 Day 1361/January 1983: 58.

9. Haddad 'Adel in *Ettela'at,* 7 Shahrivar 1362/August 1983.

10. *Sorush,* No. 175, 11 Day 1361/1 January 1983: 41.

11. See Chapter 8, pp. 155–56.

12. As we have seen, three Grand Ayatollahs have dissented. Three other *maraji'-e taqlid,* Golpayegani, Mar'ashi-Najafi, and Shirazi have tacitly accepted Khomeini's superior authority as *vali-ye amr.*

13. Madani, in *Sorush,* No. 177, 25 Day 1361/January 1983: 45.

14. Haddad 'Adel in *Ettala'at* 2 Shahrivar 1362/August 1983.

15. *Qiyam-e Iran,* December 12, 1985.

16. Arjomand 1984a.

17. *Ettala'at,* 14 Day 1362/January 1983.

18. Friday Sermon, 24 Shahrivar 1361/September 1982.

19. *Ettela'at,* 28 Aban 1362/November 1983.

20. *Sorush,* No. 187, 27 Farvardin 1362/April 1983: 51.

21. *Jomhuri-ye Eslami,* 22 Day and 3 Bahman 1366/January 1988.

22. Grand Ayatollah Golpayegani took the occasion of the reelection of President Khamene'i to challenge the legitimacy of secondary commandment and their allegedly Islamic character. "We hope that in the future . . . they will be careful and be governed by the primary commandments of Islam, not commandments which God's servants [i.e., the ruling clerics] wish to say are God's commandments (*Ettala'at,* 1 Farvardin 1365). I am grateful to Professor Shahrough Akhavi for this reference.

23. Rajaee 1983: 70.

24. Weber 1978: 820; Schacht 1950: 95, 102.

25. Coulson 1969: 58–60.

26. Schacht 1935: 222.

27. Arjomand 1984a: 209.

28. Udovitch 1985: 460.

29. Geertz 1983: 190–91.

30. *Ruznameh-ye Rasmi-ye Jomhuri-ye Eslami:* 4 and 24 Aban 1361.

31. *Ruznameh-ye Rasmi-ye Jomhuri-ye Eslami:* 24 Aban 1361.

32. *Ruznameh-ye Rasmi-ye Jomhuri-ye Eslami:* 18 Day 1361.

33. Al-e Kashif al-Ghita' 1938: 138–39.

34. Santillana 1938, 2: 592–93; Schacht 1964: 188–98.

35. Juynboll 1910: 314–15, esp. 315 n. 1.

36. Kharaqani 1922: 25–26; Sangelaji 1977 [1959/1338]): 140–43.

37. Kharaqani 1922: 25 n. 2, 252 nn. 1, 2; Sangelaji 1950: 43–46.

38. Sangelaji 1950: 46.

39. Khomeini 1981a, 2: 408.
40. Khomeini n.d. [1980]: 219–20.
41. Khomeini n.d. [1980]: 220.
42. *Ruznameh-ye Rasmi-ye Jomhuri-ye Eslami:* 4 Aban 1361: 2, 5.

Chapter 10

1. Tilly 1978; Skocpol 1979; Zimmermann 1983: 309–14.
2. Baechler 1975.
3. Tocqueville 1959: 82.
4. Eisenstadt 1978: ch. 9; Goldstone 1982: 196–97.
5. The Shah was aware of these vulnerabilities and, knowing he had cancer, began trying to make the regime more "democratic" for the succession of his son in 1978. See Pahlavi 1980.
6. Sullivan 1981: 212.
7. Tilly 1978: 20.
8. Skocpol 1979: ch. 3 and p. 286; Goldfrank 1979: 153; Zimmermann 1983: 315, 322, 336–42, 352–57.
9. Skocpol 1979: 318 n. 2. On the Cuban revolution, see Dunn 1972: ch. 8.
10. Skocpol 1982.
11. Huntington 1968: 304–6.
12. Weber 1968: 54–56 and ch. 15.
13. Hintze 1968; Badie and Birnbaum 1983: 63, 110–11.
14. Arjomand 1984a.
15. This is the kind of situation we encounter in rebellions in Castile in 1520, where Franciscan and Dominican monks figured prominently among the leaders of the Communeros. Similarly, the priest Pau Claris as the president of the Catalan Diputacio assumed the leading position in the rebellion of the summer of 1640. When the Spanish people rose against Napoleon in 1808 without any king or government, they were led by priests and monks. See Brenan 1943: 42; and Zagorin 1982, 1: 266–67.
16. Eisenstadt 1978; Baechler 1975: 139; Goldstone 1982: 194–95.
17. By 1640, the English Crown had alienated a large segment of the elite, which included, notably, the proponents of aristocratic constitutionalism and the rising local landed gentry who resisted its increasingly statist policies (Stone 1972: 30, 57, 92, 124).
18. Tocqueville 1959; Cobban 1968; Zagorin 1982.
19. Zagorin 1982, 2: 94.
20. See Chapter 5.
21. See Chapter 3.
22. It is interesting to compare the heterogeneity and lack of cohesiveness of Iran's new middle class with the same features pertaining to their Western counterpart whom Gouldner erroneously portrayed as a new class in the Marxian schema (Gouldner 1979).
23. Tilly 1975: 507–10. It is highly revealing that the period identified by Tilly as marking the transition from traditional to modern forms of collective action, the mid-nineteenth century, coincided with the *end* of the classic age of revolutions. See Tilly 1972.
24. Hexter 1979: 129–31.
25. Cobban 1964; Goldstone 1982: 201.

26. Kumar 1983.
27. Calhoun 1983: 886, 897, 908.
28. Tilly 1972, cited in Zimmermann 1983: 374–75.
29. Sewell 1980; Moore 1978: 126–27.
30. Calhoun 1983: 911.
31. Lefebvre 1954 [1933]: 250, 254; Tilly 1975: 498; Zimmermann 1983.
32. Dunn 1972: 52–53.
33. Brenan 1943: 206–11, 213 n. A. In the Second Carlist War (1870–76), monks and priests similarly led the guerrilla bands.
34. Dunn 1972: 49, 64–69; Chevalier 1967: 161–69; Lewy 1974: Ch. 16; Hennessy 1976: 280.
35. Larsen, Hagtvet, and Myklebust 1980; Hamilton 1982.
36. Davis 1973: 85–86.
37. Rudé 1959, cited in Zimmermann 1983: 387.
38. Carsten 1976; Linz 1976: 38–39; Merkl 1980: 764, 789; Lako 1980: 395–96; Hamilton 1982: esp. 444–55.
39. Durkheim 1951 [1897].
40. Merkl 1980: 760–62.
41. Hill 1975: 45–48.
42. Stone 1972: 103, 120–21.
43. Halévy 1971.
44. Merkl 1980: 757.
45. Linz 1976: 50.
46. Stone 1972: 96–97; Walzer 1965: 140–43.
47. Seton-Watson 1956: 44; Carsten 1976: 418; Linz 1976: 48–50; Linz 1980: 167; Barbu 1980: 385–87.
48. See Appendix, Table 16.
49. Barbu 1980: 392.
50. Weber 1966: 107.
51. Weber 1968: 1195.
52. Fischer 1980.
53. Abrahamian 1980: 26.
54. It was neither the first nor the last time that a social class participated in a revolution that did not further its interests. As Barrington Moore pointed out, the peasants have often been the principal victims of modernization brought about by communist governments they helped create by their participation in revolutionary movements (Moore 1966: 428–29; see also Zimmermann 1983: 339–41, 356). Similarly, the outcome of the French Revolution was not especially favorable to the *petite bourgeoisie,* the *sans-culottes,* who most vigorously participated in it (Zimmermann 1983: 387, 407).
55. Mosse 1966: 22.
56. Wallace 1956.
57. Walzer 1965: 313, 315.
58. Ranulf 1964 [1938]; Weber 1965; Deak 1965: 394; Barbu 1980.
59. See Chapter 8, p. 174. See further p. 207.
60. Elliott 1969: 42–44, 48.
61. Hexter 1979: 178.
62. Furet 1981: esp. 29, 48–49, 70–74.
63. Kamenka 1966: 126.
64. Dunn 1972: 8–11.

65. Monerot 1960 [1949]: 12.
66. Nolte 1969: 281; Baechler 1975: 10 n. 15.
67. Mosse 1966: 21.
68. Linz 1976: 16. Once the attempt to export the Islamic Revolution, temporarily checked by the setback in the Iran-Iraq war, is resumed, one may expect further resonance of the Italian fascist ideas of "an imperialism of the poor" and "proletarian imperialism" (Sternhell 1976: 334–35; Baglieri 1980: 322–23).
69. Linz 1976: 5.
70. Fletcher 1982.
71. Arjomand 1984b.
72. Sternhell 1976: 320–21, 326, 335–37.
73. Sternhell 1976: 335, 347.
74. Safi 1982: 16–18.
75. Bracher 1970: 7–13; Carsten 1976: 428.
76. Mannheim 1953.
77. Baechler 1975: 108.
78. Hennessy 1976: 258.
79. See Chapter 9.
80. Zagorin 1982, 1: 741.
81. Walzer 1965: 92–113.
82. Mousnier 1973: 50, 61; Zagorin 1982, 2: Ch. 10.
83. Stone 1972: 90.
84. Elliott 1963: 487.
85. Trevor-Roper 1972.
86. Liu 1973: 50–51, 94–97, 146–60; Zagorin 1982, 2: 166.
87. Tocqueville 1955: 13, 156.
88. Monnerot 1960; Voegelin 1968. It is interesting to note that in 1949 Monnerot described communism as "the twentieth-century Islam."
89. It was "a spectacular but transient phenomenon . . . in no sense basic to its program." Therefore, the antireligious features faded as the true political teleology of the revolution unfolded. (Tocqueville 1955: 5–7). On the vitality of religious sentiment among the insurgent masses during the French Revolution, see Soboul 1956.
90. These variations become intelligible in the light of Linz's demonstration that the organized preemption of the political space by Christian-Democratic or Catholic-conservative parties was a decisive factor in inhibiting the growth of fascism (as was the case in Spain and Belgium). Where such parties existed and had carved up electoral territories for themselves, fascism found a formidable rival. It would also tend to be anticlerical to differentiate itself from the rival religious party (as were the Belgian Rex and the Nazis vis-à-vis the Zentrum party). See Linz 1980: 16–28, 52, 84; Linz 1976: 156; Hamilton 1982: 37–41.

    Mexican fascism, the Sinarquism of the late 1930s and early 1940s, also fits Linz's pattern. The movement declined when its middle-class supporters defected to the Catholic Acción Nacional (Hennessy 1976: 280–82). However, Linz's account of cases in which fascism was not anticlerical but intensely Christian is unsatisfactory in the light of our next section (Linz 1980: 16; Linz 1976: 164, 184 n. 51). The reverse of Linz's argument is well put by Merkl: "There is ample evidence that religious decline and confrontations played a role in fascist development . . . , creating a massive reservoir of confused quasi-religious fears and longings open to exploitation by fascist demagogues" (Merkl 1980: 757).

91. Jelinek 1980.
92. Weber 1965, 1966; Nagy-Talavera 1970: 247, 266–70.
93. Hilton 1972: 12; Williams 1974: 436–40. In this typical search for "a third way," Salgado also sought to "Brazilianize" Italian fascism. Needless to say, he considered the two aspects of his project fully compatible and would declare: "My nationalism is full of God" (Ibid., 434–36).
94. Monnerot 1960: ch. 3; Berdiaev 1961; Rhodes 1980: 79.
95. Lewy 1974: ch. 7.
96. Arjomand 1984a: 269, 270.

Glossary of
Persian and Arabic Terms

Except for Imam, Islam Kurd, Turk, and *'ulama,* the transliteration system of *Iranian Studies* is followed for Persian words, and the *IJMES* system for the Arabic terms and titles.

ahkam awwaliyya   primary commandments; rules of the Sacred Law that are obligatory per se

ahkam-e hokumati   governmental ordinances of the ruling jurist

ahkam thanawiyya   secondary commandments; rules that are obligatory not per se but by virtue of being necessary for the fulfillment of a primary commandment

ahl-e menbar   people of the pulpit; preachers and reciters of *rawzeh*

akhbar   Traditions of the Prophet and the Imams; plural of *khabar* which is synonymous with *hadith*

akhbari   Traditionalist; a tendency in Shi'ism that advocates strict adherence to the *akhbar* and is opposed to *ejtehad* and rational jurisprudence

al-dawlat al-islamiyya   Islamic government

Allah-o akbar   God is Great

amir   commander; emir

anjoman   association; political society

anjomanha-ye ayalati va velayati   provincial councils set up by the legislation of the First Majles in 1907

anjomanha-ye baladi   city councils set up by the legislation of the First Majles in 1907

Aryamehr   Sun of the Aryans; title assumed by Mohammad Reza Shah in the 1960s

'Ashura   the tenth day of Moharram; anniversary of the martyrdom of Imam Hosayn in 680/61

'avamm   common people

awliya'-e dam   those with authority over the blood; the persons entitled to blood money or retaliation according to Islamic Law

awqaf (pl. of *vaqf*)   religious endowments

bakhsh   [administrative] district

ballook (*boluk*)   rural district

barat   draft on provincial revenues issued by the central government in the nineteenth century

bast   sanctuary, usually a shrine

bayyina   clear proof in Islamic Law

dadgahha-ye madani-ye khass   Special Civil Courts; set up to deal with cases of family law since the Islamic revolution

Dar al-Tabliq   House of Mission, established in Qom by Ayatollah Shari'at-madari in the 1960s

dawlat   the state, in modern Persian

divanbegi   the chief judiciary official of the Safavid state

divan-khaneh   the state's court of justice

diyanat   religiosity

diyat   compensations for injuries in Islamic Law

du'a-gu'i   praying for a benefactor

du'a-ye nodbeh   prayer of supplication to the Hidden Imam

'ebadat   acts of worship

ejtehad   endeavor; the jurist's capacity to derive the rules of the Sacred Law from its sources

elteqati   mixed; the pejorative term used by the Islamic fundamentalists to characterize the thinking of the reformists influenced by the West

'erfan   gnostic philosophy

estekhareh   bibliomancy; using the Koran to divine the auspices for contemplated action

E'tedaliyyun   moderates; one of the two parties of the Second Majles formed at Regent Naser al-Mulk's insistence in 1911

faqih   jurist; expert in Islamic jurisprudence

farangi   Frankish; European

farman   decree, issued by the ruler

fatva   injunction in Islamic Law

Feda'iyan-e Khalq   those who sacrifice themselves for the people translated as Devotees of the People; a Marxist Leninist revolutionary organization

Feda'iyan-e Islam   those who sacrifice themselves for Islam, translated as Devotees of Islam; a group of Islamic militants active after World War II

feqh   practical jurisprudence

ghassalbashi   the Safavid official in charge of the ablution of the corpses with special responsibilities in the administration of criminal law

gholam (pl., gholaman)   slave; member of the corps of royal slaves created by Shah 'Abbas in the 1590s

gholovv   exaggeration; religious extremism, usually with millenarian and anthropolatric connotations

hadith   Tradition; recorded saying or deed of the Prophet or the Imams

hakem shar'   the judge administering the Sacred Law

haram   forbidden by the Sacred Law

hesba   norms of policing according to the Sacred Law

hey'at   association

Hey'at-e Mo'talefeh-ye Eslami   The Allied Islamic Association; a clandestine organization of clerical and lay Islamic militants active in the mid-1960s

hojja(t)   proof

hokumat   government in ordinary Persian; judging in Islamic legal terminology

hoquq Allah   the rights of God; expression in Islamic Law referring to obligations with respect to God

hoquq al-nas   the rights of the people; expression in Islamic Law referring to obligations with respect to other persons

hosayniyyeh   religious center for commemoration of the martyrdom of Imam Hosayn and performance of related ceremonies

'ilm   knowledge, science

imam   leader of the Shi'ite community of believers

imamat-e mostamarr   continuous Imamate; idea put forth in the Constitution of the Islamic Republic implying that the functions of Imamate should not be interrupted by the Occultation of the Twelfth Imam

Imam-e 'asr   Imam of the Age; the hidden Twelfth Imam of the Shi'a to reappear at the end of time

imam jom'eh   leader of the Friday congregational prayer

Jahiliyya   Ignorance; the culture of pre-Islamic Arabia

jame'eh-ye tawhidi   unitarian society; society so organized as to reflect the unity of God according to contemporary Islamic ideologies

jihad   holy war

khaleseh   royal domains

khasseh   an earlier variant of *khaleseh*

khass-e shahi   another variant of *khaleseh*

khoms   a religious tax of one-fifth on certain pecuniary and other gains

komiteh   revolutionary committee

lashkar-nevis   secretary for the army

luti   a tough of the city quarter

madrasa   college for religious learning; seminary

Majles   the Iranian parliament

majles-e shura-ye eslami   Islamic Consultative Assembly; official designation of the Iranian parliament since 1980

majles-e shura-ye melli   National Consultative Assembly; official designation of the Iranian parliament from 1906 to 1980

maktabi   doctrinaire; ideologically committed Islamic activists in contemporary Iran

mamlekat   country

marja'-e taqlid   source of imitation; the highest religious authority in Shi'ite Law

marja'iyyat   position of the source of imitation

mashruteh   conditional; the term adopted from Turkish for constitutional government; limited government

mashruteh mashru'eh   religiously legitimized constitutional government; term used by the clerical opponents of the First Majles in opposition to secular constitutionalism

mostamarri   annuity

ma'sum   sinless or infallible; attribute of the Twelve Imams

mavajeb   salary

mellat   nation

mohassel   tax collector

mojadded   renovator or renewer [of each century] in the Shi'ite tradition

Mojahedin-e Khalq   Fighters for the People; a party of Islamic radicals vehemently opposed to the Shah's and the present regime

mojtahed   the practitioner of *ejtehad;* doctor of religious jurisprudence

mollabashi   the head mollah; state-appointed leader of the *'ulama* in the eighteenth century

monshi al-mamalek   secretary of the state bureau for the interior under the Safavids and the early Qajars

moqaddas   holy

moqalled   imitator; an ordinary believer who follows a *mojtahed* in Shi'ite Law

mostamarri   annuity, pension

mostawfi   financial secretary

mostawfi al-mamalek   the chief financial functionary of the state before the twentieth century

mustaz'af (pl., *mostaz'afin*)   the weak; the disinherited

na'eb   deputy; vicegerent

navamis (sing., *namus*) norms or laws

ne'mat   benefit

nezam   order; military service

nezam-e jadid   the new order; term adopted from the Ottoman Empire for the new military organization during the first decade of the nineteenth century

nezami   military

niyabat 'amma   collective or general vicegerency; collective office of the Shi'ite *'ulama* as deputies of the Hidden Imam

nokar   servant, steward

os.an   province, department; the largest administrative unit in which Iran is divided

Osuli   principled—from *Osul* (Principles [of jurisprudence]); the rationalist movement in jurisprudence which advocates *ejtehad* and became dominant in Shi'ism after 1770s

pishkesh   gift, especially to the king

qadi   judge

qanat   wells interconnected underground

qanun   law of the state

qarasuran   military corps in charge of security of roads in the nineteenth century

qari   Koranic cantor

qollar   slaves; the elite corps of royal slaves created by Shah 'Abbas I in 1590s

qollar-aqasi   the commander of the *qollar*

ra'aya   subjects

rahbari   leadership

ra'is (pl., *ru'asa*)   head, leader

rawzeh   recital of the lives and afflictions of the Imams, especially those of Imam Hosayn and his family

rawzeh-khani   religious ceremonies centering on the above recitals

riyasat-ma'abi   seeking dominion and supremacy in leadership

ruz-nameh   journal

sadr   Safavid official in charge of religious endowments and institutions

sadr-e a'zam   title of the prime minister in the Qajar period

saheb divan   master of the ruler's chancery; controller-general

sahm-e imam   share of the Imam; one half of the *khoms* put at the disposal of the *marja'-e taqlid*

sarbaz   soldier

sayyed   master; honorific title of descendants of the Prophet

Sepah-e Pasdaran-e Enqetab-e Eslami   Corps of the Guardians of the Islamic Revolution, created in 1979

setad   commission, organization

shahri   inhabitant of a city

shar'i   in accordance with the Sacred Law

shari'a   the Sacred Law

shari'at-e mohammadi   the Sacred Law of Mohammad

shaykh al-Islam   chief religious functionary of a city under the Safavids

Shi'a   the Shi'ite community

shura   consultation

tafavote-e 'amal   difference of the operation; the difference between the amount of taxes collected and the tax assessment in the nineteenth century

taghut   idol; earthly power opposed to God

takfir   excommunication, anathematization

taklif   obligation according to the Sacred Law

tas'ir   conversion of taxation in kind into cash payment

Tasu'a   the ninth day of Moharram; eve of 'Ashura

tawhid   unity of God

ta'zieh   theatrical performance of the tragedy of Karbala and the martyrdom of Imam Hosayn and his family

tofangchi   musketeer

tollab (sing., *talabeh*)   student at a *madrasa;* seminarian

tuman   unit of currency, equivalent to 1000 *dinars*

tuyul   land held conditionally on the basis of a grant on state or royal domains

'ulama   the learned; the hierocracy

ulu'l-amr   those in authority, referred to by the Koranic verse 4:59

umma   the community of believers

vajeb (Arabic var., *wajib*)   obligatory according to the Sacred Law

vajeb-e 'eyni   individually incumbent duties laid down by the Sacred Law

vajeb-e kefa'i   duties whose discharge is incumbent upon the community as a whole rather than on any specific individual

vakil al-dawleh   representative or deputy of the state; title assumed by Karim Khan Zand

vakil al-ra'aya   representative or deputy of the subjects; title assumed by Karim Khan Zand

vali   person invested with authority or office

vali-ne'mat   lord of benefit; benefactor

vaqf   religious endowment

vazifeh   stipend, pension

vazir   minister; vizier

velayat   mandate; authority to hold office

velayat-e amr   authority to command; mandate to rule

velayat-e faqih   mandate of the jurist—according to Khomeini, to rule on behalf of the Hidden Imam

vo"az (pl. of *va'ez*)   preachers

wila' al-imama   authority of [the office of] Imamate

wilaya   Arabic variant of *velayat*

zolm   oppression, tyranny

References

Abrahamian, E. (1980) "Structural Causes of the Iranian Revolution," *Middle East Research and Information Project* 87.

———— (1982) *Iran Between Two Revolutions*. Princeton: Princeton University Press.

Adamiyyat, F. (1969/1348) *Amir Kabir va Iran*. Tehran: Khvarazmi.

———— (1976/1355) *Ideologi-ye Nahzat-e Mashrutiyyat-e Iran*. Tehran: Payam.

———— and H. Nateq (1977) *Afkar-e Ejtema'i va Siyasi va Eqtesadi dar Athar-e Montashernashodeh-ye Dawran-e Qajar*. Tehran: Amir Kabir.

Ajami, F. (1986) *The Vanished Imam. Musa al-Sadr and the Shia of Lebanon*. Ithaca: Cornell University Press.

Akhavi, S. (1980) *Religion and Politics in Contemporary Iran*. Albany: State University of New York Press.

Al-e Kashif al-Ghita', J. (1938) *Usul al-Shi'a va Asluha*. Tehran.

Algar, H. (1969) *Religion and the State in Iran: 1785–1906*. Berkeley: University of California Press.

———— (1972) "The Oppositional Role of the Ulama in Twentieth-Century Iran." In N. R. Keddie, ed., *Scholars, Saints and Sufis*. Berkeley and Los Angeles: University of California Press.

Amanat, A. (1983) "Introduction" to *Cities and Trade: Consul Abbott on the Economy and Society of Iran: 1847–1866*. London: Ithaca Press.

———— (1988) "In Between the Madrasa and the Market-place: The Designation of Clerical Leadership in Modern Shi'ism." In S. A. Arjomand, ed., *Authority and Political Culture in Shi'ism*. Albany: SUNY Press.

Amirahmadi, H. (1982) "Roots of Dependent Development and Dictatorship in Iran: 1800–1920." Unpublished Cornell University Ph.D. dissertation.

Arani, Sh. [pseud.] (1980) "The Iranian Revolution: Year Zero." *Dissent*, Spring.

———— (1981) "The Toppling of Bani-Sadr." *The Nation*, July 4.

Ardant, G. (1975) "Financial Policy and Economic Infrastructure of Modern States and Nations." In C. Tilly, ed., *The Formation of National States in Western Europe*. Princeton: Princeton University Press.

Arjomand, S. A. (1981a) "Shi'ite Islam and the Revolution in Iran." *Government and Opposition*, Vol. 16, No. 3.

———— (1981b) "The *Ulama's* Traditionalist Opposition to Parliamentarianism: 1907–1909." *Middle Eastern Studies*, Vol. 17, No. 2.

———— (1982) *"À la recherche de la conscience collective:* The Ideological Impact of Durkheim in Turkey and Iran." *The American Sociologist*, Vol. 17, No. 2.

255

———— (1984a) *The Shadow of God and the Hidden Imam. Religion, Political Order and Societal Change in Shi'ite Iran from the Beginning to 1890*. Chicago: University of Chicago Press.

———— (1984b) "Traditionalism in Twentieth Century Iran." In *From Nationalism to Revolutionary Iran*. London: Macmillan and Albany: SUNY Press.

———— (1985) "Religion, Political Order and Societal Change: With Special Reference to Shi'ite Islam." *Current Perspective in Social Theory*, Vol. 6.

———— (1986) "Social Change and Movements of Revitalization in Contemporary Islam." In J. A. Beckford, ed., *New Religious Movements and Rapid Social Change*. London and Beverly Hills, Calif.: Sage Publications.

———— (1988) "Ideological Revolution in Shi'ism." In *Authority and Political Culture in Shi'ism*. Albany: SUNY Press.

Ashraf, A. (1982/1361) "Dehqanan Zamin va Enqelab." In *Kitab-e Agah*. Tehran: Agah.

———— and Banuazizi, A. (1985) "State and Class in the Iranian Revolution." *State, Economy and Culture*, Vol. 1, No. 3.

Avery, P. W. (1965) *Modern Iran*. London: Ernest Benn.

Badie, B., and Birnbaum, P. (1983) *The Sociology of the State*. Chicago: University of Chicago Press.

Baechler, J. (1975) *Revolution*. New York: Harper and Row.

Baglieri, J. (1980) "Italian Fascism and the Crisis of Liberal Hegemony: 1901–1922." In Larsen, et al., *Who Were the Fascists?*

Bahar, M. (1978/1357) *Tarikh-e Mokhtasar-e Ahzab-e Siyasi-ye Iran*. Vol. 1. Tehran: Jibi.

———— (1984/1363) *Tarikh-e Mokhtasar-e Ahzab-e Siyasi-ye Iran*. Vol. 2. Tehran: Amir Kabir.

Bakhash, Sh. (1971) "The Evolution of Qajar Bureaucracy." *Middle Eastern Studies*, Vol. 7, No. 2.

———— (1978) *Iran: Monarchy, Bureaucracy and Reform under the Qajars: 1858–1896*. London: Ithaca Press.

———— (1983) "The Failure of Reform: The Prime Ministership of Amin al-Dawla, 1897–1898." In E. Bosworth and C. Hillenbrand, eds., *Qajar Iran*. Edinburgh: Edinburgh University Press.

———— (1984) *The Reign of the Ayatollah*. New York: Basic Books.

Bamdad, M. (1968–74) *Tarikh-e Rejal-e Iran*. 6 vols. Tehran: Zavvar.

Banani, A. (1961) *The Modernization of Iran: 1921–1941*. Stanford: Stanford University Press.

Bani-Sadr, A.-H. (1979/1358) *Negahi beh Tarikh-e Siyasi-ye Iran*. Qom: Hadid.

———— (1982) *L'Ésperance trahie*. Paris: S.P.A.G.-Papyrus Éditions.

Barbu, Z. (1980) "Psycho-Historical and Sociological Perspectives on the Iron Guard, the Fascist Movement of Romania." In Larsen, et al., *Who Were the Fascists?*

Barker, E. (1944) *The Development of Public Services in Western Europe: 1660–1930*. Oxford: Oxford University Press.

Batatu, H. (1978) *The Old Social Classes and the Revolutionary Movements of Iraq. A Study of Iraq's Old Landed and Commercial Classes and of its Communists, Ba'thists, and Free Officers*. Princeton: Princeton University Press.

Bavar, M. (1945/1324) *Kuh Giluyeh va Ilat-e An*. Gachsaran: Sherkat-e Chap.

Bazargan, M. (1984/1363) *Enqelab-e Iran dar Daw Harakat*. Tehran: Naraqi.

Berdiaev, N. A. (1961) *The Russian Revolution*. Ann Arbor: University of Michigan Press.

Bharier, J. (1971) *Economic Development of Iran 1900–1970*. Oxford: Oxford University Press.

Bill, J. A. (1972) *The Politics of Iran. Groups, Classes and Modernization*. Columbus, Ohio: Merrill.

Bracher, K. D. (1970) *The German Dictatorship*. New York: Praeger Publishers.

Brenan, G. (1943) *The Spanish Labyrinth. An Account of the Social and Political Background of the Spanish Civil War*. Cambridge: Cambridge University Press.

Brière, C., and Blanchet, P. (1979) *Iran: la révolution au nom de Dieu*. Paris: Seuil.

Browne, E. G. (1910) *The Persian Revolution 1905–1909*. Cambridge University Press. [Reprint, Frank Cass, 1966.]

Brzezinski, Z. (1983) *Power and Principle*. New York: Farrar, Straus, Giroux.

Calder, N. (1982) "Accommodation and Revolution in Imami Shi'i Jurisprudence; Khumayni and the Classical Tradition." *Middle Eastern Studies*, Vol. 18, No. 1.

Calhoun, C. J. (1983) "The Radicalism of Tradition: Community Strength or Venerable Disguise and Borrowed Language?" *American Journal of Sociology*, Vol. 88, No. 5.

Carsten, J. L. (1976) "Interpretation of Fascism." In W. Laqueur, ed., *Fascism: A Reader's Guide*. Berkeley: University of California Press.

Chardin, J. (1811) *Les Voyages du Chevalier Chardin en Perse*. 10 Vols. L. Langlès, ed. Paris.

Chevalier, F. (1967) "The *Ejido* and Political Stability in Mexico." In C. Veliz, ed., *The Politics of Conformity in Latin America*. Oxford: Oxford University Press.

Cobban, A. (1964) *Social Interpretations of the French Revolution*. Cambridge: Cambridge University Press.

——— (1968) *Aspects of the French Revolution*. New York: Norton.

Constitution of the Islamic Republic of Iran. (1980) English translation by H. Algar. Berkeley: Mizan Press.

Cottam, R. W. (1964) *Nationalism in Iran*. Pittsburgh: University of Pittsburgh Press.

——— (1986) "The Iranian Revolution." In J. R. I. Cole and N. R. Keddie, eds., *Shi'ism and Social Protest*. New Haven: Yale University Press.

Coulson, N. J. (1969) *Conflict and Tension in Islamic Jurisprudence*. Chicago: University of Chicago Press.

Curzon, G. N. (1892) *Persia and the Persian Question*. 2 Vols. London: Longmans, Green and Co.

Danesh-pazhuh, M. T. (1968/1347) *"Dastur al-Moluk-e* Mirza Rafi'a and *Tadhkirat al-Moluk-e* Mirza Sami'a." *Majalleh-ye Daneshkadeh-ye Adabiyyat va 'Olum-e Ensani*, Vol. 15, Nos. 5 and 6.

Dastur al-'Amal-e Shahaneh (1907/1325). Tehran.

Davies, J. C. (1962) "Towards a Theory of Revolution." *American Sociological Review*, Vol. 27, No. 1.

Davis, N. Z. (1973) "Religious Riots in Sixteenth Century France." *Past and Present*, Vol. 59.

Dawlatabadi, Yaha (n.d.) *Hayat-e Yahya*. 4 vols. Tehran: Ebn Sina.

Deak, I. (1965) "Hungary." In H. Rogger and E. Weber, eds., *The European Right: A Historical Profile.* Berkeley: University of California Press.

Demorgny, G. (1915) *Les Institutions Financières en Perse.* Paris.

Donboli, 'Abd al-Razzaq (1825–6/1241Q) *Ma'ather-e Soltaniyyeh.* Tabriz.

Douglas, W. O. (1951) *Strange Lands and Friendly People.* New York: Harper and Brothers.

Dunn, J. (1972) *Modern Revolutions. An Introduction to the Analysis of a Political Phenomenon.* New York: Cambridge University Press.

Durkheim, E. (1951 [1897]) *Suicide: A Study in Sociology.* Glencoe, N.Y.: Free Press.

Edwards, L. P. (1927) *The Natural History of Revolution.* Chicago: University of Chicago Press.

Eisenstadt, S. N. (1978) *Revolutions and the Transformation of Societies.* New York: Wiley.

Elliott, J. (1963) *The Revolt of the Catalans. A Study in the Decline of Spain (1598–1640).* New York: Cambridge University Press.

——— (1969) "Revolution and Continuity in Early Modern Europe." *Past and Present,* Vol. 42.

Elwell-Sutton, L. P. (1941) *Modern Iran.* London: George Routledge and Sons.

Enayat, H. (1982) *Modern Islamic Political Thought.* London: Macmillan.

——— (1983) "Khumayni's Concept of the 'Guardianship of the Jurisconsult.' " In J. P. Piscataori, ed., *Islam in the Political Process.* New York: Cambridge University Press.

Eqbal Ashtiyani, 'A. (1976/1355) *Mirza Taqi Khan Amir Kabir.* I. Afshar, ed. Tehran: Tus.

———, ed. (1946/1325) [Mohammad, Kalantar of Fars]. *Ruznameh-ye Mirza Mohammad Kalantar-e Fars.* Tehran: Sherkat-e Chap.

Esfahanian, K., ed. (1970–71/1350) *Majmu'eh-ye Asnad va Madarek-e Farrokh Khan Amin al-Dawleh.* Tehran: Tehran University Press.

Estahbanati, Abu'l-Hasan (1894/1312Q) *Salsabil.* Bombay.

E'temad al-Saltaneh, Mohammad Hasan (1970/1349) *Sadr al-Tavarikh.* Tehran: Vahid.

——— (1971/1350) *Ruznameh-ye Khaterat-e E'temad al-Saltaneh.* I. Afshar, ed. Tehran: Amir Kabir.

Fakhra'i, E. (1973/1352) *Gilan dar Jonbesh-e Mashrutiyyat.* Tehran: Jibi.

Farid al-Mulk Hamadani, M. A. (1975/1354) *Khaterat-e Farid.* M. Farid, ed. Tehran: Zavvar.

Fasa'i, Hasan (1972) *History of Persia under Qajar Rule.* H. Busse, ed. and tr. New York: Columbia University Press.

Fischer, M. M. J. (1980) *Iran. From Religious Dispute to Revolution.* Cambridge, Mass.: Harvard University Press.

Fletcher, W. M. (1982) *The Search for a New Order. Intellectuals and Fascism in Prewar Japan.* Chapel Hill: University of North Carolina Press.

Floor, W. (1980) "The Revolutionary Character of the Iranian Ulama: Wishful Thinking or Reality?" *International Journal of Middle East Studies,* Vol. 12, No. 4.

——— (1983) "Change and Development in the Judicial System of Qajar Iran (1800–1925)." In E. Bosworth and C. Hillenbrand, eds., *Qajar Iran.* Edinburgh: Edinburgh University Press.

Fraser, J. B. (1834) *An Historical and Descriptive Account of Persia.* Edinburgh.

Furet, F. (1981) *Interpreting the French Revolution*. E. Foster, trans. New York: Cambridge University Press.

Garthwaite, G. R. (1983) *Khans and Shahs. A Documentary Analysis of the Bakhtiari in Iran*. New York: Cambridge University Press.

Geertz, C. (1983) *Local Knowledge*. New York: Basic Books.

Gellner, E. (1981) *Muslim Society*. New York: Cambridge University Press.

Gilbar, G. G. (1977) "The Big Merchants (*tujjar*) and the Persian Constitutional Revolution of 1906." *Asian and African Studies*, Vol. 11, No. 3.

Goldfrank, W. L. (1979) "Theories of Revolution and Revolution Without Theory: The Case of Mexico." *Theory and Society*, Vol. 7, No. 3.

Goldstone, J. A. (1982) "The Comparative and Historical Study of Revolutions." *Annual Review of Sociology*, Vol. 8.

Gouldner, A. (1979) *The Future of Intellectuals and the Rise of the New Class*. New York: Seabury.

Graham, R. (1978) *Iran: The Illusion of Power*. London: Croom Helm.

Green, J. D. (1982) *Revolution in Iran. The Politics of Countermobilization*. New York: Praeger.

Grignaschi, M. (1966) "Quelques spécimens de la littérature sassanide conservés dans les bibliothèques d'Istanbul." *Journal Asiatique*, Vol. 254.

Haddad, Y. Y. (1983) "The Qur'anic Justification for Revolution: The View of Sayyid Qutb." *The Middle Eastern Journal*, Vol. 37, No. 1.

Hairi, A. H. (1977) *Shi'ism and Constitutionalism in Iran, A Study of the role played by the Persian Residents of Iraq in Iranian Politics*. Leiden: E. J. Brill.

Hajj-Sayyed-Javadi, A. A. (1979/1358) *Daftarha-ye Enqelab*. Tehran: Entesharat-e Jonbesh baraye Azadi.

Halévy, E. (1971) *The Birth of Methodism in England*. Trans. with Intro. by Bernard Semmel. Chicago: University of Chicago Press.

Halliday, F. (1979) *Iran: Dictatorship and Development*. London: Penguin.

Hamilton, R. F. (1982) *Who Voted for Hitler?* Princeton: Princeton University Press.

Hearing Committee on Foreign Affairs, House of Representatives (1982) *General Huyser's Mission to Iran, January 1979*. Washington, D.C.: U.S. Government Reports.

Hedayat, Mokhber al-Saltaneh (1965/1344) *Khatarat va Khatarat*, Tehran: Zavvar.

——— (1984/1363) *Gozaresh-e Iran. Qajar va Mashrutiyyat*. Tehran: Entesharat-e Noqreh.

Hennessy, A. (1976) "Fascism and Populism in Latin America." In W. Laqueur, ed., *Fascism: A Reader's Guide*. Berkeley: University of California Press.

Hexter, J. H. (1979) *Reappraisals in History*. 2d ed. Chicago: University of Chicago Press.

Hickman, W. F. (1982) *Ravaged and Reborn: The Iranian Army*. Washington, D.C.: Brookings Institution.

Hill, C. (1975) *The World Turned Upside Down*. London: Penguin.

al-Hilli, al Muhaqqiq, Abu'l-Qasim Najm al-Din (1864/1383Q) *al-Mukhtasar al-Nafi'*. Najaf.

Hilton, S. (1972) "Ação Integralista Brasiliera: Fascism in Brazil, 1932–1938." *Luso-Brazilian Review*, Vol. 9, No. 2.

Hintze, O. (1968) "The State in Historical Perspective." In R. Bendix, et al.,

State and Society. A Reader in Comparative Political Sociology. Berkeley: University of California Press.

———— (1975) *The Historical Essays of Otto Hintze.* F. Gilbert, ed. New York: Oxford University Press.

Hobbes, T. (1963 [1840]) *Behemoth: The History of the Causes of the Civil Wars of England.* W. Molesworth, ed. New York: Burt Franklin.

———— (1914 [1651]) *Leviathan.* London: Everyman's Library.

Hobsbawm, E. J. (1957) "Methodism and the Threat of Revolution in Britain." *History Today,* February.

———— (1959) *Primitive Rebels. Studies in Archaic Forms of Social Movement in 19th and 20th Centuries.* New York: Norton.

Hooglund, E. (1982) *Land and Revolution in Iran, 1960–1980.* Austin: University of Texas Press.

Hourcade, B. (1980) "Géographie de la révolution iranienne." *Hérodote,* No. 18, April–June.

Huntington, S. P. (1968) *Political Order in Changing Societies.* New Haven: Yale University Press.

Huyser, R. E. (1986) *Mission to Tehran.* New York: Harper & Row.

Iranian Army (Setad-e Bozorg Arteshtaran) (n.d.) *Tarikh-e Artesh-e Novin-e Iran.* Vol. 1. Tehran: Chapkhaneh-ye Artesh.

Issawi, C. (1971) *The Economic History of Iran 1800–1914.* Chicago: University of Chicago Press.

———— (1977) "Population and Resources in the Ottoman Empire and Iran." In T. Naff and R. Owen, eds., *Studies in Eighteenth Century Islamic History.* Carbondale, Ill.: Southern Illinois University Press.

Ivanov, M. S. (1976/1355) *Tarikh-e Novin-e Iran.* Trans. into Persian by H. Tizabi and H. Qa'em-panah. Tudeh Party Publications.

Jamalzadeh, S. M. A. (1916–17/1335) *Ganj-e Shayegan.* Berlin: Kaveh.

Jelinek, Y. (1980) "Clergy and Fascism: The Hlinka Party in Slovakia and the Croatian Ustasha Movement." In Larsen, et al., *Who Were the Fascists?*

Juynboll, W. (1910) *Handbuch des islamischen Gesetzes.* Leiden and Leipzig.

Kamenka, E. (1966) "The Concept of Political Revolution." In C. J. Friedrich, ed., *Revolution. Nomos VIII.* New York: Atherton.

Kanun-e Vokala' (1965/1344) *Tahavvolat-e Bist va Panj Saleh-ye Vekalat-e Dadgostari.* Tehran.

Kasravi, A. (1967/1346 and 1976/1355) *Tarikh-e Masruteh-ye Iran.* 2 Vols. Tehran: Amir Kabir.

Kazemi, F. (1980) *Poverty and Revolution in Iran.* New York: New York University Press.

Keddie, N. R. (1966) *Religion and Rebellion in Iran. The Tobacco Protests of 1891–92.* London: Frank Cass.

———— (1969) "Iranian Politics 1900–1905: Background to Revolution I, II, and III." *Middle Eastern Studies,* Vol. 5, Nos. 1–3.

———— (1972) "The Roots of the Ulama's Power in Modern Iran." In *Scholars, Saints and Sufis.* Berkeley and Los Angeles: University of California Press.

———— (1981) *Roots of Revolution.* New Haven: Yale University Press.

Kermani, Nazem al-Eslam, M. (1970/1349) *Tarikh-e Bidari-ye Iraniyan,* Intro. Vol. and Pts. I and II. A.-A. Sa'idi Sirjani, ed. Tehran: Bonyad-e Farhang-e Iran.

Khal'atbari, A.-S. (1983/1362) *Sepahdar-e Tonekaboni.* Tafazzoli, M., ed. Tehran: Novin.

Khalkali, 'Emad al-'Ulama', A. (1907/1325) *Resaleh-ye Ma'na-ye Mashruteh.* Tehran.

Kharaqani, Sayyed Asadollah (1922/1341Q) *Resaleh-ye Moqaddaseh-ye Qaza va Shahadat va Muhakemat-e 'Ebadi-ye Eslami.* Tehran: Matba'eh-ye Azad.

Khomeini, Sayyed Ruhollah. (1971/1391Q) *Hukumat-e Islami.* Najaf and Tehran.

―――― (1977/1355) *Ava-ye Enqelab* (Selections from speeches), n.p.

―――― (n.d. [1980]) *Resaleh-ye Novin.* Trans. and Commentaries by A.-K. Biazar-Shirazi. Tehran.

―――― (1981a/1401Q) *Tahrir al-Wasila.* 3d ed., 2 Vols. Beirut: Dar al-Ta'aruf li'l-Matbu'at.

―――― (1981b/1360) *Misbah al-Hidaya ila'l-Khilafa wa'l-Wilaya.* S. A. Fahri, trans. Tehran: Payam-e Azadi.

―――― (1981c) *Islam and Revolution. Writings and Declarations of Imam Khomeini.* Translated and annotated by H. Algar. Berkeley, Calif.: Mizan Press.

Klein, I. (1980) "Prospero's Magic: Imperialism and Nationalism in Iran 1909–11." *Journal of Asian History,* Vol. 14, No. 1.

Kumar, K. (1983) "Class and Political Action in Nineteenth Century England. Theoretical and Comparative Perspectives." *European Journal of Sociology,* Vol. 24, No. 1.

Lacko, M. (1980) "The Social Roots of Hungarian Fascism: The Arrow Cross." In Larsen, et al., *Who Were the Fascists?*

Lambton, A. K. S. (1953) *Landlord and Peasant in Persia.* New York: Oxford University Press.

―――― (1958) "Secret Societies and the Persian Revolution of 1905–6." *St. Antony's Papers,* Vol. 4; *Middle Eastern Affairs,* No. 1.

―――― (1963) "Persian Political Societies 1906–11." *Middle Eastern Affairs,* No. 3; *St. Antony Papers,* Vol. 16.

―――― (1965a) "Dustur, Iran." *Encyclopedia of Islam.* 2d Ed., Vol. 2. Leiden: E. J. Brill.

―――― (1965b) "The Tobacco Régie: Prelude to Revolution." *Studia Islamica,* Vol. 22.

―――― (1969) *The Persian Land Reform.* New York: Oxford University Press.

―――― (1970) "The Persian 'Ulama and Constitutional Reform." In *Le Shi'isme imamite* (Colloque de Strasbourg). Paris: Presses Universitaires de France.

―――― (1974) "Some New Trends in Islamic Political Thought in Late Eighteenth and Early Nineteenth Century Persia." *Studia Islamica,* Vol. 40.

―――― (1977) "The Tribal Resurgence and the Decline of the Bureaucracy in the Eighteenth Century." In T. Naff and R. Owen, eds., *Studies in Eighteenth Century Islamic History.* Carbondale, Ill.: Southern Illinois University Press.

―――― (1979) "Ilat." *Encyclopedia of Islam.* 2d Ed., Vol. 3. Leiden: E. J. Brill.

―――― (1984) "Land and Revolution in Iran (review article)." *Iranian Studies,* Vol. 17, No. 1.

Larsen, S. U., Hagtvet, B., and Myklebust, J. P. (1980) *Who Were the Fascists? Social Roots of European Fascism.* Oslo: Universitetsforlaget.

Lefebvre, G. (1954 [1933]) "La Révolution française et les paysans." In *Études sur la Révolution française.* Paris: Presses Universitaires de France.

Lerman, E. (1981) "Mawdudi's Concept of Islam." *Middle Eastern Studies,* Vol. 17, No. 4.

Lewis, B. (1973) *Islam in History.* London: Alcove Press.

Lewy, G. (1974) *Religion and Revolution.* New York: Oxford University Press.

Linz, J. J. (1976) "Some Notes towards a Comparative Study of Fascism in Sociological Historical Perspective." In W. Laqueur, ed., *Fascism: A Reader's Guide.* Berkeley: University of California Press.

——— (1980) "Political Space and Fascism as a Late-Comer." In Larson, et al. *Who Were the Fascists?*

Liu, T. (1973) *Discord in Zion: The Puritan Divines and the Puritan Revolution 1640–1660.* The Hague: Nijhoff.

Looney, R. E. (1982) *Economic Origins of the Iranian Revolution.* New York: Pergamon Press.

Lorenz, J. H. (1974) "Modernization and Political Change in Nineteenth-Century Iran: The Role of Amir Kabir." Unpublished Princeton University Ph.D. dissertation.

McDaniel, R. A. (1974) *The Shuster Mission and the Persian Constitutional Revolution.* Minneapolis: Bibliotheca Islamica.

Madani, J. (1983/1362) "Hoquq-e Asasi dar Jomhuri-ye Eslami." *Sorush,* Nos. 165–187. Aban 1362–Farvardin 1363 (Series of Articles).

Majles-e Shura-ye Eslami (1983/1361) *Ashna'i ba Majles-e Shura-ye Eslami (Karnameh-ye 1360).* 2d ed. Tehran.

Makki, H. (1945/1324) *Tarikh-e Bist Saleh-ye Iran.* Tehran.

Malekzadeh, M. (n.d.) *Tarikh-e Enqelab-e Mashrutiyyat-e Iran.* 6 Vols. Tehran: Ebn Sina.

Mandelstam, N. (1983) *Hope Against Hope.* New York: Atheneum.

Mannheim, K. (1953) "Conservative Thought." In *Essays on Sociology and Psychology.* London: Routledge and Kegan Paul.

Mawdudi, A. A. (1947) *Process of Islamic Revolution.* Pathankot, Punjab: Maktab-e Jamaat-e Islami.

Meredith, C. (1971) "Early Qajar administration: an analysis of its development and functions." *Iranian Studies,* Vol. 4, Nos. 2–3.

Merkl, P. H. (1980) "Comparing the Fascist Movements." In Larsen, et al., *Who Were the Fascists?*

Meskub, Sh. (1981) *Melliyat va Zaban.* [Paris]: Entesharat-e Ferdowsi.

Minorski, V., ed. (1943) *Tadhkirat al-Muluk. A Manual of Safavid Administration (circa 1137/1725).* London: Luzac and Co.

Mitchell, B. R. (1981) *European Historical Statistics, 1750–1975.* 2d Ed. New York: Facts on File.

———, and Dean P. (1962) *Abstract of British Historical Statistics.* Cambridge: Cambridge University Press,

Mochaver, F. (1937) *L'Evolution des finances iraniennes.* Paris: Librarie Technique et Economique.

Molkara, 'Abbas Mirza (1976/1355) *Sharh-e Hal-e 'Abbas Mirza Molkara.* Nava'i, 'A., ed. Tehran: Babak.

Monnerot, J. (1960 [1949]) *Sociology and Psychology of Communism.* Boston: Beacon Press.

Moore, B. (1966) *Social Origins of Dictatorship and Democracy: Lord and Peasant in the Making of the Modern World.* Boston: Beacon Press.

————— (1978) *Injustice: The Social Basis of Obedience and Revolt.* White Plains, N.Y.: M. E. Sharpe.

Moshiri, M., ed. (1969/1348) Mohammad Hashem Rostam al-Hokama'. *Rostam al-Tavarikh,* Tehran.

Mosse, G. L. (1966) "The Genesis of Fascism." *Journal of Contemporary History,* Vol. 1, No. 1.

Motahhari, M. (1970–71/1349) *Khadamat-e Motaqabel-e Eslam va Iran.* Tehran: Anjoman-e Eslami-ye Mohandesin.

Mottahedeh, R. P. (1980) *Loyalty and Leadership in an Early Islamic Society.* Princeton: Princeton University Press.

Mousnier, R. (1973) *Social Hierarchies, 1450 to the Present.* P. Evans, trans. New York: Schocken.

Mozakerat-e Majles-e Shura-ye Melli. Tehran: Ruznameh-ye Rasmi-ye Keshvar-e Shahanshahi-ye Iran. 1946 onward.

Musgrave, R. A., and Peacock, A. T. (1958) *Classics in the Theory of Public Finance.* London: Macmillan.

Nagy-Talavera, N. M. (1970) *The Green Shirts and the Others. A History of Fascism in Hungary and Rumania.* Stanford: Hoover Institution Press.

Najafi, M. T. (1979–80/1358) *Tafsir-e Asan.* Tehran: Eslamiyyeh.

Najafi-Quchani (1972/1351) *Siyahat-e Sharq ya Zendeginameh-ye Aqa Najafi Quchani.* R. 'A. Shakeri, ed. Mashhad: Tus.

Naraqi, H. (1976/1355) *Kashan dar Jonbesh-e Mashruteh-ye Iran.* Tehran: Mash'al-e Azadi.

Naseri, A. A. (1972/1351) *Emamat va Shafa'at.* Tehran.

Nava'i, A-H. (1977/1355) *Dawlatha-ye Iran as Aghaz-e Mashrutiyyat ta Ultimatum.* Tehran: Babak.

————— (1978/1357) *Fath-e Tehran. Gushehha-yi z-Tarikh-e Mashrutiyyat.* Tehran: Babak.

Nolte, E. (1969) *Three Faces of Fascism.* New York: New American Library.

Nuri, Fazlollah, ed. (1893) *Su'al va Javab.* Bombay.

Oberling, P. (1974) *The Qashqa'i Nomads of Fars.* Paris: Mouton.

Olfat, M. B. (n.d.) Unpublished Memoirs.

Pahlavi, M. R. (1980) *Answer to History.* New York: Stein and Day.

Parsons, A. (1984) *The Pride and the Fall. Iran 1974–1979.* London: Jonathan Cape.

Perry, J. R. (1979) *Karim Khan Zand. A History of Iran, 1747–1779.* Chicago: University of Chicago Press.

Qarabaghi, 'A. (n.d.) *Haqayaq dar bareh-ya bohran-e Iran.* Paris: Soheil.

Qutb, Sayyid (1956/1335) *Dar Sayeh-ye Qur'an.* Trans. into Persian by 'A. Aram. Tehran: 'Elmi.

————— (1973 [1944]) *'Adalat-e Ejtema'i dar Islam.* Persian trans. of *al-'Adalat al-Ijtima'iyya fi'l-Islam* (1944) by M. H. Gerami and S. H. Khosrawshahi. Qum: Resalat.

Rabino, H. L. (1973/1352) *Mashruteh-ye Gilan az Yaddashha-ye Rabino.* M. Rawshan, ed. Rasht: 'Ata'i.

Rajaee, F. (1983) *Islamic Values and World View.* New York: University Press of America.

[Raja'i, M.-'A.] (1983/1362) *Shahid Raja'i, Osweh-ye Sabr va Esteqamat.* [Tehran]: Bonyad-e Shahid.

Ranulf, S. (1964 [1938]) *Moral Indignation and Middle Class Psychology*. New York: Schocken.

Razavi, H. and Vakil, F. (1984) *The Political Environment of Economic Planning in Iran. 1971–1983*. Boulder and London: Westview Press.

Razmara, 'A. (1941a/1320) *Joghrafiya-ye Nezami-ye Iran: Kurdestan*. Tehran.

———— (1941b/1320) *Joghrafiya-ye Nezami-ye Iran: Lurestan*. Tehran.

Rezvani, M. E. (1977) "Ruznameh-ye Shaykh Fazlollah-e Nuri." *Tarikh*, Vol. 1, No. 2.

Rhodes, J. M. (1980) *The Hitler Movement: A Modern Millenarian Revolution*. Stanford: Hoover Institution Press.

Richard, Y. (1982) "Base idéologique du conflit entre Mosaddeq et l'ayatollah Kashani." In J. P. Digard, ed., *Le Cuisinier et le philosophe*. Paris: Maisonneuve et Larose.

Röhrborn, K. M. (1966) *Provinzen und Zentralgewalt Persiens im 16. und 17. Jahrhundert*. Berlin: de Gruyter.

Rose, G. F. (1984) "The Post-Revolutionary Purge of Iran's Armed Forces: A Revisionist Assessment." *Iranian Studies*, Vol. 17, Nos. 2–3.

Rudé, G. (1959) *The Crowd in the French Revolution*. Oxford: Clarendon Press.

Ruhani (Ziyarati), S. H. (1982a/1360) *Barrasi va Tahlili az Nahzat-e Emam Khomeini*. Tehran: Entesharat-e Rah-e Emam.

———— (1982b/1361) *Shari'at-madari dar Dadgah-e Tarikh*. [Tehran].

Rustow, D. A. (1970) "The Political Impact of the West." In P. M. Holt, A. K. S. Lambton, and B. Lewis, eds., *The Cambridge History of Islam*, Vol. 1. Cambridge: Cambridge University Press.

Sadeqipur, A. (1968/1346) *Majmu'eh-ye Sokhanraniha-ye A'la Hazrat-e Faqid Reza Shah-e Kabir*. Tehran.

Safa'i, E. (1966/1345) *Naser al-Molk Qaragozlu, Rahbaran-e Mashruteh*. Vol. 2, No. 3. Tehran: Sharq.

———— (1975/1353) *Kudeta-ye 1299 va Athar-e An*. Tehran.

Safi, L. (1982/1361) *Nezam-e Emamat va Rahbari*. Tehran: Bonyad-e Be'that.

Salnameh-ye Amari-ye Keshvar (Statistical Yearbook). Tehran: Sazeman-e Barnameh va Budjeh.

Sangelaji, M. (1950/1329) *A'in-e Dadrasi dar Eslam*. Tehran.

Santillana, D. (1938) *Instituzioni di diritto musulmano malichita con riquardo anche al-Sistema Sciafita*. 2 Vols. Rome.

Savory, R. (1980) *Iran Under the Safavids*. New York: Cambridge University Press.

———— (1985) "Allahverdi Khan," *Encyclopaedia Iranica*. Vol. 1. London: Routledge & Kegan Paul.

Schacht, J. (1935) "Zur soziologischen Betrachtung des Islamischen Rechts." *Der Islam*, Vol. 22.

Sepehr, Mirza Mohammad 'Ali (1965–66/1344) *Nasekh al-Tavarikh*. M. B. Behbudi, ed. 4 vols. (the Qajar section). Tehran: Eslamiyyeh.

Seton-Watson, H. (1956) *The East European Revolution*. 3d Ed. New York: Praeger.

Sharif-Razi, M. (1973–74 to 1979) *Ganjineh-ye Daneshmandan*. 8 Vols. Tehran: Eslamiyyeh.

Shari'ati, A. (1971) *Entezar, Madhhab-e E'teraz*. Tehran: Abu Dharr.

Sheikholeslami, A. R. (1971) "The Sale of Offices in Qajar Iran, 1858–1896." *Iranian Studies*, Vol. 4, Nos. 2–3.

Sheil, [Lady] (1856) *Glimpses of Life and Manners in Persia*. London: John Murray.

Shuster, W. M. (1912) *The Strangling of Persia*. New York: The Century Co.

Sick, G. (1985) *All Fall Down. America's Tragic Encounter with Iran*. New York: Random House.

Skocpol, T. (1979) *States and Social Revolutions*. London and New York: Cambridge University Press.

——— (1982) "Rentier State and Shi'a Islam in the Iranian Revolution." *Theory and Society*.

Soboul, A. (1956) "Sentiment religieux et cultes populaires pendant la révolution. Saintes patriotes et martyrs de la liberté." *Archives de sociologie des religions*, Vol. 1, No. 2.

Sorel, G. (1950) *Reflections on Violence*. New York: Free Press.

Sotoudeh, H. (1936) *L'Évolution economique de l'Iran et ses problèmes*. Paris: Libraire Technique et Economique.

Stempel, J. D. (1981) *Inside the Iranian Revolution*. Bloomington: Indiana University Press.

Sternhell, Z. (1976) "Fascist Ideology." In W. Laqueur, ed., *Fascism. A Reader's Guide*. University of California Press.

Stone, L. (1972) *The Causes of the English Revolution*. London: Routledge and Kegan Paul.

Sullivan, W. H. (1981) *Mission to Iran*. New York: W. W. Norton.

Tabataba'i, M. H. (n.d.) *al-Mizan fi Tafsir al-Qur'an*. Beirut: al-Ilmi.

al-Tabrisi, Fazl ibn Hasan (1970–71/1249) *Tafsir Majma'al-Bayan*. A. Beheshti, ed. and trans. Qom.

Tapper, R. (1983) "Introduction" to *The Conflict of Tribe and State in Iran and Afghanistan*. London: Croom Helm.

Tapu Defteri, No. 904, 1727/1140.

Tavakkoli, A. (1949/1328) "Chand Estekhareh az Mohammad 'Ali Shah ba Javabha-ye Anha." *Yadegar*, Vol. 5, Nos. 8 and 9.

Tellenbach, G. (1970) *Church, State and Christian Society at the Time of the Investiture Contest*. R. F. Bennett, trans. New York: Harper.

Thaqafi Tehrani, M. (1978/1398Q) *Ravan-e Javid dar Tafsir-e Qur'an-e Majid*. 2d Ed. Tehran: Burhan.

Theqat al-Eslam Tabrizi (1976/1355) *Majmu'eh-ye Athar-e Qalami*. N. Fathi, ed. Tehran: Anjoman-e Athar-e Melli.

Tilly, C. (1972) "How Protest Modernized in France 1845–1855." In W. O. Aydelotte, A. G. Bogue, and R. Fogel, eds., *The Dimensions of Quantitative Research in History*. Princeton: Princeton University Press.

——— (1975) "Revolutions and Collective Violence." In F. I. Greenstein and N. W. Polsby, eds., *Handbook of Political Science*. Vol. 3: *Macropolitical Theory*. Reading, etc.: Addison-Wesley.

——— (1978) *From Mobilization to Revolution*. Reading, etc.: Addison Wesley.

Tocqueville, A. de (1955) *The Old Regime and the French Revolution*. New trans. by S. Gilbert New York: Doubleday.

——— (1959) *The European Revolution and Correspondence with Gobineauu*. J. Lukacs, ed. and trans. New York: Doubleday Anchor.

Trevor-Roper, H. (1972) "The Fast Sermons of the Long Parliament." In *Religion, the Reformation and Social Change*. 2d Ed. London: Macmillan.

al-Tusi, Muhammad ibn Hasan (1957) *Tafsir al-Tibyan*. A. B. Tehrani, ed. Najaf: al-Matba'at al-'Ilmiyya.

Udovitch, A. (1985) "Islamic Law and the Social Context of Exchange in the Medieval Middle East." *History and Anthropology*, Vol. 1, No. 2.

Ullmann, W. (1949) *Medieval Papalism: The Political Theories of the Medieval Canonists*. London: Methuen.

Vance, C. (1983) *Hard Choices*. New York: Simon and Schuster.

Vijuyeh, M. B. (1976/1355) *Tarikh-e Enqelab-e Azarbaijan va Balva-ye Tabriz*. A. Katebi, ed. Tehran: Simorgh.

Voegelin E. (1968) *Science, Politics and Gnosticism: Two Essays*. Chicago: Regnery Gateway.

Vothuq, 'A. (n.d.) *Dastan-e Zendegi. Khaterati az Panjah Sal Tarikh-e Mu'aser*. Tehran.

Wallace, A. F. C. (1956) "Revitalization Movements: Some Theoretical Considerations for their Comparative Study." *American Anthropologist*, Vol. 58.

Walzer, M. (1965) *The Revolution of the Saints. A Study in the Origins of Radical Politics*. Cambridge: Harvard University Press.

Weber, E. (1965) "Rumania." In H. Rogger and E. Weber, eds., *The European Right: A Historical Profile*. Berkeley: University of California Press.

———— (1966) "The Man of the Archangel." *Journal of Contemporary History*, Vol. 1, No. 1.

Weber, M. (1978) *Economy and Society*. G. Roth and C. Wittich, eds. Berkeley: University of California Press.

Williams, M. T. (1974) "Integralism and the Brazilian Catholic Church." *Hispanic American Historical Review*, Vol. 54, No. 3.

Yaddashtha-ye Khosraw Avval Anushirvan-e Dadgar (1932/1310) Rahimzadeh Safavi, ed. Tehran: Matba'eh-ye Majles.

Yazdi, E. (1984/1363) *Akharin Talashha da akharin ruzha*. Tehran: Zafar.

———— (n.d.) *Barrasi-ye Jonbeshha-ye Eslami ya Mo'arrefi-ye Chehrehha-ye Nashenakhteh-ye Ruhaniyyat-e Mo'aser* (N.p.).

Yekrangiyan, Mir Hosayn (1957/1336) *Golgun-Kafanan. Gusheh-yi az Tarikh-e Nezami-ye* Iran. Tehran: 'Elmi.

Zabih, S. (1982) *Iran Since the Revolution*. Baltimore: Johns Hopkins Press.

Zagorin P. (1982) *Rebels and Rulers, 1500–1660*. 2 Vols. Cambridge University Press.

Zahir al-Dawleh (1972/1350) *Khaterat va Asnad-e Zahir al-Dawleh*. I. Afshar, ed. Tehran: Jibi.

Zarrabi, 'Abdo'l-Rahim, the Kalantar (1950/1335) *Tarikh-e Kashan*. I. Afshar, ed. Tehran: Ebn Sina.

Zill al-Soltan, Mas'ud Mirza (1907/1325) *Tarikh-e Sargozashat-e Mas'udi*. Tehran.

Zimmermann, E. (1983) *Political Violence Crises and Revolution. Theories and Research*. Cambridge, Mass.: Shenkman.

Zonis, M. (1983) "Iran: A Theory of Revolution from Accounts of the Revolution." *World Politics*, Vol. 35, No. 4.

Index

/